STUDENT ATLAS OF

World Geography

Eighth Edition

Christopher J. Sutton
Western Illinois University

STUDENT ATLAS OF WORLD GEOGRAPHY, EIGHTH EDITION

Published by McGraw-Hill, a business unit of The McGraw-Hill Companies, Inc., 1221 Avenue of the Americas, New York, NY 10020. Copyright ©2014 by The McGraw-Hill Companies, Inc. All rights reserved. Printed in the United States of America. Previous edition(s) ©2012, 2010, and 2008. No part of this publication may be reproduced or distributed in any form or by any means, or stored in a database or retrieval system, without the prior written consent of The McGraw-Hill Companies, Inc., including, but not limited to, in any network or other electronic storage or transmission, or broadcast for distance learning.

Some ancillaries, including electronic and print components, may not be available to customers outside the United States.

This book is printed on acid-free paper.

Student Atlas® is a registered trademark of the McGraw-Hill Companies, Inc.
Student Atlas is published by the **Contemporary Learning Series** group within the McGraw-Hill Higher Education division.

2 3 4 5 6 7 8 9 LKV 21 20 19

MHID 0-07-352767-X
ISBN 978-0-07-352767-3
ISSN 1531-0221

Acquisitions Editor: *Joan L. McNamara*
Marketing Director: *Adam Kloza*
Marketing Manager: *Nathan Edwards*
Senior Developmental Editor: *Debra A. Henricks*
Lead Project Manager: *Jane Mohr*
Buyer: *Jennifer Pickel*
Cover Designer: *Studio Montage, St. Louis, MO.*
Cover Image: © *Christopher J. Sutton*
Senior Content Licensing Specialist: *Shirley Lanners*
Media Project Manager: *Sridevi Palani*

Compositor: Laserwords Private Limited

A Note to the Student

The study of geography has become an increasingly important part of the curriculum in secondary schools and institutions of higher education over the last decade. This trend, a most welcome one from the standpoint of geographers, has begun to address the massive problem of "geographic illiteracy" that has characterized the United States, almost alone among the world's developed nations. When a number of international comparative studies on world geography were undertaken, beginning in the 1970s, it became apparent that most American students fell far short of their counterparts in Europe, Russia, Canada, Australia, and Japan in their abilities to recognize geographic locations, to identify countries or regions on maps, or to explain the significance of such key geographic phenomena such as population distribution, economic or urban location, or the availability of natural resources. Indeed, many American students could not even locate the United States on world maps, let alone countries such as France, or Indonesia, or Nigeria. This atlas, and the texts it is intended to accompany, is a small part of the process of attempting to increase the geographic literacy of American students. As the true meaning of "the global community" becomes more apparent, such an increase in geographic awareness is not only important but necessary.

The maps in the *Student Atlas of World Geography* are designed to introduce you to the patterns or "spatial distribution" of the wide variety of human and physical features of the earth's surface and to help you understand the relationships between these patterns. We call such relationships "spatial correlation" and whenever you compare the patterns made by two or more phenomena that exist at or near the earth's surface—the distribution of human population and the types of climate, for example—you are engaging in spatial correlation. At the very outset of your study of this atlas, you should be aware of some limitations of the data used to create the maps. In some instances, there may be data missing. In such cases, the cause may represent the failure of a country to report information to a central international body (such as the United Nations or the World Bank), or it may mean that the shifting of political boundaries and changed responsibility for reporting data have caused some countries (for example, those countries that made up the former Soviet Union or the former Yugoslavia) to delay their reports. It is always our aim to use the most up-to-date data that is possible. Subsequent editions of this atlas will have increased data on countries such as Serbia, Montenegro, or Kosovo when it becomes available. In the meantime, as events continue to restructure our world, it's an exciting time to be a student of world geography!

You will find your study of this atlas more productive if you study the maps and tables on the following pages in the context of the five distinct themes that have been developed as part of the increasing awareness of the importance of geographic education:

1. *Location: Where Is It?* This theme offers a starting point from which you discover the precise location of places in both absolute terms (the latitude and longitude of a place) and in relative terms (the location of a place in relation to the location of other places). When you think of location, you should automatically think of both forms. Knowing something about absolute location will help you to understand a variety of features of physical geography, because such key elements are so closely related to their position on the earth. But it is equally important to think of location in relative terms. The location of places in relation to other places is often more important as a determinant of social, economic, and cultural characteristics than the factors of physical geography.

2. *Place: What Is It Like?* This theme investigates the political, economic, cultural, environmental, and other characteristics that give a place its identity. You should seek to understand the similarities and differences of places by exploring their basic characteristics. Why are some places with similar environmental characteristics so very different in economic, cultural, social, and political ways? Why are other places with such different environmental characteristics so seemingly alike in terms of their institutions, their economies, and their cultures?

3. *Human/Environment Interactions: How Is the Landscape Shaped?* This theme illustrates the ways in which people respond to and modify their environments. Certainly the environment is an important factor in influencing human activities and behavior. But the characteristics of the environment do not exert a controlling influence over human activities; they only provide a set of alternatives from which different cultures, in different times, make their choices. Observe the relationship between the basic elements of physical geography such as climate and terrain and the host of ways in which humans have used the land surfaces of the world.

4. *Movement: How Do People Stay in Touch?* This theme examines the transportation and communications systems that link people and places. Movement or "spatial interaction" is the chief mechanism for the spread of ideas and innovations from one place to another. It is spatial interaction that validates the old cliché, "the world is getting smaller." We find McDonald's restaurants in Tokyo and Honda automobiles in New York City because of spatial interaction. Advanced transportation and communications systems have transformed the world into which your parents were born. And the world your children will be born into will be very different from your world. None of this would happen without the force of movement or spatial interaction.

5. *Regions: Worlds Within a World*. This theme helps to organize knowledge about the land and its people. The world consists of a mosaic of "regions" or areas that are somehow different and distinctive from other areas. The region of Anglo-America (the United States and Canada) is, for example, different enough from the region of Western Europe that geographers clearly identify them as two unique and separate areas. Yet despite their differences, Anglo-Americans and Europeans share a number of similarities: common cultural backgrounds, comparable economic patterns, shared religious traditions, and even some shared physical environmental characteristics. Conversely, although the regions of Anglo-America and Eastern Asia are also easily distinguished as distinctive units of the earth's surface, they have a greater number of shared physical environmental characteristics. But those who live in Anglo-America and Eastern Asia have fewer similarities and more differences between them than is the case with Anglo-America and Western Europe: different cultural traditions, different institutions, different linguistic and religious patterns. An understanding of both the differences and similarities between regions such as Anglo-America and Europe on the one hand, or Anglo-America and Eastern Asia on the other, will help you to understand the world around you. At the very least, an understanding of regional similarities and differences will help you to interpret what you read on the front page of your daily newspaper or view on the evening news report on your television set.

Not all of these themes will be immediately apparent on each of the maps and tables in this atlas. But if you study the contents of *Student Atlas of World Geography*, along with the reading of your text and think about the five themes, maps and tables and text will complement one another and improve your understanding of global geography.

Christopher J. Sutton

About the Author

Christopher J. Sutton is professor of geography at Western Illinois University. Born in Virginia and raised in Illinois, he received his bachelor's degree (1988) and master's degree (1991) in Geography from Western Illinois University. In 1995 he earned his Ph.D. in Geography from the University of Denver. He is the author of numerous research articles, educational materials, and co-authors the *Student Atlas of World Politics*. A broadly trained geographer, his areas of interest include cartographic design, cultural geography, and urban transportation. After teaching at Northwestern State University of Louisiana for three years, Dr. Sutton returned to Western Illinois University in 1998, serving as chair of the Department of Geography from 2002 to 2007. Dr. Sutton has served as president of the Illinois Geographical Society and in 2012 was recipient of the Society's Distinguished Geographer award.

Academic Advisory Board

Members of the Academic Advisory Board review maps for content, accuracy, and usefulness. We think you will find their careful consideration reflected in this edition.

Acknowledgments

The Student Atlas of World Geography was first conceived by Dr. John L. Allen, emeritus professor and chair of both the University of Connecticut and University of Wyoming departments of geography. A prolific author, he has authored numerous books, book chapters, and articles on Lewis and Clark and the exploration of the American West. Dr. Allen developed a series of highly popular atlases including the Student Atlas of World Politics, Atlas of Economic Development, Atlas of Environmental Issues, Atlas of Anthropology, and Atlas of World Events. He also served for more than two decades as editor of Annual Editions: Environment. Dr. Allen received the Outstanding Teaching Award from the University of Connecticut and was named the 2010 Fellow of the Society for the History of Discoveries. The current edition of Student Atlas of World Geography builds on the legacy established by Dr. Allen. Thousands of students have gained a greater understanding of the complex nature of our world through this atlas and it is with great appreciation that I thank Dr. Allen for recognizing of the importance of an atlas comprised mostly of thematic maps and his dedication to allow this valuable work to continue.

Christopher J. Sutton

Nozar Alaolmolki
Hiram College

Barbara Batterson-Rossi
Palomar College

A. Steele Becker
University of Nebraska at Kearney

Koop Berry
Walsh University

Mark Bjelland
Gustavus Adolphus College

Daniel A. Bunye
South Plains College

Winifred F. Caponigri
Holy Cross College

Femi Ferreira
Hutchinson Community College

Eric J. Fournier
Samford University

William J. Frazier
Columbus State College

Hari P. Garbharran
Middle Tennessee State University

Baher Gosheh
Edinboro University of Pennsylvania

Donald Hagan
Northwest Missouri State University

Robert Janiskee
University of South Carolina

David C. Johnson
University of Louisiana

Effie Jones
Crichton College

Cub Kahn
Marythurst University

Artimus Keiffer
Franklin College

Leonard E. Lancette
Mercer University

Donald W. Lovejoy
Palm Beach Atlantic College

Mark Maschhoff
Harris-Stowe State College

Richard Matthews
University of South Carolina

Madolia Mills
University of Colorado–Colorado Springs

Robert Mulcahy
Providence College

Otto H. Muller
Alfred University

Judith Otto
Framingham State University

J. Henry Owusu
University of Northern Iowa

Steven Parkansky
Morehead State University

William Preston
California Polytechnic State University, San Luis Obispo

Neil Reid
The University of Toledo

Amber Ruskell
Southeastern Community College

A. L. Rydant
Keene State College

Deborah Berman Santana
Mills College

Steven Slakey
University of La Verne

Rolf Sternberg
Montclair State University

Richard Ulack
University of Kentucky

David Woo
California State University, Haywood

Donald J. Zeigler
Old Dominion University

Table of Contents

Unit IV   Global Economic Patterns   67

Unit V   Global Patterns of Environmental Disturbance   93

Unit VI   Global Political Patterns   123

Unit VII   World Regions   161

Introduction: How to Read an Atlas

An atlas is a book containing maps which are "models" of the real world. By the term "model" we mean exactly what you think of when you think of a model: a representation of reality that is generalized, usually considerably smaller than the original, and with certain features emphasized, depending on the purpose of the model. A model of a car does not contain all of the parts of the original but it may contain enough parts that it is recognizable as a car and can be used to study principles of automotive design or maintenance. A car model designed for racing, on the other hand, may contain fewer parts but would have the mobility of a real automobile. Car models come in a wide variety of types containing almost anything you can think of relative to automobiles that doesn't require the presence of a full-size car. Because geographers deal with the real world, virtually all of the printed or published studies of that world require models. Unlike a mechanic in an automotive shop, we can't roll our study subject into the shop, take it apart, and put it back together. We must use models. In other words, we must generalize our subject, and the way we do that is by using maps. Some maps are designed to show specific geographic phenomena, such as the climates of the world or the relative rates of population growth for the world's countries. We call these maps "thematic maps" and Units I through VI of this atlas contain maps of this type. Other maps are designed to show the geographic location of towns and cities and rivers and lakes and mountain ranges and so on. These are called "reference maps" and they make up many of the maps in Unit VII. All of these maps, whether thematic or reference, are models of the real world that selectively emphasize the features that we want to show on the map.

In order to read maps effectively—in other words, in order to understand the models of the world presented in the following pages—it is important for you to know certain things about maps: how they are made using what are called *projections*; how the level of mathematical proportion of the map or what geographers call *scale* affects what you see; and how geographers use *generalization* techniques such as simplification and symbols where it would be impossible to draw a small version of the real world feature. In this brief introduction, then, we'll explain to you three of the most important elements of map interpretation: projection, scale, and generalization.

Map Projections

Perhaps the most basic problem in *cartography*, or the art and science of map-making, is the fact that the subject of maps—the earth's surface—is what is called by mathematicians "a non-developable surface." Because the world is a sphere (or nearly so—it's actually slightly flattened at the poles and bulges a tiny bit at the equator), it is impossible to flatten out the world or any part of its curved surface without producing some kind of distortion. This "near sphere" is represented by a geographic grid or coordinate system of lines of latitude or *parallels* that run east and west and are used to measure distance north and south on the globe, and lines of longitude or *meridians* that run north and south and are used to measure distance east and west. All the lines of longitude are half circles of equal length and they all converge at the poles. These meridians are numbered from 0 degrees (Prime or Greenwich Meridian) east and west to 180 degrees. The meridian of 0 degrees and the meridian of 180 degrees are halves of the same "great circle" or line representing a plane that bisects the globe into two equal hemispheres. All lines of longitude are halves of great circles. All the lines of latitude are complete circles that are parallel to one another and are spaced equidistant on the meridians. The circumference of these circles lessens as you move north or south from the equator. Parallels of latitude are numbered from 0 degrees at the equator north and south to 90 degrees at the North

The Coordinate System

and South poles. The only line of latitude that is a great circle is the equator, which equally divides the world into a northern and southern hemisphere. In the real world, all these grid lines of latitude and longitude intersect at right angles. The problem for cartographers is to convert this spherical or curved grid into a geometrical shape that is "developable"; that is, it can be flattened (such as a cylinder or cone) or is already flat (a plane). The reason the results of the conversion process are called "projections" is that we imagine a world globe (or some part of it) that is made up of wires running north–south and east–west to represent the grid lines of latitude and longitude and other wires or even solid curved plates to represent the coastlines of continents or the continents themselves. We then imagine a light source at some location inside or outside the wire globe that can "project" or cast shadows of the wires representing grid lines onto a developable surface. Sometimes the basic geometric principles of projection may be modified by other mathematical principles to yield projections that are not truly geometric but have certain desirable features. We call these types of projections "arbitrary." The three most basic types of projections are named according to the type of developable surface: cylindrical, conic, or azimuthal (plane). Each type has certain characteristic features: they may be *equal area* projections in which the size of each area on the map is a direct proportional representation of that same area in the real world but shapes are distorted; they may be *conformal* projections in which area may be distorted but shapes are shown correctly; or they may be *compromise* projections in which both shape and area are distorted but the overall picture presented is fairly close to reality. It is important to remember that all maps distort the geographic grid and continental outlines in characteristic ways. The only representation of the world that does not distort either shape or area is a globe. You can see why we must use projections—can you imagine an atlas that you would have to carry back and forth across campus that would be made up entirely of globes?

Cylindrical Projections

Cylindrical projections are drawn as if the geographic grid were projected onto a cylinder. Cylindrical projections have the advantage of having all lines of latitude as true parallels or straight lines. This makes these projections quite useful for showing geographic relationships in which latitude or distance north–south is important (many physical features, such as climate, are influenced by latitude). Unfortunately, most cylindrical-type projections distort area significantly. One of the most famous is the Mercator projection shown on the next page.

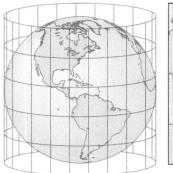 

Cylindrical Projection: The Mercator Projection

This projection makes areas disproportionately large as you move toward the pole, making Greenland, which is actually about one-seventh the size of South America, appear to be as large as the southern continent. But the Mercator projection has the quality of conformality: landmasses on the map are true in shape and thus all coastlines on the map intersect lines of latitude and longitude at the proper angles. This makes the Mercator projection, named after its inventor, a sixteenth-century Dutch cartographer, ideal for its original purpose as a tool for navigation—but not a good projection for attempting to show some geographical feature in which areal relationship is important. Unfortunately, the Mercator projection has often been used for wall maps for schoolrooms and the consequence is that generations of American school children have been "tricked" into thinking that Greenland is actually larger than South America. Much better cylindrical-type projections are those like the Robinson projection used in this atlas that is neither equal area nor conformal but a compromise that portrays the real world much as it actually looks, enough so that we can use it for areal comparisons.

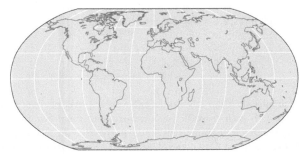

The Robinson Projection

Conic Projections

Conic projections are those that are imagined as being projected onto a cone that is tangent to the globe along a standard parallel, or a series of cones tangent along several parallels or even intersecting the

Conic Projection of Europe

globe. Conic projections usually show latitude as curved lines and longitude as straight lines. They are good projections for midlatitude regions with east–west extents or for areas with north–south extent, like the preceding map of Europe, and may be either conformal, equal area, or compromise, depending on how they are constructed. Many of the regional maps in the last map section of this atlas are conic projections.

Azimuthal Projections

Azimuthal projections are those that are imagined as being projected onto a plane or flat surface. They are named for one of their essential properties. An "azimuth" is a line of compass bearing and azimuthal projections have the property of yielding true compass directions from the center of the map. This makes azimuthal maps useful for navigation purposes, particularly air navigation. But, because they distort area and shape so greatly, they are seldom used for maps designed to show geographic relationships. When they are used as illustrative rather than navigation maps, it is often in the "polar case" projection shown above where the plane has been made tangent to the globe at the North Pole.

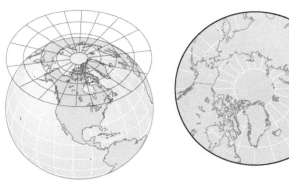

Azimuthal Projection of the North Polar Region

Map Scale

Because maps are models of the real world, it follows that they are not the same size as the real world or any portion of it. Every map, then, is subject to generalization, which is another way of saying that maps are drawn to certain scales. The term *scale* refers to the mathematical quality of *proportional representation,* and is expressed as a ratio between an area of the real world or the distance between places on the real world and the same area or distance on the map. We show map scale on maps in three different ways. Sometimes we simply use the proportion and write what is called a *natural scale* or representative fraction: for example, we might show on a map the mathematical proportion of 1:62,500. A map at this scale is one that is one sixty-two thousand five-hundredth the size of the same area in the real world. Other times we convert the proportion to a written description that approximates the relationship between distance on the map and distance in the real world. Because there are nearly 62,500 inches in a mile, we would refer to a map having a natural scale of 1:62,500 as having an "inch-mile" scale of "1 inch represents 1 mile." If we draw a line one inch long on this map, that line represents a distance of approximately one mile in the real world. Finally, we usually use a graphic or linear scale: a bar or line, often graduated into miles or kilometers, that shows graphically the proportional representation. A graphic scale for our 1:62,500 map might be about five inches long, divided into five equal units clearly labeled as "1 mile," "2 miles," and so on. Our examples on the following page show all three kinds of scales.

The most important thing to keep in mind about scale, and the reason why knowing map scale is important to being able to read a map correctly, is the relationship between proportional representation and generalization. A map that fills a page but shows the whole world is

Map 1 Small Scale Map of the United States

Map 2 Map of the Northeast

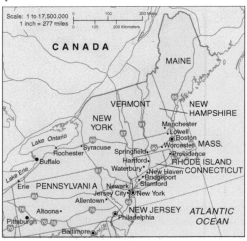

Map 3 Map of Southeastern New England

Map 4 Large Scale Map of Boston, MA

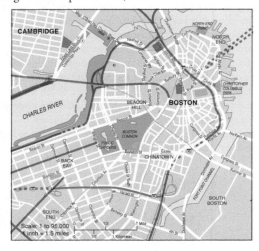

much more highly generalized than a map that fills a page but shows a single city. On the world map, the city may appear as a dot. On the city map, streets and other features may be clearly seen. We call the first map, the world map, a *small scale* map because the proportional representation is a small number. A page-size map showing the whole world may be drawn at a scale of 1:150,000,000. That is a very small number indeed—hence the term *small scale* map even though the area shown is large. Conversely, the second map, a city map, may be drawn at a scale of 1:250,000. That is still a very small number but it is a great deal larger than 1:150,000,000! And so we'd refer to the city map as a *large scale* map, even though it shows only a small area. On our world map, geographical features are generalized greatly and many features can't even be shown at all. On the city map, much less generalization occurs—we can show specific features that we couldn't on the world map—but generalization still takes place. The general rule is that the smaller the map scale, the greater the degree of generalization; the larger the map scale, the less the degree of generalization. The only map that would not generalize would be a map at a scale of 1:1 and that map wouldn't be very handy to use. Examine the relationship between scale and generalization in the four maps on this page.

Generalization on Maps

A review of the four maps on this page should give you some indication of how cartographers generalize on maps. One thing that you should have noticed is that the first map, that of the United States, is

much simpler than the other three and that the level of *simplification* decreases with each map. When a cartographer simplifies map data, information that is not important for the purposes of the map is just left off. For example, on Map 1 the objective may have been to show cities with more than 1 million in population. To do that clearly and effectively, it is not necessary to show and label rivers and lakes. The map has been simplified by leaving those items out. Map 4, on the other hand, is more complex and shows and labels geographic features that are important to the character of the city of Boston; therefore, the Charles River is clearly indicated on the map.

Another type of generalization is *classification*. Map 1 shows cities with more than 1 million in population. Map 2 shows cities of several different sizes and a different symbol is used for each size classification or category. Many of the thematic maps used in this atlas rely on classification to show data. A thematic map showing population growth rates (see Map 26) will use different colors to show growth rates in different classification levels or what are sometimes called *class intervals*. Thus, there will be one color applied to all countries with population growth rates between 1.0 percent and 1.4 percent, another color applied to all countries with population growth rates between 1.5 percent and 2.1 percent, and so on. Classification is necessary because it is impossible to find enough symbols or colors to represent precise values. Classification may also be used for qualitative data, such as the national or regional origin of migrating populations. Cartographers show both quantitative and qualitative classification levels or class intervals in important sections of maps called *legends*.

These legends, as in the samples that follow, make it possible for the reader of the map to interpret the patterns shown.

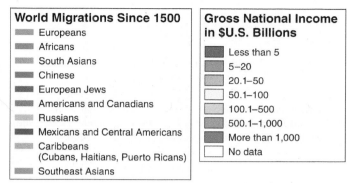

Map Legends from Maps 25 and 48

A third technique of generalization is *symbolization* and we've already noted several different kinds of symbols: those used to represent cities on the preceding maps, or the colors used to indicate population growth levels on Map 26. One general category of map symbols is quantitative in nature and this category can further be divided into a number of different types. For example, the symbols showing city size on Maps 1 and 2 on the preceding page can be categorized as *ordinal* in that they show relative differences in quantities (the size of cities). A cartographer might also use lines of different widths to express the quantities of movement of people or goods between two or more points as on Map 25.

The color symbols used to show rates of population growth can be categorized as *interval* in that they express certain levels of a mathematical quantity (the percentage of population growth). Interval symbols are often used to show physical geographic characteristics such as inches of precipitation, degrees of temperature, or elevation above sea level. The following sample, for example, shows precipitation (from Map 3a).

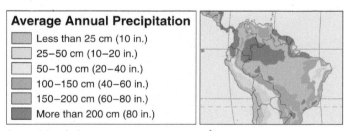

Interval Symbols

Still another type of mathematical symbolization is the *ratio* in which sets of mathematical quantities are compared: the number of persons per square mile (population density) or the growth in gross national product per capita (per person). The map below shows Gross National Income per capita (from Map 49).

Ratio Symbols

Finally, there are a vast number of cartographic symbols that are not mathematical but show differences in the kind of information being portrayed. These symbols are called *nominal* and they range from the simplest differences such as land and water to more complex differences such as those between different types of vegetation. Shapes or patterns or colors or iconographic drawings may all be used as nominal symbols on maps. The following map (from Map 8) uses color to show the distribution of soil types.

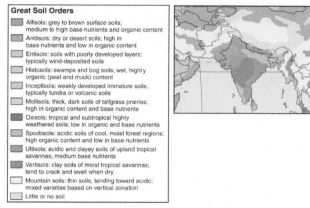

Nominal Symbols

The final technique of generalization is what cartographers refer to as *interpolation*. Here, the maker of a map may actually show more information on the map than is actually supplied by the original data. In understanding the process of interpolation it is necessary for you to visualize the quantitative data shown on maps as being three dimensional: x values provide geographic location along a north–south axis of the map; y values provide geographic location along the east–west axis of the map; and z values are those values of whatever data (for example, temperature) are being shown on the map at specific points. We all can imagine a real three-dimensional surface in which the x and y values are directions and the z values are the heights of mountains and the depths of valleys. On a topographic map showing a real three-dimensional surface, contour lines are used to connect points of equal elevation above sea level. These contour lines are not measured directly; they are estimated by interpolation on the basis of the elevation points that are provided.

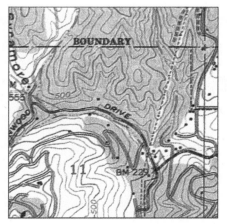

Interpolation

It is harder to imagine the statistical surface of a temperature map in which the *x* and *y* values are directions and the *z* values represent degrees of temperature at precise points. But that is just what cartographers do. And to obtain the values between two or more specific points where *z* values exist, they interpolate based on a class interval they have decided is appropriate and use *isolines* (which are statistical equivalents of a contour line) to show increases or decreases in value. The diagram to the right shows an example of an interpolation process. Occasionally interpolation is referred to as *induction*. By whatever name, it is one of the most difficult parts of the cartographic process.

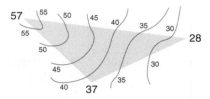

Degrees of Temperature (Celsius)
Interval = 5 degrees

And you thought all you had to do to read an atlas was look at the maps! You've now learned that it is a bit more involved than that. As you read and study this atlas, keep in mind the principles of projection and scale and generalization (including simplification, classification, symbolization, and interpolation) and you'll do just fine. Good luck and enjoy your study of the world of maps as well as maps of the world!

Unit I

Global Physical Patterns

Map 1 World Political Boundaries

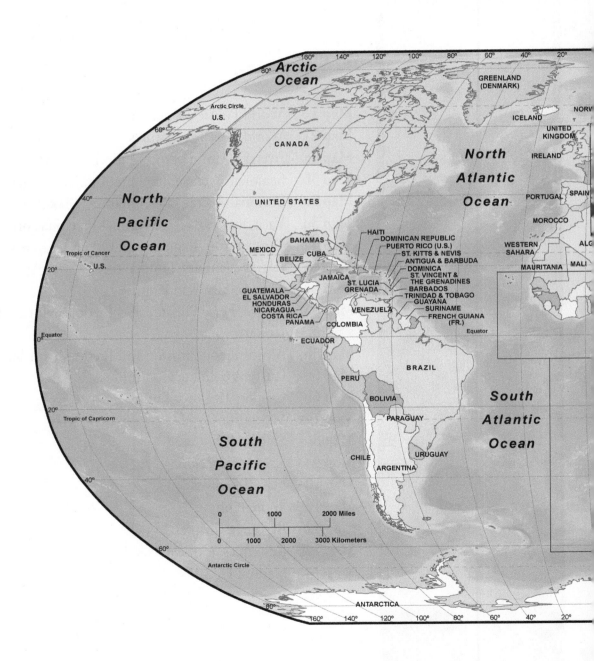

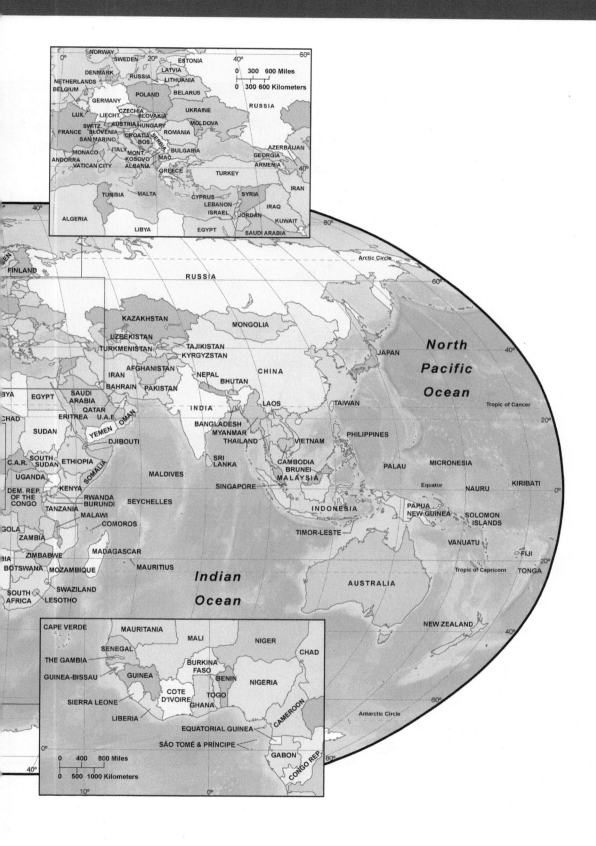

Map 2   World Physical Features

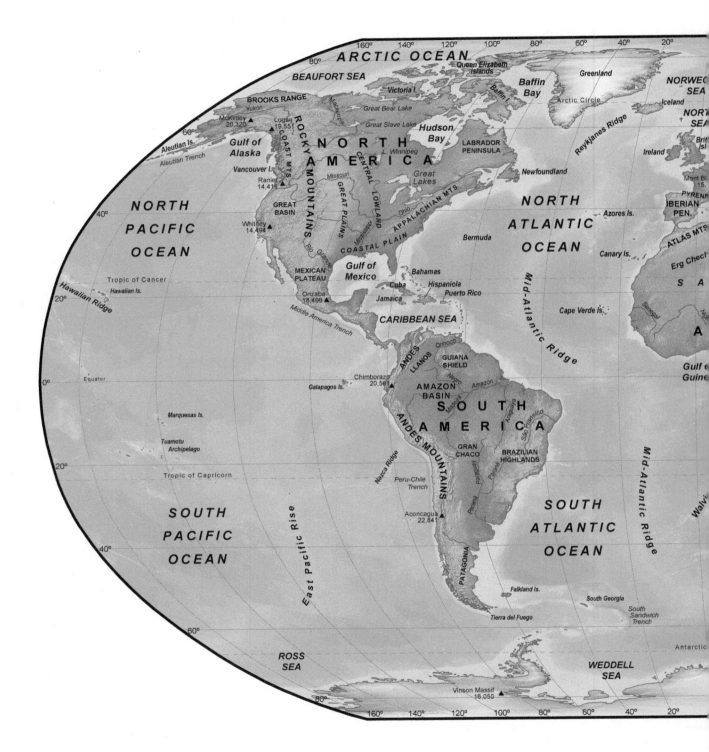

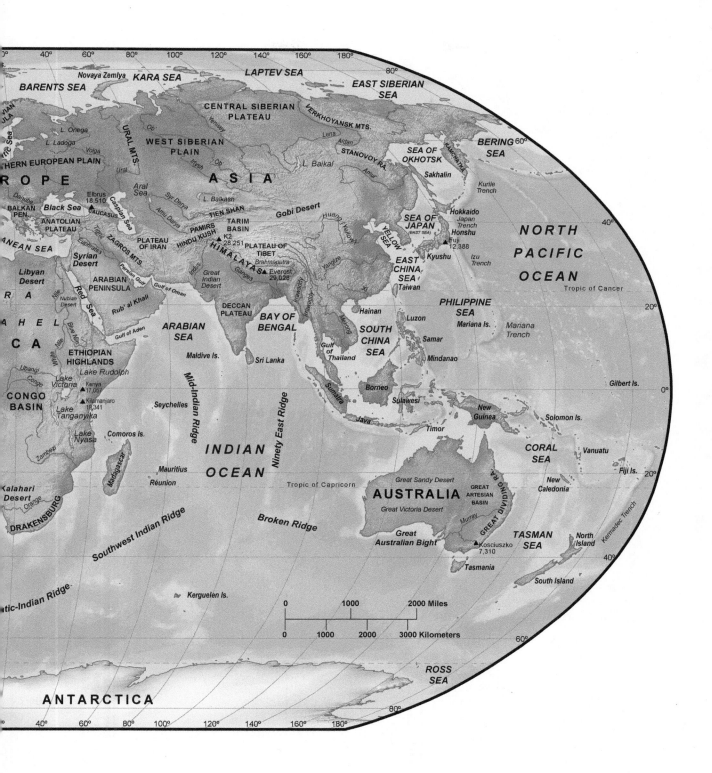

Map 3a   Average Annual Precipitation

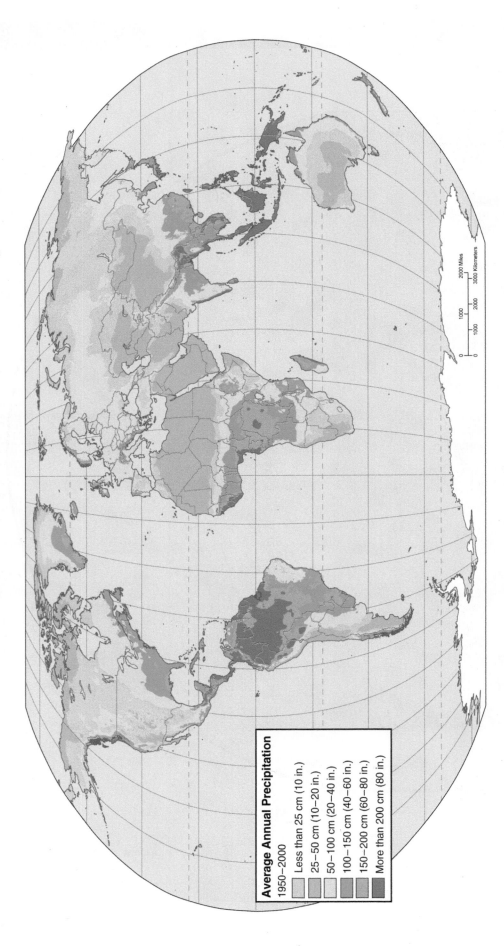

Average Annual Precipitation
1950–2000

- Less than 25 cm (10 in.)
- 25–50 cm (10–20 in.)
- 50–100 cm (20–40 in.)
- 100–150 cm (40–60 in.)
- 150–200 cm (60–80 in.)
- More than 200 cm (80 in.)

The two most important physical geographic variables are precipitation and temperature, the essential elements of weather and climate. Precipitation is a conditioner of both soil type and vegetation. More than any other single environmental element, it influences where people do or do not live. Water is the most precious resource available to humans, and water availability is largely a function of precipitation. Water availability is also a function of several precipitation variables that do not appear on this map: the seasonal distribution of precipitation (is precipitation or drought concentrated in a particular season?), the ratio between precipitation and temperature (how

much of the water that comes to the earth in the form of precipitation is lost through mechanisms such as evaporation and transpiration that are a function of temperature?), and the annual variability of precipitation (how much do annual precipitation totals for a place or region tend to vary from the "normal" or average precipitation?). In order to obtain a complete understanding of precipitation, these variables should be examined along with the more general data presented on this map. The study of precipitation and other climatic elements is the concern of the branch of physical geography called "climatology."

Map 3b   Seasonal Average Precipitation, November through April

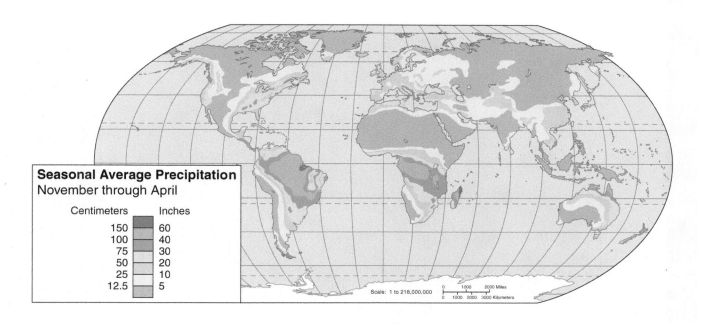

Seasonal Average Precipitation
November through April

Centimeters		Inches
150		60
100		40
75		30
50		20
25		10
12.5		5

Scale: 1 to 218,000,000

0   1000   2000 Miles
0   1000   2000   3000 Kilometers

Map 3c   Seasonal Average Precipitation, May through October

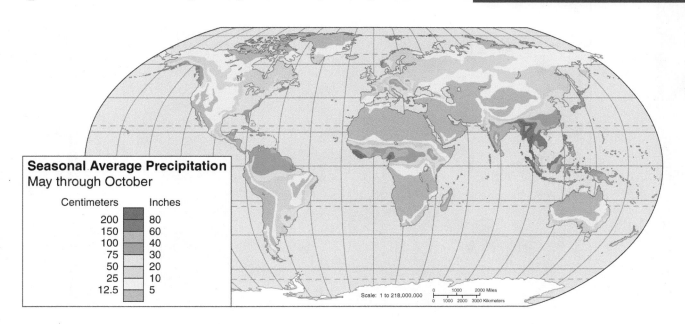

Seasonal Average Precipitation
May through October

Centimeters		Inches
200		80
150		60
100		40
75		30
50		20
25		10
12.5		5

Scale: 1 to 218,000,000

0   1000   2000 Miles
0   1000   2000   3000 Kilometers

Seasonal average precipitation is nearly as important as annual precipitation totals in determining the habitability of an area. Critical factors are such things as whether precipitation coincides with the growing season and thus facilitates agriculture or during the winter when it is less effective in aiding plant growth, and whether precipitation occurs during summer with its higher water loss through evaporation and transpiration or during the winter when more of it can go into storage. Several of the world's great climate zones have pronounced seasonal precipitation rhythms. The tropical and subtropical savanna grasslands have a long winter dry season and abundant precipitation in the summer. The Mediterranean climate is the only major climate with a marked dry season during the summer, making agriculture possible only through irrigation or other adjustments to cope with drought during the period of plant growth. And the great monsoon climates of South and Southeast Asia have their winter dry season and summer rain that have conditioned the development of Asian agriculture and the rhythms of Asian life.

Map 3d Variation in Average Annual Precipitation

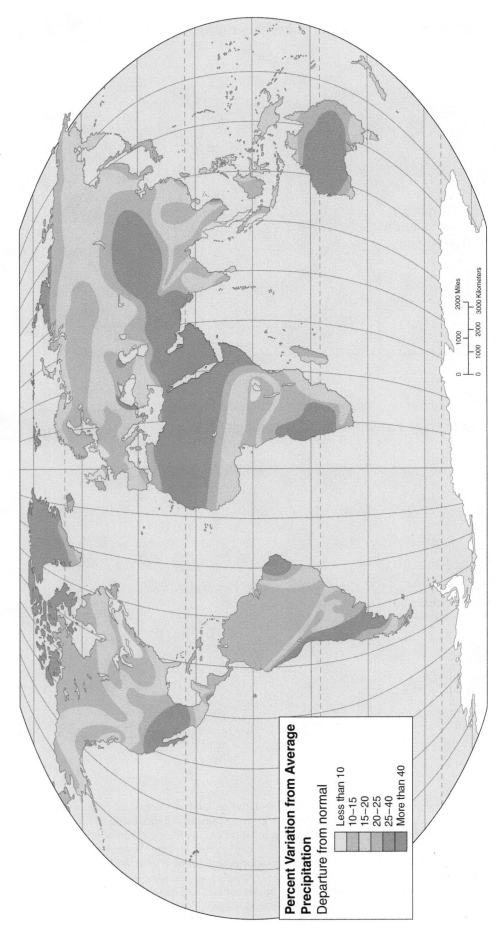

Percent Variation from Average Precipitation

Departure from normal

- Less than 10
- 10–15
- 15–20
- 20–25
- 25–40
- More than 40

0	1000	2000 Miles	
0	1000	2000	3000 Kilometers

Although annual precipitation totals and seasonal distribution of precipitation are important variables, the variability of precipitation from one year to the next may be even more critical. You will note from the map that there is a general spatial correlation between the world's drylands and the amount of annual variation in precipitation. Generally, the drier the climate, the more likely it is that there will be considerable differences in rainfall and/or snowfall from one year to the next. We might determine that the average precipitation of the mid-Sahara is 2 inches per year. What this really means is that a particular location in the Sahara during one year might receive 0.5", during the next year 3.5", and during a third year 2". If you add these together and divide by the number of years, the "average" precipitation is 2" per year. The significance of this is that much of the world's crucial agricultural output of cereals (grains) comes from dryland climates (the Great Plains of the United States, the Pampas of Argentina, the steppes of Ukraine and Russia, for example), and variations in annual rainfall totals can have significant impacts on levels of grain production and, therefore, important consequences for both economic and political processes.

Map 4a   Temperature Regions and Ocean Currents

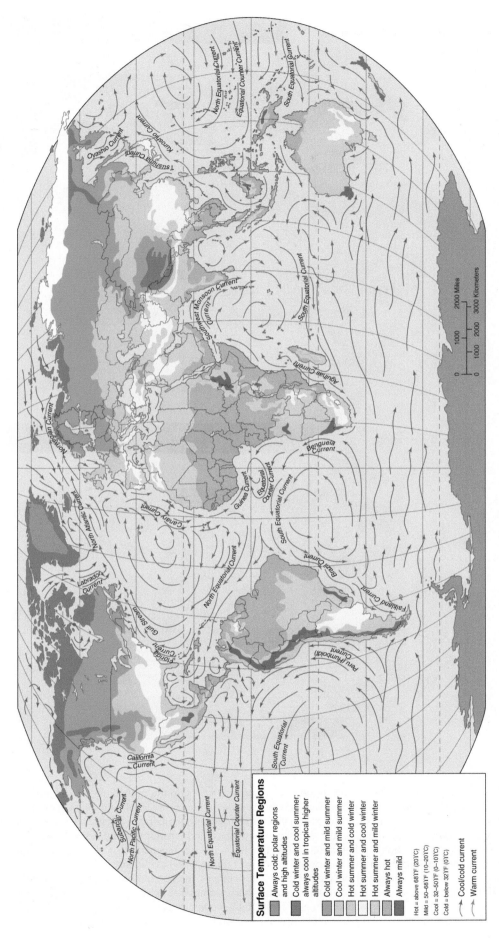

Surface Temperature Regions

- Always cold: polar regions and high altitudes
- Cold winter and cool summer; always cool in tropical higher altitudes
- Cold winter and mild summer
- Cool winter and mild summer
- Hot summer and cold winter
- Hot summer and cool winter
- Hot summer and mild winter
- Always hot
- Always mild

Hot = above 68°F (20°C)
Mild = 50–68°F (10–20°C)
Cool = 32–50°F (0–10°C)
Cold = below 32°F (0°C)

⟶ Cool/cold current
⟶ Warm current

Along with precipitation, temperature is one of the two most important environmental variables, defining the climate conditions so essential for the distribution of the human activities as agriculture and the distribution of the human population. The seasonal rhythm of temperature, including such measures as the average annual temperature range (difference between the average temperature of the warmest month and that of the coldest month), is an additional variable not shown on the map but,

like the seasonality of precipitation, should be a part of any comprehensive study of climate. The ocean currents illustrated exert a significant influence over the climate of adjacent regions and are the most important mechanism for redistributing surplus heat from the equatorial region into middle and high latitudes. Physical geographers known as "climatologists" study the phenomenon of temperature and related climatic characteristics.

Map 4b Average January Temperature

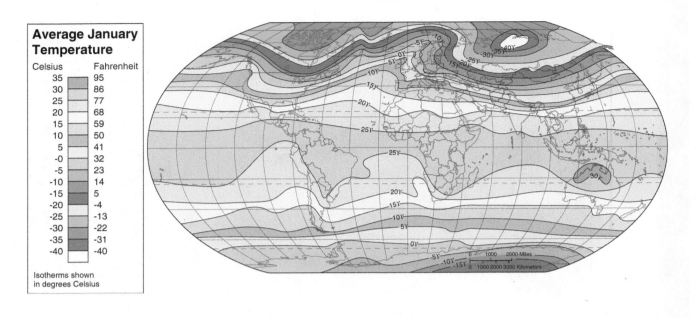

Average January Temperature

Celsius	Fahrenheit
35	95
30	86
25	77
20	68
15	59
10	50
5	41
-0	32
-5	23
-10	14
-15	5
-20	-4
-25	-13
-30	-22
-35	-31
-40	-40

Isotherms shown in degrees Celsius

Map 4c Average July Temperature

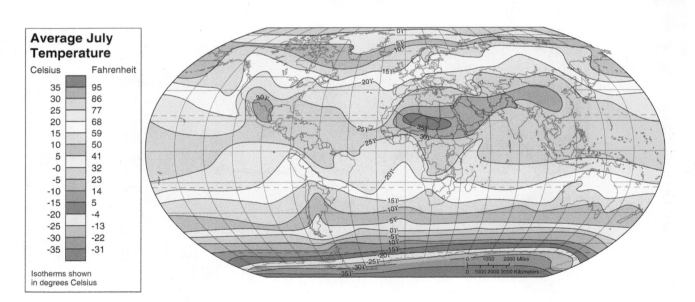

Average July Temperature

Celsius	Fahrenheit
35	95
30	86
25	77
20	68
15	59
10	50
5	41
-0	32
-5	23
-10	14
-15	5
-20	-4
-25	-13
-30	-22
-35	-31

Isotherms shown in degrees Celsius

Where moisture availability tends to mark the seasons in the tropics and subtropics, in the midlatitudes, seasons are marked by temperature. Temperature is determined by latitudinal transition, by altitude or elevation above sea level, and by location of a place relative to the world's landmasses and oceans. The most important of these controls is latitude, and temperatures generally become lower with increasing latitude. Proximity to water, however, tends to moderate temperature extremes, and "maritime" climates influenced by the oceans will be warmer in the winter and cooler in the summer than continental climates in the same general latitude. Maritime climates will also show smaller temperature ranges, the difference between January and July temperatures, whereas climates of the continental interiors, far from the moderating influences of the oceans, will tend to have greater temperature ranges. In the Northern Hemisphere, where there are both large landmasses and oceans, the range is great. But in the Southern Hemisphere, dominated by water and, hence, by the more moderate maritime air masses, the temperature range is comparatively small. Significant temperature departures from the "normal" produced by latitude may also be the result of elevation. With exceptions, lower temperatures produced by topography are difficult to see on maps of this scale.

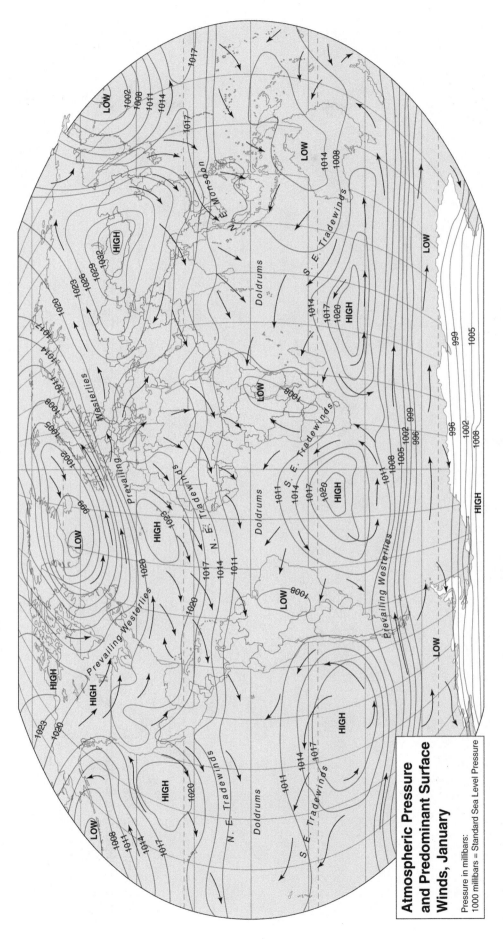

**Atmospheric Pressure
and Predominant Surface
Winds, January**

Pressure in millibars:
1000 millibars = Standard Sea Level Pressure

Map 5b Atmospheric Pressure and Predominant Surface Winds, July

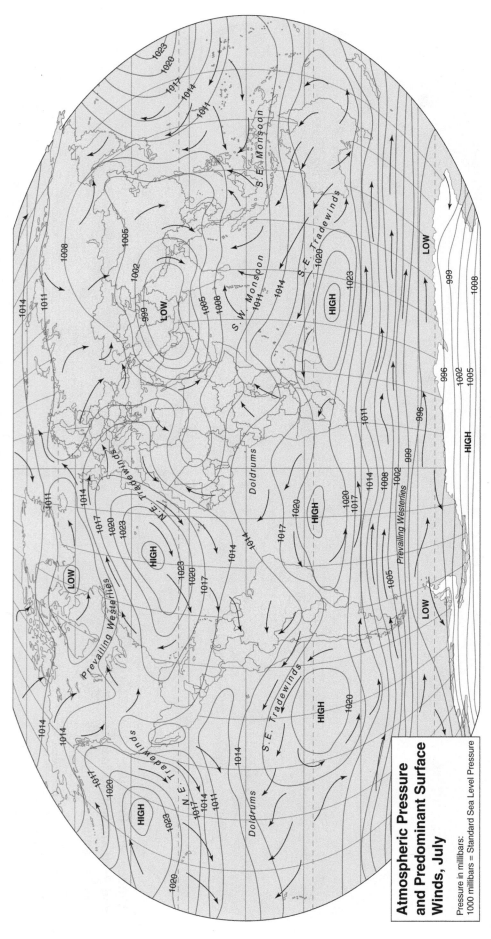

Atmospheric Pressure and Predominant Surface Winds, July

Pressure in millibars:
1000 millibars = Standard Sea Level Pressure

Atmospheric pressure, or the density of air, is a function largely of air temperature: the colder the air, the denser and heavier it is, hence the higher its pressure; the warmer the air, the lighter and less stable it is, hence the lower its pressure. Global pressure systems are the alternating low and high pressure systems that, from the equator north and south, include: the equatorial low (sometimes called the intertropical convergence) centered on the equator for much of the year; the subtropical highs with their centers near the 30th parallel of north and south latitude; the subpolar lows or polar fronts centered near the 60th parallel of north and south latitude; and the polar highs near the north and south poles. Air flows from high pressure to low pressure regions, and this air flow constitutes the earth's major surface winds such as the tropical tradewinds and the prevailing westerlies. This flow of air is one of the chief mechanisms by which surplus heat energy from the equatorial region is redistributed to higher latitudes. It is also the primary conditioner of the world's major precipitation belts, with rainfall and snowfall associated primarily with lower atmospheric pressure conditions.

Map 6a   Current Climate Conditions

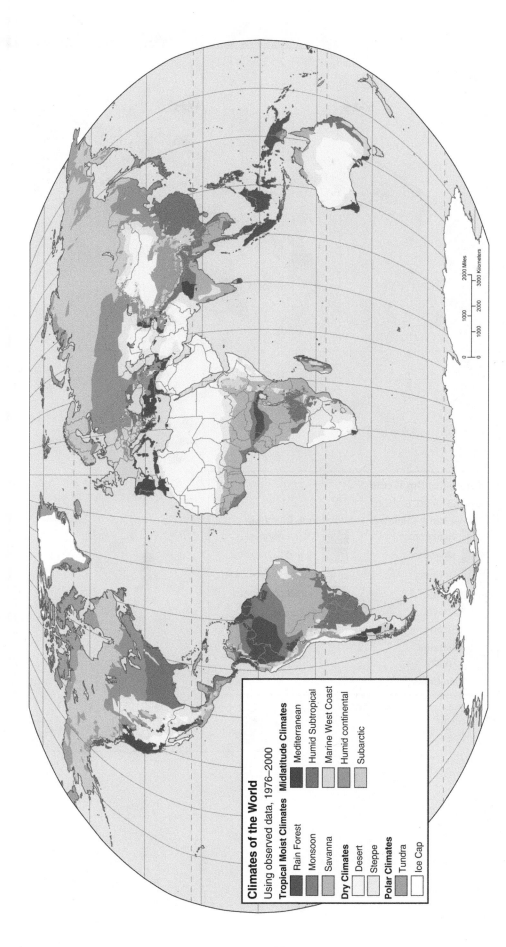

Climates of the World

Using observed data, 1976–2000

Tropical Moist Climates
- Rain Forest
- Monsoon
- Savanna

Dry Climates
- Desert
- Steppe

Midlatitude Climates
- Mediterranean
- Humid Subtropical
- Marine West Coast
- Humid continental
- Subarctic

Polar Climates
- Tundra
- Ice Cap

0   1000   2000 Miles
0   1000   2000   3000 Kilometers

Of the world's many patterns of physical geography, climate or the long-term average of weather conditions such as temperature and precipitation is the most important. It is climate that conditions the distribution of natural vegetation and the types of soils that will exist in an area. Climate also influences the availability of our most crucial resource: water. From an economic standpoint, the world's most important activity is agriculture; no other element of physical geography is more important for agriculture than climate. Ultimately, it is agricultural production that determines where the bulk of human beings live, and therefore, climate is a basic determinant of the distribution of human populations as well. The study of climates or "climatology" is one of the most important branches of physical geography.

The climate classification system shown on this map represents current climate conditions and is based on that developed by Wladimir Köppen. To establish his climate regions, Köppen used the climatic parameters of *precipitation, temperature,* and *evapotranspiration* as they impacted certain kinds of major vegetative associations. Hence the names for many of the climate regions are also the names of vegetative regions.

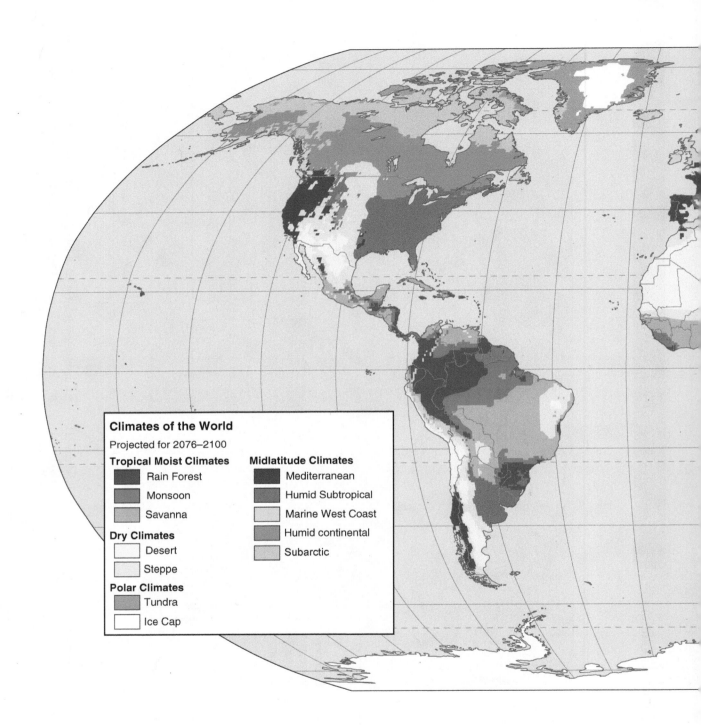

Climates of the World

Projected for 2076–2100

Tropical Moist Climates

- Rain Forest
- Monsoon
- Savanna

Dry Climates

- Desert
- Steppe

Polar Climates

- Tundra
- Ice Cap

Midlatitude Climates

- Mediterranean
- Humid Subtropical
- Marine West Coast
- Humid continental
- Subarctic

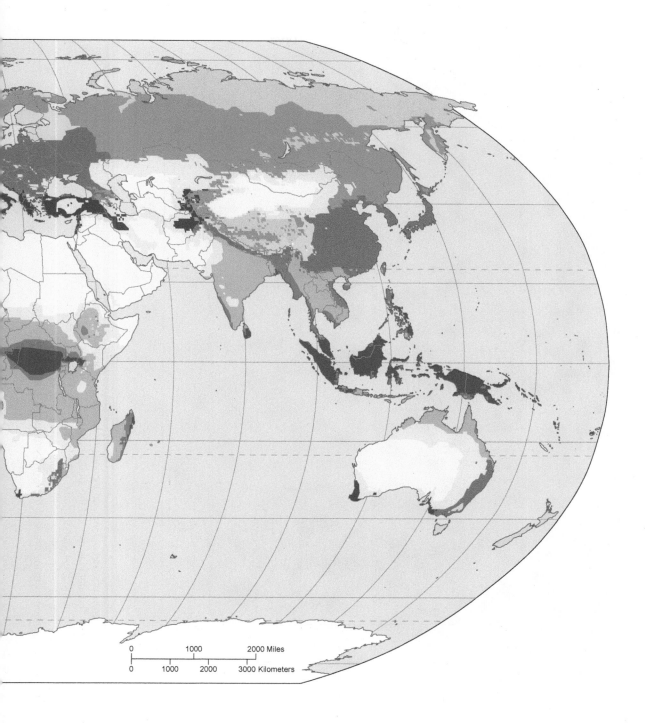

0   1000   2000 Miles

0   1000   2000   3000 Kilometers

Map 7 Vegetation Types

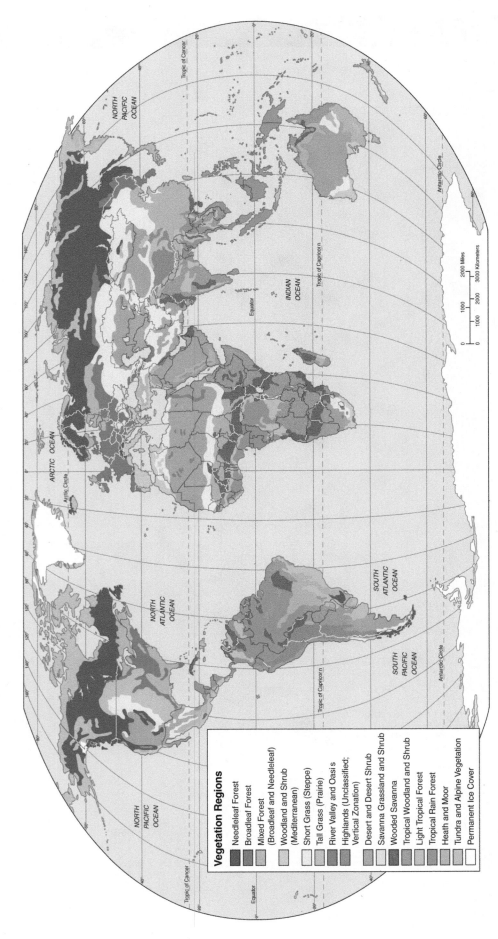

Vegetation Regions

- Needleleaf Forest
- Broadleaf Forest
- Mixed Forest (Broadleaf and Needleleaf)
- Woodland and Shrub (Mediterranean)
- Short Grass (Steppe)
- Tall Grass (Prairie)
- River Valley and Oasis
- Highlands (Unclassified; Vertical Zonation)
- Desert and Desert Shrub
- Savanna Grassland and Shrub
- Wooded Savanna
- Tropical Woodland and Shrub
- Light Tropical Forest
- Tropical Rain Forest
- Heath and Moor
- Tundra and Alpine Vegetation
- Permanent Ice Cover

Vegetation is the most visible consequence of the distribution of temperature and precipitation. The global pattern of vegetative types or "habitat classes" and the global pattern of climate are closely related and make up one of the great global spatial correlations. But not all vegetation types are the consequence of temperature and precipitation or other climatic variables. Many types of vegetation in many areas of the world are the consequence of human activities, particularly the grazing of domesticated livestock, burning, and forest clearance. This map shows the pattern of natural or "potential" vegetation, or vegetation as it might be expected to exist without significant human influences, rather than the actual vegetation that results from a combination of environmental and human factors.

-16-

Map 8 Soil Orders

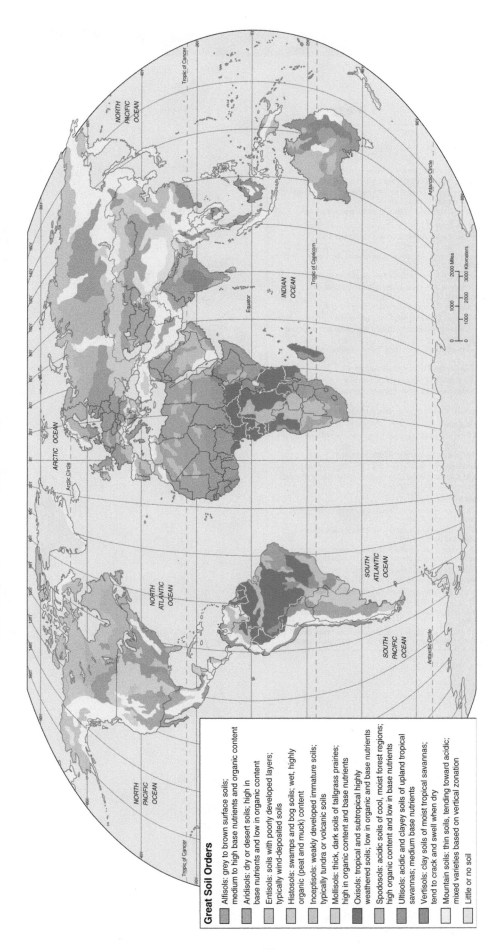

Great Soil Orders

- Alfisols: grey to brown surface soils; medium to high base nutrients and organic content
- Aridisols: dry or desert soils; high in base nutrients and low in organic content
- Entisols: soils with poorly developed layers; typically wind-deposited soils
- Histosols: swamps and bog soils; wet, highly organic (peat and muck) content
- Inceptisols: weakly developed immature soils; typically tundra or volcanic soils
- Mollisols: thick, dark soils of tallgrass prairies; high in organic content and base nutrients
- Oxisols: tropical and subtropical highly weathered soils; low in organic and base nutrients
- Spodosols: acidic soils of cool, moist forest regions; high organic content and low in base nutrients
- Ultisols: acidic and clayey soils of upland tropical savannas; medium base nutrients
- Vertisols: clay soils of moist tropical savannas; tend to crack and swell when dry
- Mountain soils: thin soils, tending toward acidic; mixed varieties based on vertical zonation
- Little or no soil

The characteristics of soil are one of the three primary physical geographic factors, along with climate and vegetation, that determine the habitability of regions for humans. In particular, soils influence the kinds of agricultural uses to which land is put. Because soils support the plants that are the primary producers of all food in the terrestrial food chain, their characteristics are crucial to the health and stability of ecosystems. Two types of soil are shown on this map: zonal soils, the characteristics of which are based on climatic patterns; and azonal soils, such as alluvial (water-deposited) or aeolian (wind-deposited)

soils, the characteristics of which are derived from forces other than climate. However, many of the azonal soils, particularly those dependent upon drainage conditions, appear over areas too small to be readily shown on a map of this scale. Thus, almost none of the world's swamp or bog soils appear on this map. People who study the geographic characteristics of soils are most often called "soil scientists," a discipline closely related to that branch of physical geography called "geomorphology."

-17-

Map 9 Ecological Regions

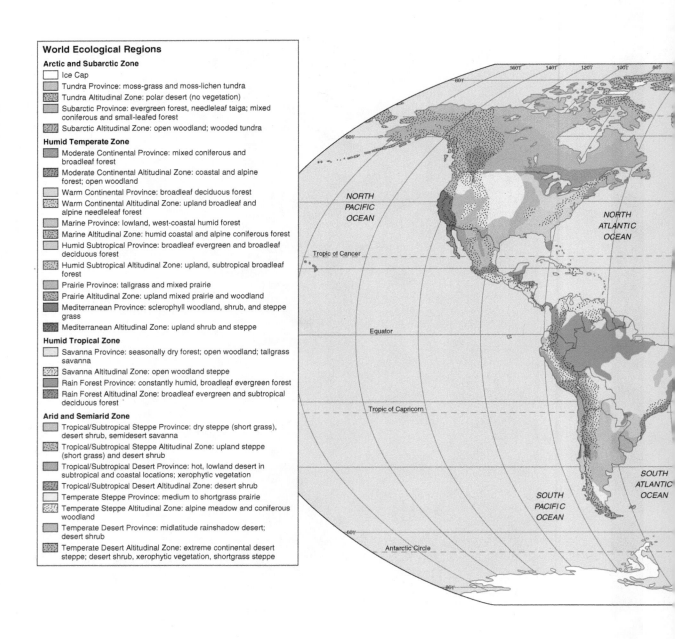

World Ecological Regions

Arctic and Subarctic Zone
- Ice Cap
- Tundra Province: moss-grass and moss-lichen tundra
- Tundra Altitudinal Zone: polar desert (no vegetation)
- Subarctic Province: evergreen forest, needleleaf taiga; mixed coniferous and small-leafed forest
- Subarctic Altitudinal Zone: open woodland; wooded tundra

Humid Temperate Zone
- Moderate Continental Province: mixed coniferous and broadleaf forest
- Moderate Continental Altitudinal Zone: coastal and alpine forest; open woodland
- Warm Continental Province: broadleaf deciduous forest
- Warm Continental Altitudinal Zone: upland broadleaf and alpine needleleaf forest
- Marine Province: lowland, west-coastal humid forest
- Marine Altitudinal Zone: humid coastal and alpine coniferous forest
- Humid Subtropical Province: broadleaf evergreen and broadleaf deciduous forest
- Humid Subtropical Altitudinal Zone: upland, subtropical broadleaf forest
- Prairie Province: tallgrass and mixed prairie
- Prairie Altitudinal Zone: upland mixed prairie and woodland
- Mediterranean Province: sclerophyll woodland, shrub, and steppe grass
- Mediterranean Altitudinal Zone: upland shrub and steppe

Humid Tropical Zone
- Savanna Province: seasonally dry forest; open woodland; tallgrass savanna
- Savanna Altitudinal Zone: open woodland steppe
- Rain Forest Province: constantly humid, broadleaf evergreen forest
- Rain Forest Altitudinal Zone: broadleaf evergreen and subtropical deciduous forest

Arid and Semiarid Zone
- Tropical/Subtropical Steppe Province: dry steppe (short grass), desert shrub, semidesert savanna
- Tropical/Subtropical Steppe Altitudinal Zone: upland steppe (short grass) and desert shrub
- Tropical/Subtropical Desert Province: hot, lowland desert in subtropical and coastal locations; xerophytic vegetation
- Tropical/Subtropical Desert Altitudinal Zone: desert shrub
- Temperate Steppe Province: medium to shortgrass prairie
- Temperate Steppe Altitudinal Zone: alpine meadow and coniferous woodland
- Temperate Desert Province: midlatitude rainshadow desert; desert shrub
- Temperate Desert Altitudinal Zone: extreme continental desert steppe; desert shrub, xerophytic vegetation, shortgrass steppe

Ecological regions are distinctive areas within which unique sets of organisms and environments are found. We call the study of the relationships between organisms and their environmental surroundings "ecology." Within each of the ecological regions portrayed on the map, a particular combination of vegetation, wildlife, soil, water, climate, and terrain defines that region's habitability, or ability to support life, including human life. Like climate and landforms, ecological relationships are crucial to the existence of agriculture, the

-18-

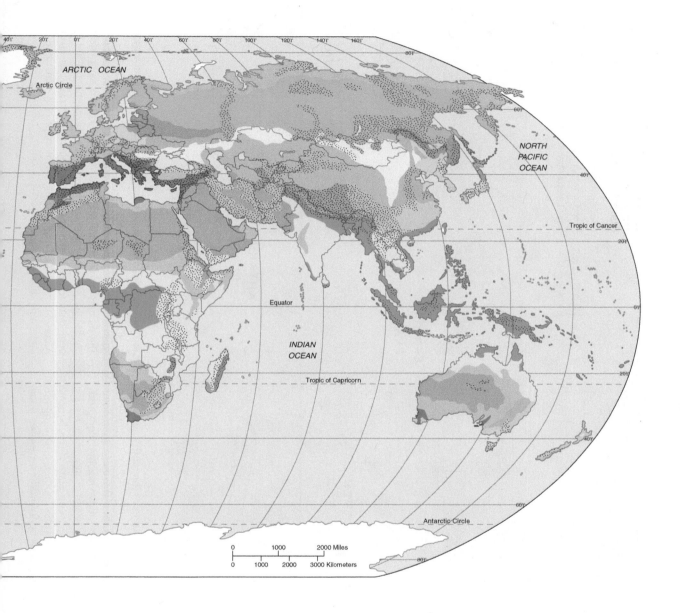

most basic of our economic activities, and important for many other kinds of economic activity as well. Biogeographers are especially concerned with the concept of ecological regions because such regions so clearly depend upon the geographic distribution of plants and animals in their environmental settings.

Map 10   Plate Tectonics

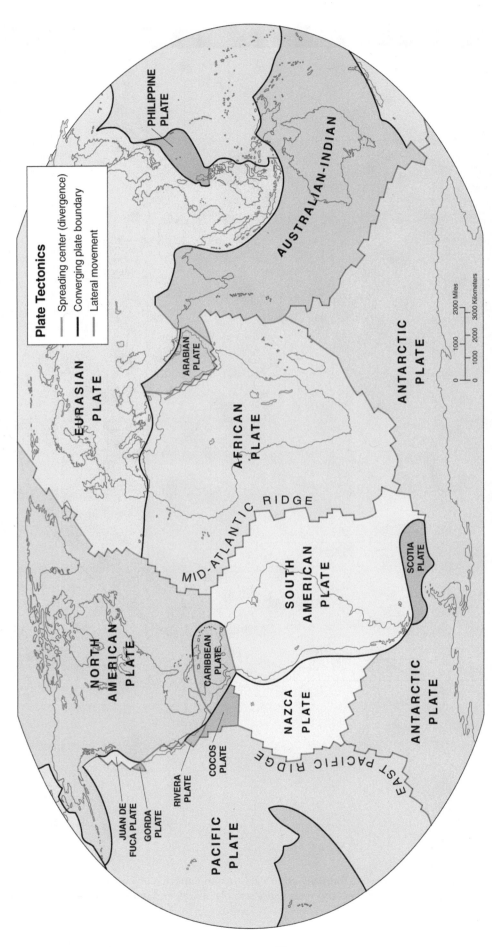

Plate Tectonics

— Spreading center (divergence)
— Converging plate boundary
— Lateral movement

PHILIPPINE PLATE

AUSTRALIAN-INDIAN

EURASIAN PLATE

ARABIAN PLATE

AFRICAN PLATE

ANTARCTIC PLATE

NORTH AMERICAN PLATE

MID-ATLANTIC RIDGE

SOUTH AMERICAN PLATE

SCOTIA PLATE

CARIBBEAN PLATE

JUAN DE FUCA PLATE

GORDA PLATE

RIVERA PLATE

COCOS PLATE

NAZCA PLATE

PACIFIC PLATE

EAST PACIFIC RIDGE

ANTARCTIC PLATE

0   1000   2000   3000 Kilometers
0   1000   2000 Miles

An understanding of the forces that shape the primary features of the earth's surface—the continents and ocean basins—requires a view of the earth's crust as fragments or "lithospheric plates" that shift position relative to one another. There are three dominant types of plate movement: *convergence*, in which plates move together, compressing former ocean floor or continental rocks together to produce mountain ranges, or producing mountain ranges through volcanic activity if one plate slides beneath another; *divergence*, in which the plates move away from one another, producing rifts in the earth's crust through which molten material wells up to produce new sea floors and mid-oceanic ridges; and *lateral shift*, in which plates move horizontally relative to one another, causing significant earthquake activity. All the major forms of these types of shifts are extremely slow and take place over long periods of geologic time. The movement of crustal plates, or what is known as "plate tectonics," is responsible for the present shape and location of the continents but is also the driving force behind some much shorter-term earth phenomena such as earthquakes and volcanoes. A comparison of the map of plates with maps of hazards and terrain will reveal some interesting relationships.

Map 11   Topography

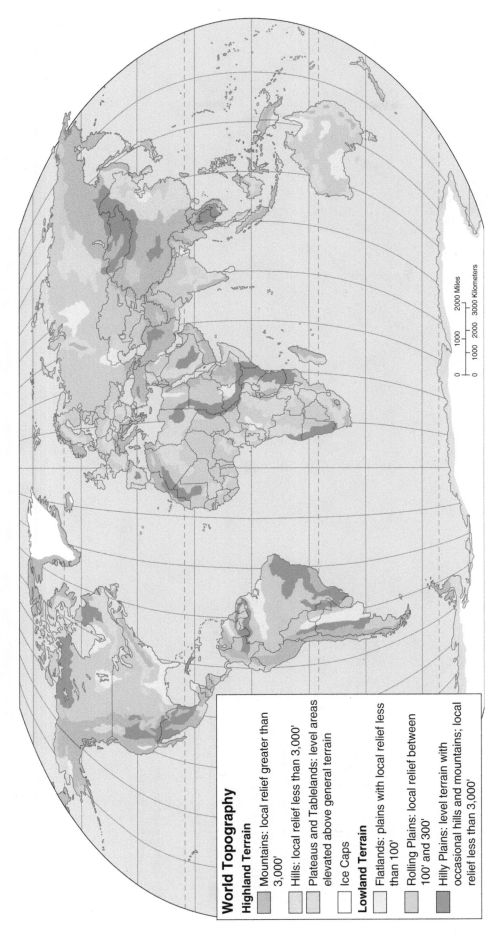

World Topography

Highland Terrain

- Mountains: local relief greater than 3,000'
- Hills: local relief less than 3,000'
- Plateaus and Tablelands: level areas elevated above general terrain
- Ice Caps

Lowland Terrain

- Flatlands: plains with local relief less than 100'
- Rolling Plains: local relief between 100' and 300'
- Hilly Plains: level terrain with occasional hills and mountains; local relief less than 3,000'

0   1000   2000 Miles

0   1000   2000   3000 Kilometers

Topography or terrain, also called "landforms," is second only to climate as a conditioner of human activity, particularly agriculture but also the location of cities and industry. A comparison of this map of mountains, valleys, plains, plateaus, and other features of the earth's surface with a map of land use (Map 16) shows that most of the world's productive agricultural zones are located in lowland and relatively level regions. Where large regions of agricultural productivity are found, we also tend to find urban concentrations and, with cities, we find industry. There is also a good spatial correlation between the map of topography and the map showing the distribution and density of the human population (Map 15). Normally, the world's major landforms

are the result of extremely gradual primary geologic activity such as the long-term movement of crustal plates. This activity occurs over hundreds of millions of years. Also important is the more rapid (but still slow by human standards) geomorphological or erosional activity of water, wind, glacial ice, and waves, tides, and currents. Some landforms may be produced by abrupt or "cataclysmic" events such as a major volcanic eruption or a meteor strike, but such events are relatively rare and their effects are usually too minor to show up on a map of this scale. The study of the processes that shape topography is known as "geomorphology" and is an important branch of physical geography.

Map 12a   Resources: Mineral Fuels

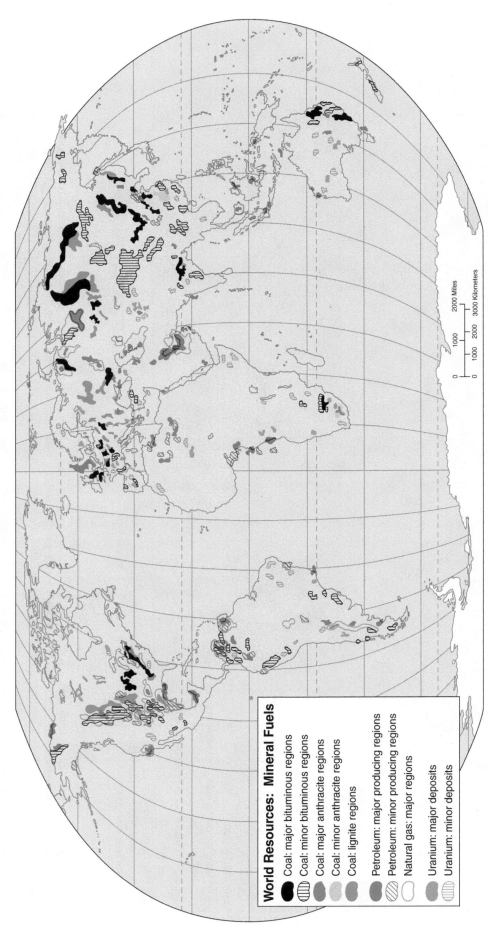

World Resources: Mineral Fuels

- Coal: major bituminous regions
- Coal: minor bituminous regions
- Coal: major anthracite regions
- Coal: minor anthracite regions
- Coal: lignite regions

- Petroleum: major producing regions
- Petroleum: minor producing regions
- Natural gas: major regions

- Uranium: major deposits
- Uranium: minor deposits

0   1000   2000 Miles
0   1000   2000   3000 Kilometers

The extraction and transportation of mineral fuels rank with agriculture and forestry as "primary" human activities that impact the environment on a global scale. Nearly all of the most highly publicized environmental disasters of recent decades—the Gulf of Mexico oil spill or the Chernobyl nuclear accident, for example—have involved mineral fuels that were being stored, transported, or used. And the continuing extraction of mineral fuels such as oil, natural gas, coal, and uranium produces high levels of atmospheric, soil, and water pollution. The location of mineral fuels tells us a great deal about where

environmental degradation is likely to be occurring or to occur in the future. One need only look at the levels of atmospheric pollution and vegetative disruption in central and eastern Europe to recognize the damaging consequences of heavy reliance on coal as a domestic and industrial fuel. The location of mineral fuels also tells us something about existing or potential levels of economic development with those countries possessing abundant reserves of mineral fuels having more of a chance to maintain or attain higher levels of prosperity.

Map 12b   Resources: Critical Metals

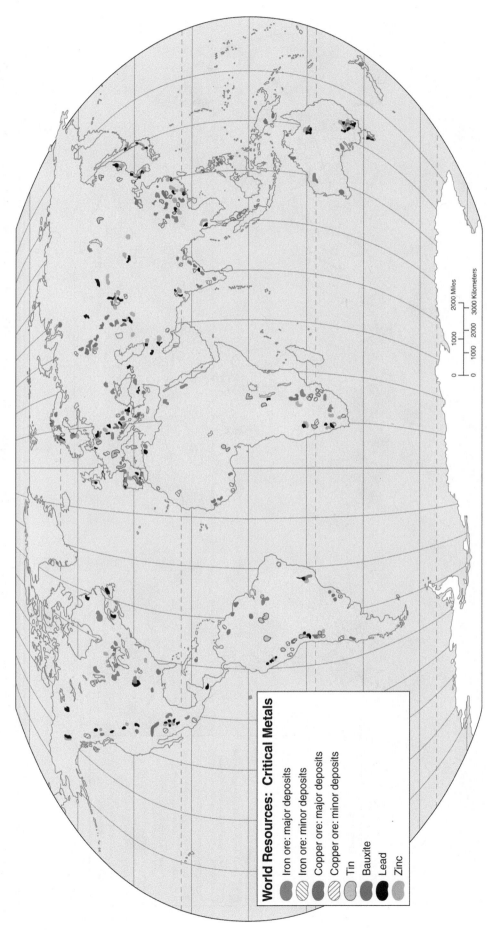

World Resources: Critical Metals

- Iron ore: major deposits
- Iron ore: minor deposits
- Copper ore: major deposits
- Copper ore: minor deposits
- Tin
- Bauxite
- Lead
- Zinc

0   1000   2000 Miles
0   1000   2000   3000 Kilometers

The location of deposits of critical metals such as iron, copper, tin, and others is an important determinant of the location of mining activities, like mineral fuel extraction, a "primary" economic activity. Also like mineral fuel extraction, mining for critical metallic ores makes significant environmental impact, particularly on vegetation, soils, and water resources. Some of the world's most dramatic examples of human modification of environments are located in areas of metallic ore extraction: the open pit copper mining areas of Arizona and Utah, for example. Environmental impact aside, those countries with significant critical metal deposits tend to stand a better chance of reaching higher levels of economic

development, as long as they can extract and market the ores themselves rather than having the extraction process controlled by outside concerns. The average Bolivian, for example, does not benefit greatly from the fact that his/her country is an important producer of tin and other metals. Bolivia is a "colonial dependency" country and the wealth generated by metallic ore production there tends to flow out of the country to Europe and North America. On the other hand, another South American country, Brazil, is paying for much of its own current economic development by utilizing its reserves of iron and other metals and more of the wealth from the extraction of those resources stays within the country.

Map 13   Natural Hazards

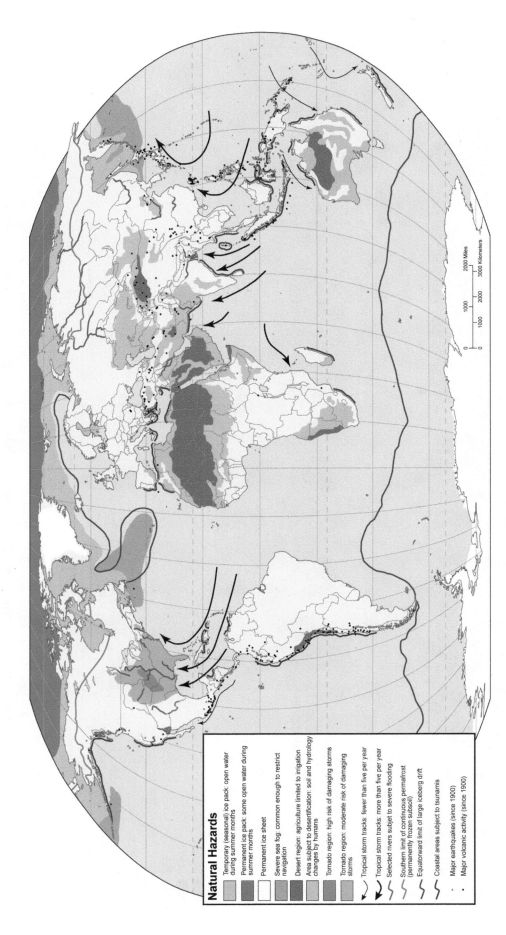

Natural Hazards

- Temporary (seasonal) ice pack: open water during summer months
- Permanent ice pack: some open water during summer months
- Permanent ice sheet
- Severe sea fog: common enough to restrict navigation
- Desert region: agriculture limited to irrigation
- Area subject to desertification: soil and hydrology changes by humans
- Tornado region: high risk of damaging storms
- Tornado region: moderate risk of damaging storms
- Tropical storm tracks: fewer than five per year
- Tropical storm tracks: more than five per year
- Selected rivers subject to severe flooding
- Southern limit of continuous permafrost (permanently frozen subsoil)
- Equatorward limit of large iceberg drift
- Coastal areas subject to tsunamis
- Major earthquakes (since 1900)
- Major volcanic activity (since 1900)

0   1000   2000 Miles
0   1000   2000   3000 Kilometers

Unlike other elements of physical geography, most natural hazards are unpredictable. However, there are certain regions where the probability of the occurrence of a particular natural hazard is high. This map shows regions affected by major natural hazards at rates that are higher than the global norm. The presence of persistent natural hazards may influence the types of modifications that people make in the environment and certainly influence the styles of housing and other elements of cultural geography. Natural hazards may also undermine the utility of an area for economic purposes and some scholars suggest that regions of environmental instability may be regions of political instability as well. The study of natural hazards has become an important activity for "resource geographers" whose areas of interest overlap both human and physical fields of geography.

-24-

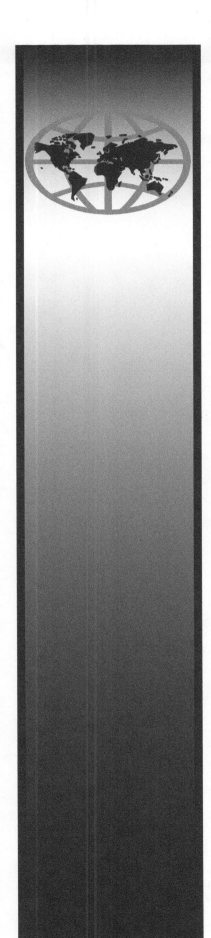

Unit II

Global Human Patterns

Map 14 Past Population Distributions and Densities

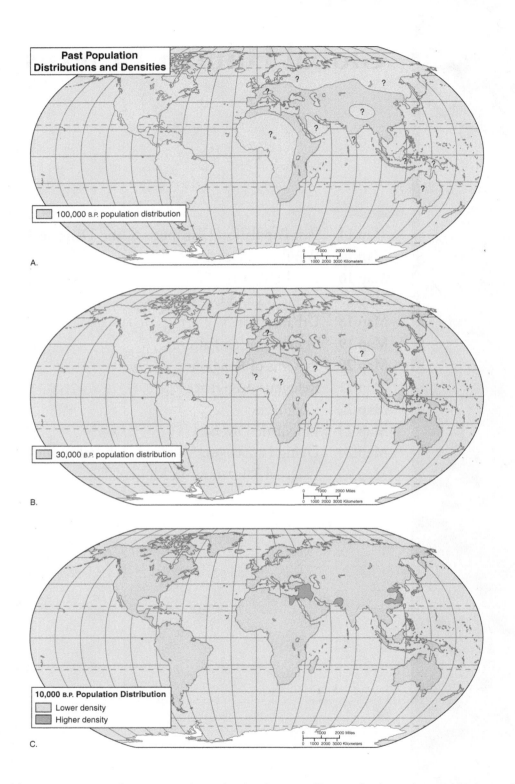

The map of the world at 100,000 B.P. (Before Present) shows the distributions of hominids who at that time had spread from their probable origin in Africa into parts of the Old World. At 30,000 B.P. few places in the Old World remained uninhabited. All the people then were hunters and gatherers who subsisted on wild foods. The environment probably was not at its carrying capacity for human populations until about 15,000 B.P.

By 10,000 B.P. hunting and gathering people had spread throughout the world. Plant and animal domestication (farming and pastoralism) had begun in some parts of the Old World perhaps as a response to behavioral changes necessitated by the population, which

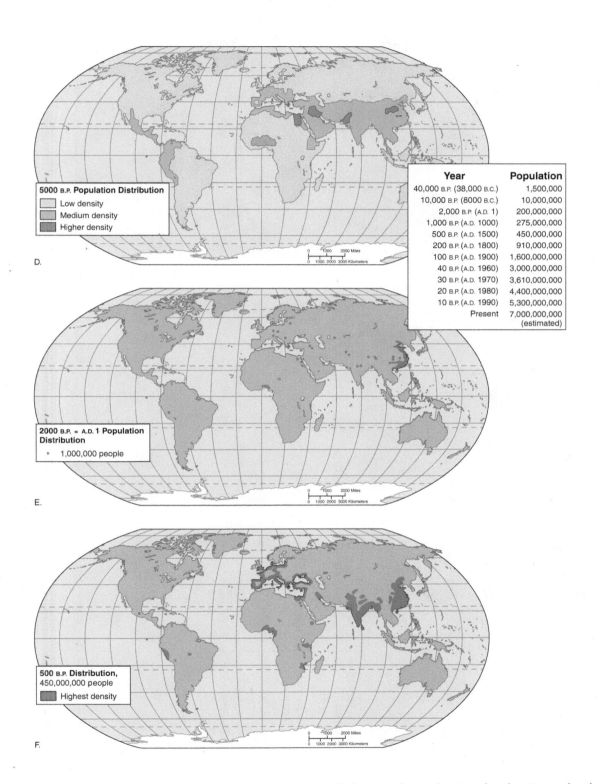

5000 B.P. Population Distribution
- Low density
- Medium density
- Higher density

D.

Year	Population
40,000 B.P. (38,000 B.C.)	1,500,000
10,000 B.P. (8000 B.C.)	10,000,000
2,000 B.P. (A.D. 1)	200,000,000
1,000 B.P. (A.D. 1000)	275,000,000
500 B.P. (A.D. 1500)	450,000,000
200 B.P. (A.D. 1800)	910,000,000
100 B.P. (A.D. 1900)	1,600,000,000
40 B.P. (A.D. 1960)	3,000,000,000
30 B.P. (A.D. 1970)	3,610,000,000
20 B.P. (A.D. 1980)	4,400,000,000
10 B.P. (A.D. 1990)	5,300,000,000
Present	7,000,000,000 (estimated)

2000 B.P. = A.D. 1 Population Distribution
- • 1,000,000 people

E.

500 B.P. Distribution, 450,000,000 people
- Highest density

F.

now exceeded the environmental carrying capacity. Farming supports higher population densities than hunting and gathering. Urban civilization with cities dependent on their hinterlands had developed by 5000 B.P. The maps of A.D. 1 and A.D. 1500 approximate actual population density on a scale of one dot to every million people. A chart provides total world population figures for different time periods. Compare these maps to the contemporary population density map on the following page.

Map 15 Population Density

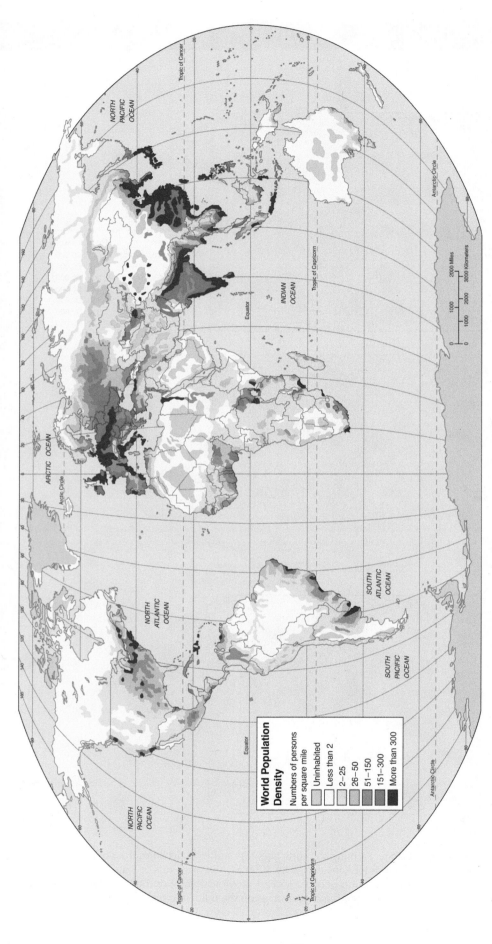

World Population Density

Numbers of persons per square mile

- Uninhabited
- Less than 2
- 2–25
- 26–50
- 51–150
- 151–300
- More than 300

No feature of human activity is more reflective of environmental conditions than where people live. In the areas of densest populations, a mixture of natural and human factors has combined to allow maximum food production, maximum urbanization, and maximum centralization of economic activities. Three great concentrations of human population appear on the map—East Asia, South Asia, and Europe—with a fourth, lesser concentration in eastern North America. One of these great population clusters—South Asia—is still growing rapidly and is expected to become even more densely populated during the twenty-first century. The other concentrations are likely to remain about as they now appear. In Europe and North America, this is the result of economic development that has caused population growth to level off during the last century. In East Asia, population has also begun to grow more slowly. In the case of Japan and the Koreas, this is the consequence of economic development; in the case of China, it is the consequence of government intervention in the form of strict family planning. The areas of future high density (in addition to those already existing) are likely to be in Middle and South America and Africa, where population growth rates are well above the world average.

Map 16   Land Use, A.D. 1500

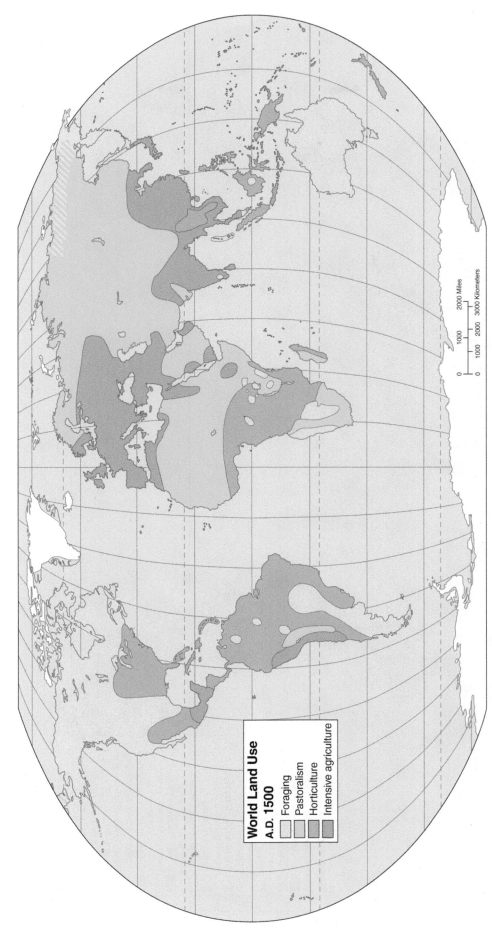

World Land Use
A.D. 1500

Foraging
Pastoralism
Horticulture
Intensive agriculture

0   1000   2000 Miles
0   1000   2000   3000 Kilometers

Europeans began to explore the world in the late 1400s. They encountered many independent people with self-sustaining economies at that time. Foraging people practiced hunting and gathering, utilizing the wild forms of plants and animals in their environments. Horticultural people practiced a simple form of agriculture using hoes or digging sticks as their basic tools. They sometimes cleared their land by burning and then planted crops. Pastoralists herded animals as their basic subsistence pattern. Complex state-level societies, such as the Mongols, had pastoralism as their base. Intensive agriculturalists based their societies on complicated irrigation systems and/or the plow and draft animals. Wheat and rice were two kinds of crops that supported large populations. In many of the areas of intensive agriculture—particularly in MesoAmerica, Europe, Southwest Asia, South Asia, and East Asia—complex patterns of market economies had begun to develop well before the fifteenth century and the beginnings of European expansion.

Map 17 Modern Land Use Patterns

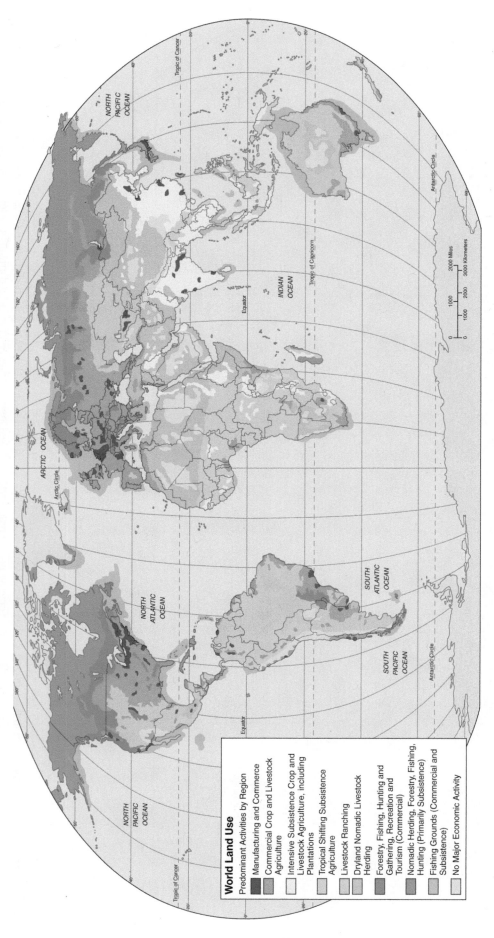

World Land Use

Predominant Activities by Region

- Manufacturing and Commerce
- Commercial Crop and Livestock Agriculture
- Intensive Subsistence Crop and Livestock Agriculture, including Plantations
- Tropical Shifting Subsistence Agriculture
- Livestock Ranching
- Dryland Nomadic Livestock Herding
- Forestry, Fishing, Hunting and Gathering, Recreation and Tourism (Commercial)
- Nomadic Herding, Forestry, Fishing, Hunting (Primarily Subsistence)
- Fishing Grounds (Commercial and Subsistence)
- No Major Economic Activity

Many of the major land use patterns of the world (such as urbanization, industry, and transportation) are relatively small in area and are not easily seen on maps, but the most important uses people make of the earth's surface have more far-reaching effects. This map illustrates, in particular, the variations in primary land uses (such as agriculture) for the entire world. Note the differences between land use patterns in the more developed countries of the middle latitude zones and the less developed countries of the tropics.

Map 18 Urbanization

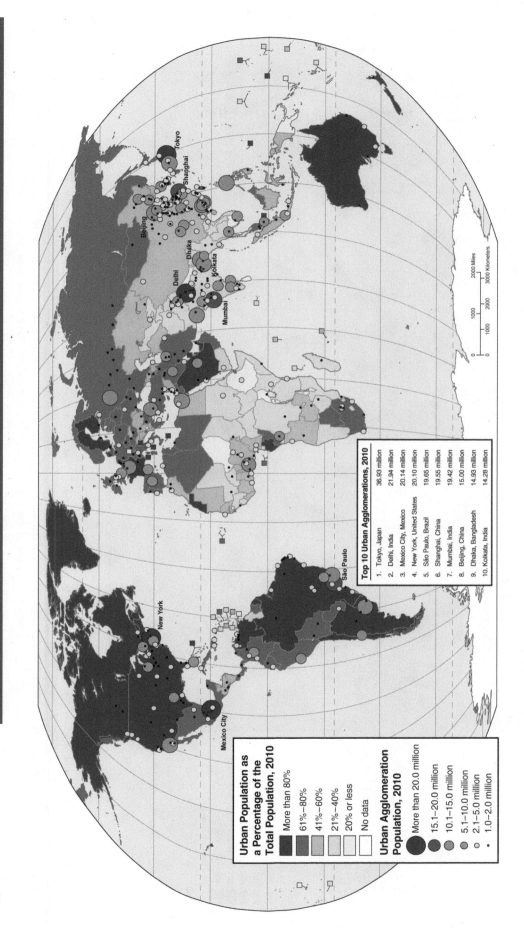

Urban Population as a Percentage of the Total Population, 2010

- More than 80%
- 61%–80%
- 41%–60%
- 21%–40%
- 20% or less
- No data

Urban Agglomeration Population, 2010

- More than 20.0 million
- 15.1–20.0 million
- 10.1–15.0 million
- 5.1–10.0 million
- 2.1–5.0 million
- 1.0–2.0 million

Top 10 Urban Agglomerations, 2010	
1. Tokyo, Japan	36.93 million
2. Delhi, India	21.94 million
3. Mexico City, Mexico	20.14 million
4. New York, United States	20.10 million
5. São Paulo, Brazil	19.65 million
6. Shanghai, China	19.55 million
7. Mumbai, India	19.42 million
8. Beijing, China	15.00 million
9. Dhaka, Bangladesh	14.93 million
10. Kolkata, India	14.28 million

The degree to which a region's population is concentrated in urban areas is a major indicator of a number of things: the level of economic development, and the problems associated with human concentrations. Urban dwellers are rapidly becoming the norm among the world's people and rates of urbanization are increasing worldwide, with the greatest increases in urbanization taking place in developing regions. Whether in developed or developing countries, those who live in cities exert an influence on the environment, politics, economics, and social systems that goes far beyond the confines of the city itself. Acting as the focal points for the flow of goods and ideas, cities draw resources and people not just from their immediate hinterland but from the entire world. This process creates far-reaching impacts as resources are extracted, converted through industrial processes, and transported over great distances to metropolitan regions, and as ideas spread or *diffuse* along with the movements of people to cities and the flow of communication from them. The significance of urbanization can be most clearly seen, perhaps, in North America where, in spite of vast areas of relatively unpopulated land, more than 90 percent of the population lives in urban areas.

Map 19 Transportation Patterns

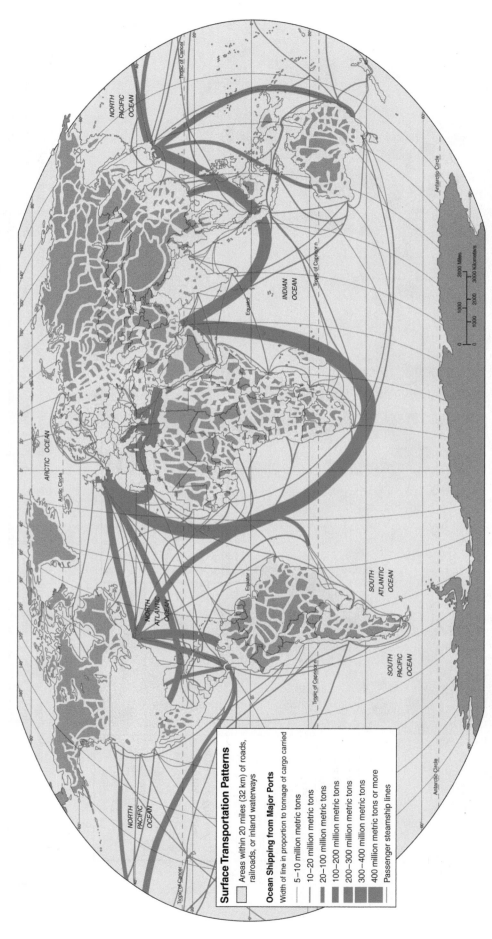

Surface Transportation Patterns

☐ Areas within 20 miles (32 km) of roads, railroads, or inland waterways

Ocean Shipping from Major Ports

Width of line in proportion to tonnage of cargo carried

— 5–10 million metric tons
— 10–20 million metric tons
— 20–100 million metric tons
— 100–200 million metric tons
— 200–300 million metric tons
— 300–400 million metric tons
— 400 million metric tons or more
— Passenger steamship lines

As a form of land use, transportation is second only to agriculture in its coverage of the earth's surface and is one of the clearest examples in the human world of a *network*, a linked system of lines allowing flows from one place to another. The global transportation network and its related communication web are responsible for most of the *spatial interaction*, or movement of goods, people, and ideas between places. As the chief mechanism of spatial interaction, transportation is linked firmly with the concept of a shrinking world and the development of a global community and economy. Because transportation systems require significant modification of the earth's surface, transportation is also responsible for massive alterations in the quantity and quality of water, for major soil degradations and erosion, and (indirectly) for the air pollution that emanates from vehicles utilizing the transportation system. In addition, as improved transportation technology draws together places on the earth that were formerly remote, it allows people to impact environments a great distance away from where they live.

-32-

Map 20 Patterns of Religion

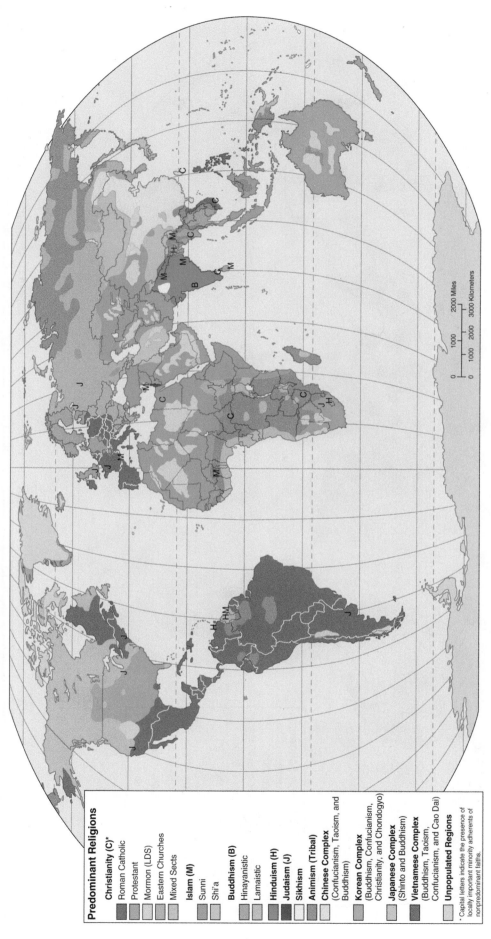

Predominant Religions

Christianity (C)*
Roman Catholic
Protestant
Mormon (LDS)
Eastern Churches
Mixed Sects

Islam (M)
Sunni
Shi'a

Buddhism (B)
Hinayanistic
Lamaistic

Hinduism (H)

Judaism (J)

Sikhism

Animism (Tribal)

Chinese Complex
(Confucianism, Taoism, and Buddhism)

Korean Complex
(Buddhism, Confucianism, Christianity, and Chondogyo)

Japanese Complex
(Shinto and Buddhism)

Vietnamese Complex
(Buddhism, Taoism, Confucianism, and Cao Dai)

Unpopulated Regions

* Capital letters indicate the presence of locally important minority adherents of nonpredominant faiths.

0   1000   2000 Miles
0   1000   2000   3000 Kilometers

Religious adherence is one of the fundamental defining characteristics of human culture, the style of life adopted by a people and passed from one generation to the next. Because of the importance of religion for culture, a depiction of the spatial distribution of religions is as close as we can come to a map of cultural patterns. More than just a set of behavioral patterns having to do with worship and ceremony, religion is a vital conditioner of the ways that people deal with one another, with their institutions, and with the environments they occupy. In many areas of the world, the ways in which people make a living, the patterns of occupation that they create on the land, and the impacts that they make on ecosystems are the direct consequences of their adherence to a religious faith. An examination of the map in the context of international and intranational conflict will also show that tension between countries and the internal stability of states is also a function of the spatial distribution of religion.

Map 21   Religious Adherence

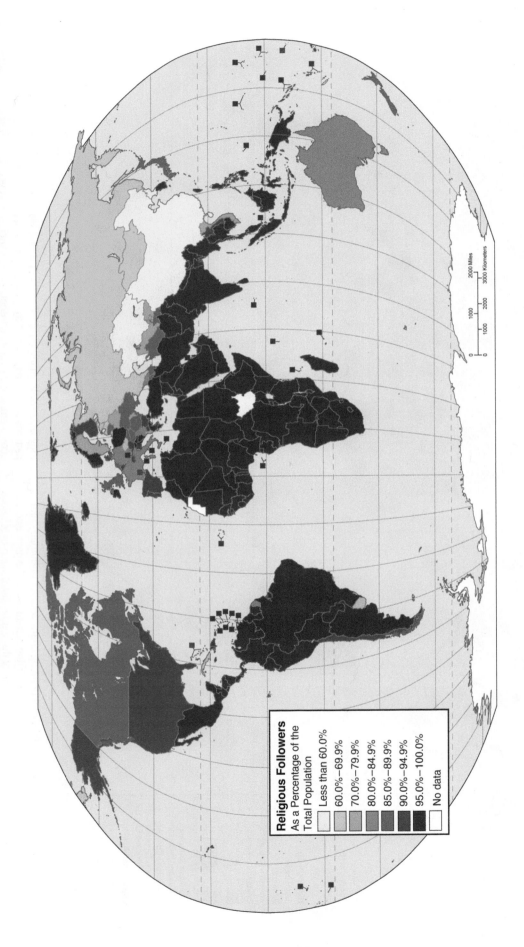

Religious Followers
As a Percentage of the
Total Population

- Less than 60.0%
- 60.0%–69.9%
- 70.0%–79.9%
- 80.0%–84.9%
- 85.0%–89.9%
- 90.0%–94.9%
- 95.0%–100.0%
- No data

Throughout much of the past century, the numbers of proclaimed religious adherents declined as a percentage of populations as the consequence of the emergence of Communism in what became the U.S.S.R. and in China. Communism enforced atheism as a state "religion," and it is impossible to know how many Orthodox Christians actually existed in Russia or how many Christians, Muslims, Buddhists, Taoists, and Confucianists,, existed in China. With the downfall of the old Soviet Union and the reemergence of an Orthodox Christian Russia, percentages of religious adherents in that country have increased tremendously. And as the European Union has opened its doors to "guest workers" and others from Southwest Asia and Africa, the number of Muslims has grown considerably in most of western Europe while Islam has reestablished itself as a religion among many inhabitants of the Balkan peninsula. It is worth noting that whereas past shifts in regional percentages of religious adherents had often been the result of missionary activity (the growing number of Christians in Africa during the twentieth century, for example), the recent changes have come about through political and economic change.

Map 22   Languages

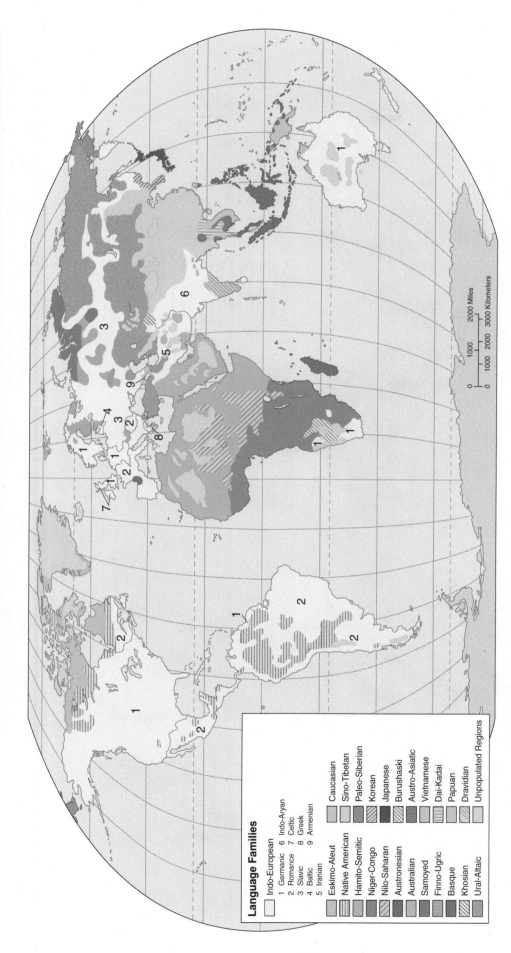

Language Families

Indo-European
1 Germanic 6 Indo-Aryan
2 Romance 7 Celtic
3 Slavic   8 Greek
4 Baltic   9 Armenian
5 Iranian

Eskimo-Aleut
Native American
Hamito-Semitic
Niger-Congo
Nilo-Saharan
Austronesian
Australian
Samoyed
Finno-Ugric
Basque
Khoisan
Ural-Altaic

Caucasian
Sino-Tibetan
Paleo-Siberian
Korean
Japanese
Burushaski
Austro-Asiatic
Vietnamese
Dai-Kadai
Papuan
Dravidian
Unpopulated Regions

0   1000   2000 Miles
0 1000 2000 3000 Kilometers

Language, like religion, is an important identifying characteristic of culture. Indeed, it is perhaps the most durable of all those identifying characteristics or *cultural traits*: language, religion, institutions, material technologies, and ways of making a living. After centuries of exposure to other languages or even conquest by speakers of other languages, the speakers of a specific tongue will often retain their own linguistic identity. Language helps us to locate areas of potential conflict, particularly in regions where two or more languages overlap. Many, if not most, of the world's conflict zones are also areas of linguistic diversity. Knowing the distribution of languages helps us to understand some of the reasons behind important current events: for example, linguistic

identity differences played an important part in the disintegration of the Soviet Union in the early 1990s; and in areas emerging from recent colonial rule, such as Africa, the participants in conflicts over territory and power are often defined in terms of linguistic groups. Language distributions also help us to comprehend the nature of the human past by providing clues that enable us to chart the course of human migrations, as shown in the distribution of Indo-European, Austronesian, or Hamito-Semitic languages. Finally, because languages have a great deal to do with the way people perceive and understand the world around them, linguistic patterns help to explain the global variations in the ways that people interact.

-35-

Map 23   Linguistic Diversity

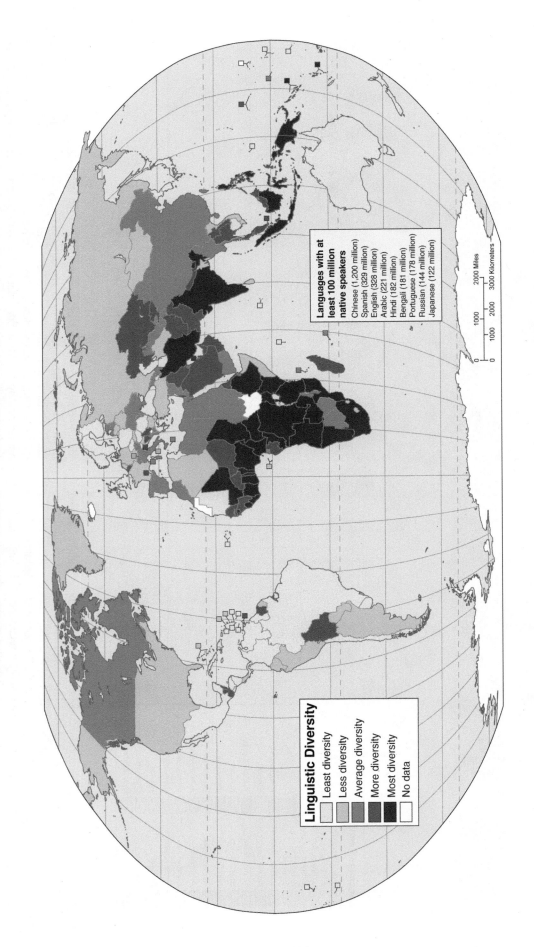

Linguistic Diversity

- Least diversity
- Less diversity
- Average diversity
- More diversity
- Most diversity
- No data

Languages with at least 100 million native speakers

Chinese (1,200 million)
Spanish (329 million)
English (328 million)
Arabic (221 million)
Hindi (182 million)
Bengali (181 million)
Portuguese (178 million)
Russian (144 million)
Japanese (122 million)

0   1000   2000   3000 Kilometers
0   1000   2000 Miles

Of the more than 6,000 languages spoken in the world today half the world's population speaks one of approximately one dozen languages. Additionally, fewer than 100 are official languages—those designated by a country as the language of government, commerce, education, and information. In this map, diversity is measured using Greenberg's Diversity Index, which measures the likelihood that a person speaking one native language will encounter a person speaking a different native language. Although a country's linguistic diversity is obviously influenced by the absolute number of languages present in a country, it is also affected by the population size, connectivity, and the degree to which political boundaries conform to the territory of ethnolinguistic groups. An example of the latter phenomenon is clearly seen in Africa where political boundaries reflect the era of European colonialism and not necessarily the geography of indigenous populations. Using this index, Papua New Guinea ranks as the most linguistically diverse country in the world and North Korea and Vatican City are the least diverse.

Map 24 Dying Languages: A Loss of Cultural Wealth

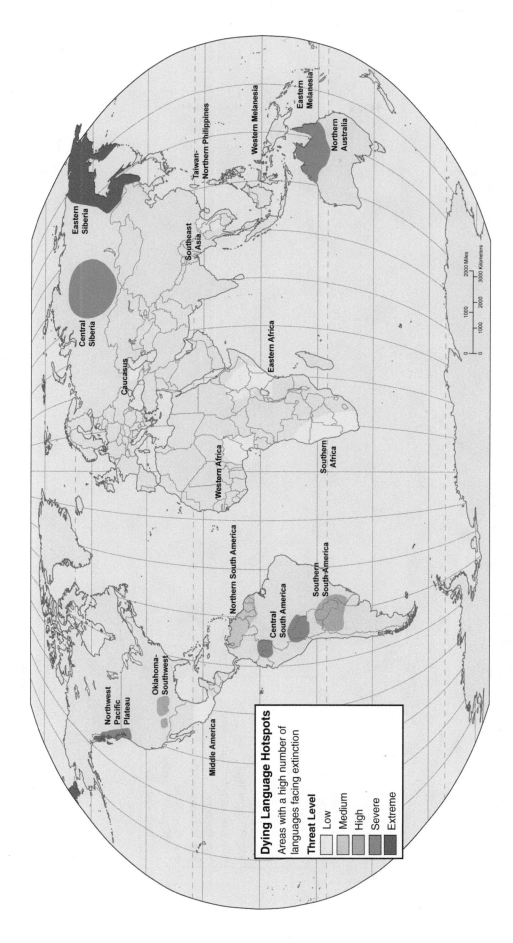

Dying Language Hotspots

Areas with a high number of languages facing extinction

Threat Level

- Low
- Medium
- High
- Severe
- Extreme

Eastern Siberia

Central Siberia

Caucasus

Taiwan- Northern Philippines

Southeast Asia

Western Melanesia

Eastern Melanesia

Northern Australia

Western Africa

Eastern Africa

Southern Africa

Northwest Pacific Plateau

Oklahoma- Southwest

Middle America

Northern South America

Central South America

Southern South-America

The number of languages spoken in the world of the sixteenth century was probably in excess of 10,000, possibly half of those in Africa. After four centuries of colonialism, migration, and economic and political change, less than half of the languages spoken in 1500 still exist—and many of these are in jeopardy of becoming extinct. Language is an essential marker of culture and when languages disappear, the cultures that used them tend to disappear as well, becoming assimilated—at least linguistically—into the dominant cultures of their regions. Thus, in North America, languages such as Mojave or Wichita are currently used by fewer than 10 native speakers. If those speakers die before passing their languages on, the languages become "dead" and so do many significant elements of the cultures associated with them. On many American Indian reservations there are currently important efforts to regain or restore native tongues. But these movements are active among the larger American Indian populations; although Northern Arapahoe may well be preserved, the South Haida of the Pacific Northwest may not.

-37-

Map 25   External Migrations in Modern Times

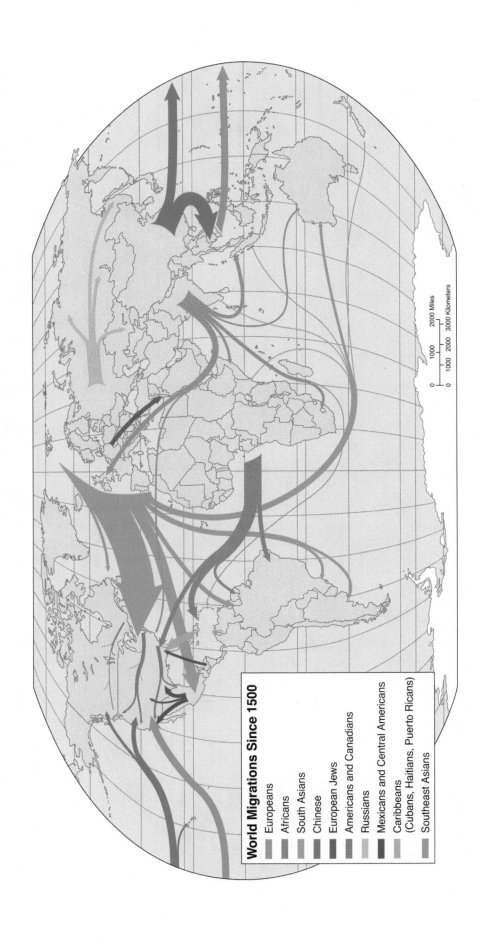

World Migrations Since 1500

- Europeans
- Africans
- South Asians
- Chinese
- European Jews
- Americans and Canadians
- Russians
- Mexicans and Central Americans
- Caribbeans (Cubans, Haitians, Puerto Ricans)
- Southeast Asians

0   1000   2000 Miles
0   1000   2000   3000 Kilometers

Migration has had a significant effect on world geography, contributing to cultural change and development, to the diffusion of ideas and innovations, and to the complex mixture of people and cultures found in the world today. *Internal migration* occurs within the boundaries of a country; *external migration* is movement from one country or region to another. Over the last 50 years, the most important migrations in the world have been internal, largely the rural-to-urban migration that has been responsible for the recent rise of global urbanization. Prior to the mid-twentieth century, three types of external migrations were most important: *voluntary*, most often in search of better economic conditions and opportunities; *involuntary* or *forced*, involving people who have been driven from their homelands by war, political unrest, or environmental disasters, or who have been transported as slaves or prisoners; and *imposed*, not entirely forced but which conditions make highly advisable. Human migrations in recorded history have been responsible for major changes in the patterns of languages, religions, ethnic composition, and economies. Particularly during the last 500 years, migrations of both the voluntary and involuntary or forced type have literally reshaped the human face of the earth.

Unit III

Global Demographic Patterns

Map 26 Population Growth Rates

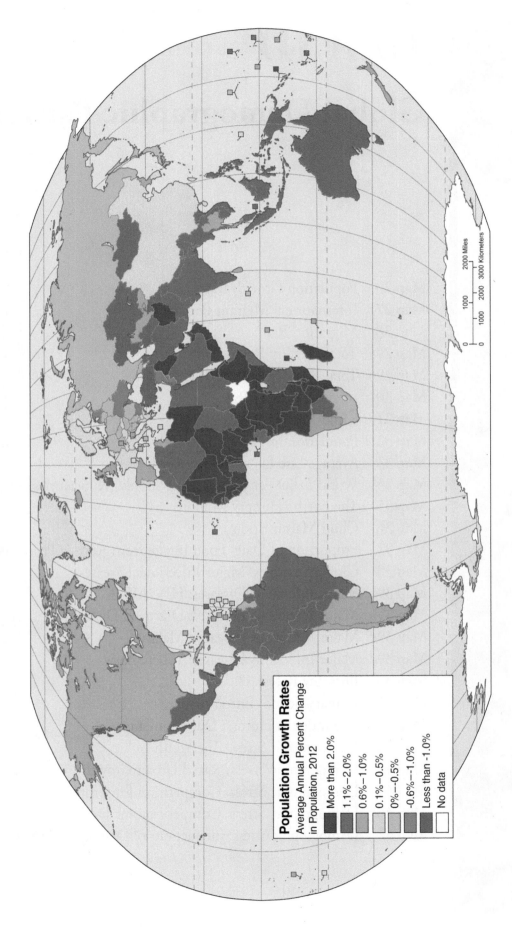

Population Growth Rates
Average Annual Percent Change
in Population, 2012

- More than 2.0%
- 1.1%–2.0%
- 0.6%–1.0%
- 0.1%–0.5%
- 0%–0.5%
- -0.6%–-1.0%
- Less than -1.0%
- No data

Of all the statistical measurements of human population, that of the rate of population growth is the most important. The growth rate of a population is a combination of natural change (births and deaths), in-migration, and out-migration; it is obtained by adding the number of births to the number of immigrants during a year and subtracting from that total the sum of deaths and emigrants for the same year. For a specific country, this figure will determine many things about the country's future ability to feed, house, educate, and provide medical services to its citizens. Some of the countries with the largest populations (such as India) also have high growth rates. Because these countries tend to be in developing regions, the combination of high population and high growth rates poses special problems for political stability and continuing economic development; the combination also carries heightened risks for environmental degradation. Many people believe that the rapidly expanding world population is a potential crisis that may cause environmental and human disaster by the middle of the twenty-first century.

-40-

Map 27   International Migrant Populations

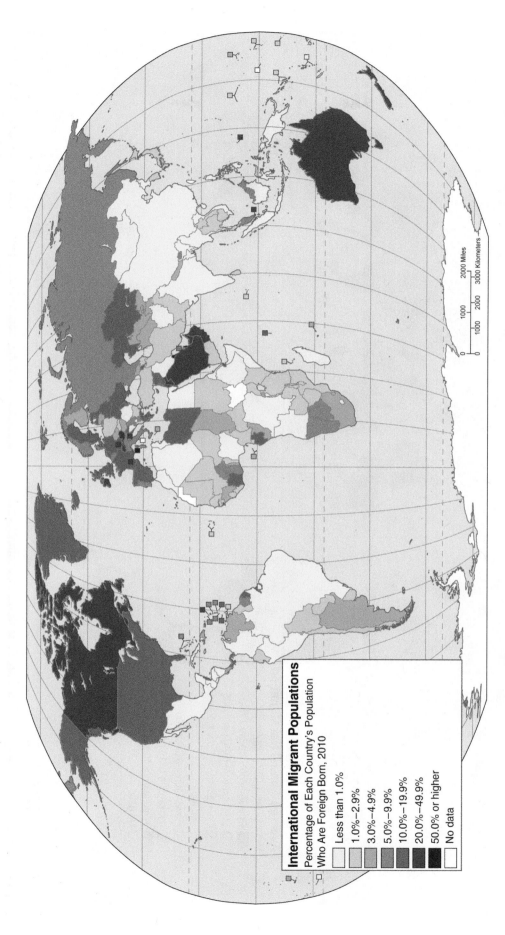

Migration—the movement from one place to another—takes a number of different forms. It may be a move within a country from an old job to a new one. It may also mean a migration, either forced or voluntary, from one country to another. The map here depicts international migration: migration between countries. Migration is distinguished from a refugee movement in that migrants are not defined as refugees granted a humanitarian and temporary protection status under international law. The Middle Americans who leave the

Central American countries, Mexico, or the Caribbean for the United States plan to live in the United States permanently, although still retaining cultural and family ties to their native country. This map clearly shows that those countries viewed as having the most favorable opportunities for improvement in personal living conditions are those with the highest numbers of in-migrants; those countries that are overcrowded, with little economically upward mobility, or international conflict tend to be those with the greatest number of out-migrants.

International Migrant Populations
Percentage of Each Country's Population
Who Are Foreign Born, 2010

- Less than 1.0%
- 1.0%–2.9%
- 3.0%–4.9%
- 5.0%–9.9%
- 10.0%–19.9%
- 20.0%–49.9%
- 50.0% or higher
- No data

2000 Miles
3000 Kilometers
0   1000   2000
0   1000   2000

Map **28** Migration Rates

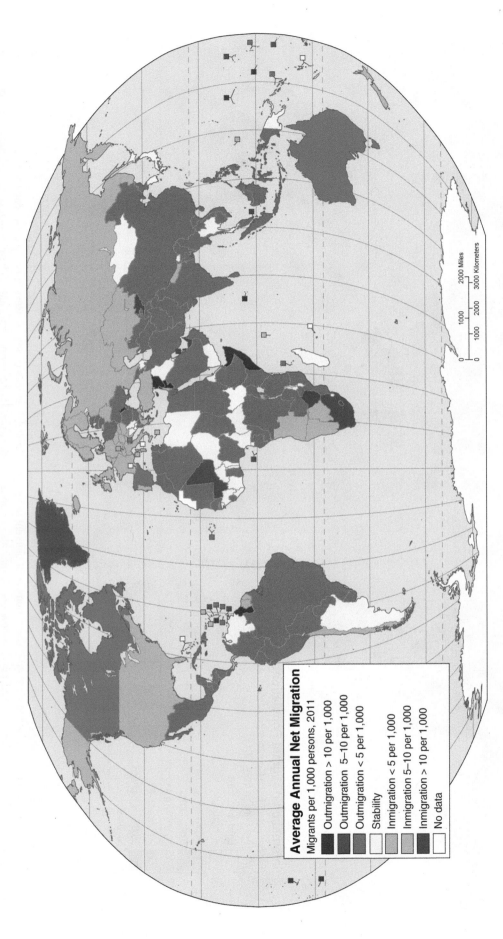

Average Annual Net Migration

Migrants per 1,000 persons, 2011

- Outmigration > 10 per 1,000
- Outmigration 5–10 per 1,000
- Outmigration < 5 per 1,000
- Stability
- Immigration < 5 per 1,000
- Immigration 5–10 per 1,000
- Immigration > 10 per 1,000
- No data

At the most fundamental level, international migration occurs because of two factors: *push* factors and *pull* factors. Push factors are those things that compel a person to leave his or her country. Push factors are many, varied, and are perceived in a negative light. Frequently cited push factors include political stresses (e.g., fear of persecution, war, or internal conflict), economic stresses (e.g., lack of employment opportunities), social stresses (e.g., lack of educational opportunities or adequate health care), and environmental stresses (e.g., natural disasters such as a major

earthquake or hurricane/typhoon, desertification). Conversely, pull factors are those attributes of another country that a person finds attractive. As might be expected, they are the opposite of push factors. Thus, international migration is spurred by the perception that things are *bad* in one's own country and *better* in another. Both push and pull factors need be present. A person likely will not migrate to another country if it is perceived that conditions in that country are as bad or worse than in that person's home country.

Map 29a   Total Fertility Rates, 1985

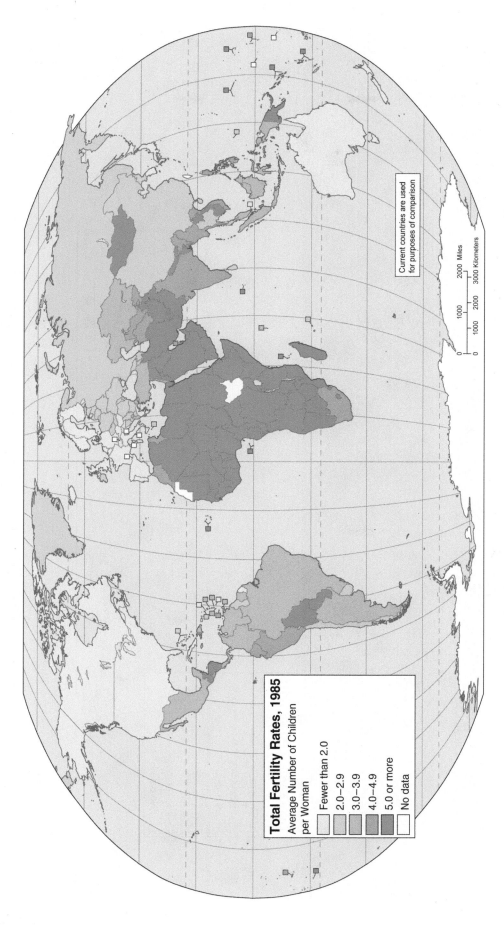

Total Fertility Rates, 1985
Average Number of Children per Woman

- Fewer than 2.0
- 2.0–2.9
- 3.0–3.9
- 4.0–4.9
- 5.0 or more
- No data

Current countries are used for purposes of comparison

0   1000   2000   2000 Miles
0   1000   2000   3000 Kilometers

Fertility rates measure the number of children that might be expected to be given birth by a single woman during her child-bearing years. It is often a better predictor of potential population growth than such measures as birth rate. More important, it is a major indicator of the "demographic transition" or shift from a population characterized by large families and short lives to one characterized by small families and longer lives. For the last century, the first type of population profile (large families and short lives) has been evident in the lesser developed parts of the world whereas the developed nations had populations with the smaller family and longer life span profile. What these comparative maps show us is that over the last quarter century, the same kind of demographic transition begun in Europe and America in the nineteenth century is now beginning in Middle and South America, in Africa, and in Asia. There are still areas of concern—African fertility rates are, for example, still far too high. But particularly in South America, Southeast Asia, and East Asia, fertility rates have dropped dramatically, indicating the potential for much slower population growth in the next generation.

-43-

Map 29b Total Fertility Rates, 2011

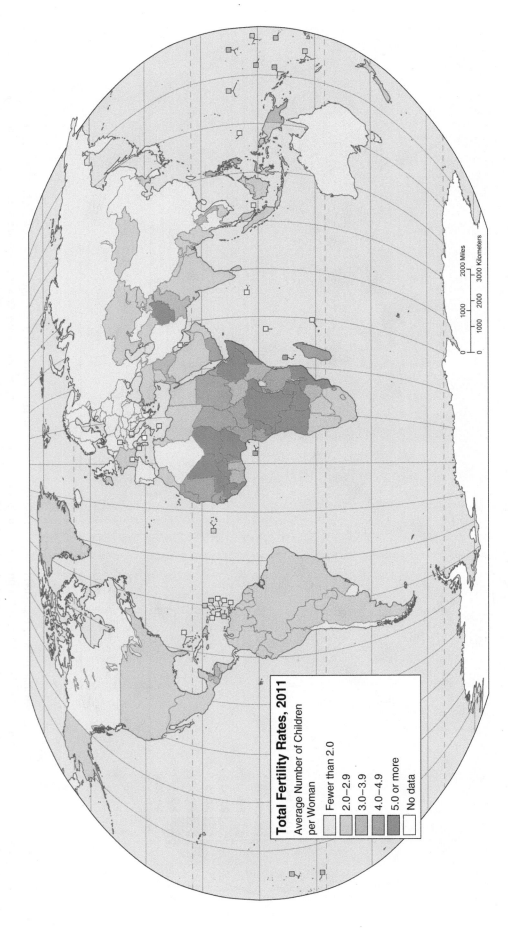

Total Fertility Rates, 2011

Average Number of Children
per Woman

- Fewer than 2.0
- 2.0–2.9
- 3.0–3.9
- 4.0–4.9
- 5.0 or more
- No data

1000   2000 Miles

0   1000   2000   3000 Kilometers

Map 30   Infant Mortality Rate

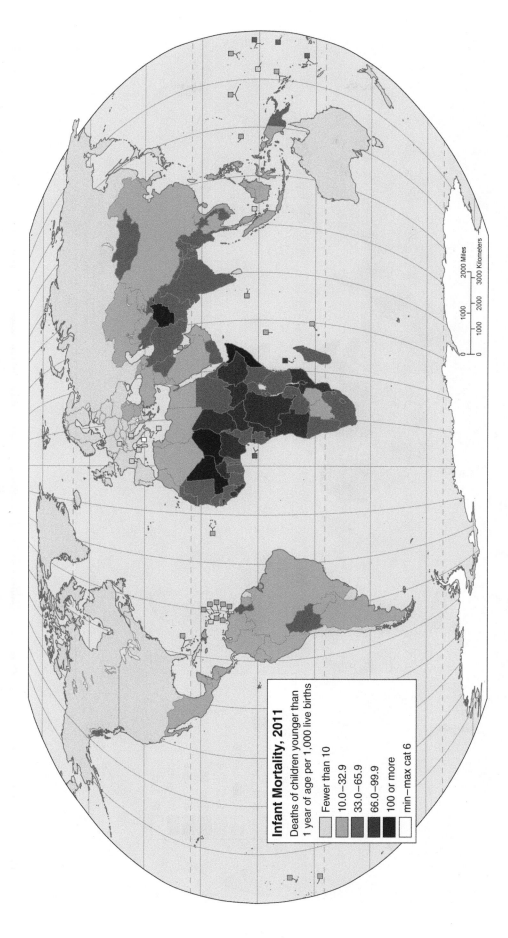

Infant Mortality, 2011

Deaths of children younger than
1 year of age per 1,000 live births

- Fewer than 10
- 10.0–32.9
- 33.0–65.9
- 66.0–99.9
- 100 or more
- min–max cat 6

0   1000   2000 Miles

0   1000   2000   3000 Kilometers

Infant mortality rates are calculated by dividing the number of children born in a given year who die before their first birthday by the total number of children born that year and then multiplying by 1,000; this shows how many infants have died for every 1,000 births. Infant mortality rates are prime indicators of economic development. In highly developed economies, with advanced medical technologies, sufficient diets, and adequate public sanitation, infant mortality rates tend to be quite low. By contrast, in less developed countries, with the disadvantages of poor diet, limited access to medical technology, and the other problems of poverty, infant mortality rates tend to be high. Although worldwide infant mortality has decreased significantly during the last two decades, many regions of the world still experience infant mortality above the 10 percent level (100 deaths per 1,000 live births). Such infant mortality rates not only represent human tragedy at its most basic level, but also are powerful inhibiting factors for the future of human development. Comparing infant mortality rates in the midlatitudes and the tropics shows that children in most African countries are more than 10 times as likely to die within a year of birth as children in European countries.

-45-

Map 31 Child Mortality Rate

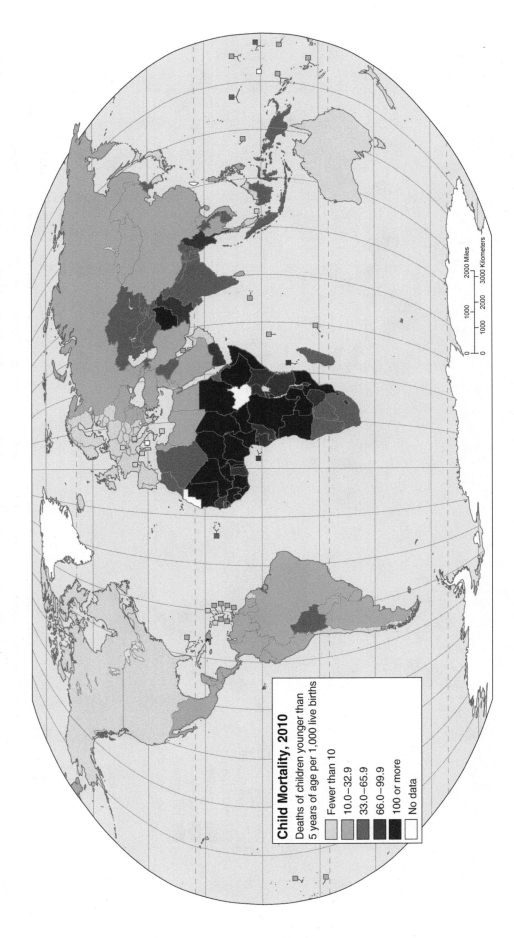

Child Mortality, 2010

Deaths of children younger than
5 years of age per 1,000 live births

- Fewer than 10
- 10.0–32.9
- 33.0–65.9
- 66.0–99.9
- 100 or more
- No data

Child mortality rates are calculated by determining the probability that a child born in a specified year will die before reaching age 5, using current age-specific mortality rates for a population. The major sources of mortality rates are vital registration systems and estimates made from surveys and/or census reports. Along with infant mortality and average life-expectancy rates, child mortality rates, according to the World Bank, "are probably the best general indicators of a community's current health status and are often cited as overall measures of a population's welfare or quality of life." Where infant mortality often reflects health care conditions, child mortality is usually a reflection of the inadequacy of nutrition, leading to early deaths from nutritionally related diseases. In some less developed countries in Africa and Asia, child mortality is also an indicator of the widespread presence of infectious diseases such as malaria, tuberculosis, and HIV/AIDS.

Map 32   Population by Age Group

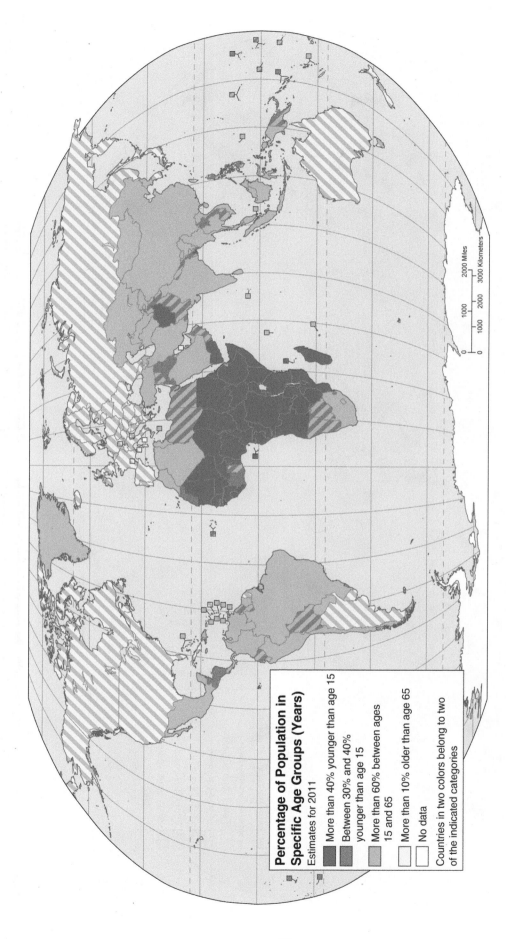

Percentage of Population in Specific Age Groups (Years)

Estimates for 2011

- More than 40% younger than age 15
- Between 30% and 40% younger than age 15
- More than 60% between ages 15 and 65
- More than 10% older than age 65
- No data
- Countries in two colors belong to two of the indicated categories

0   1000   2000   3000 Kilometers
0   1000   2000 Miles

Of all the measurements that illustrate the dynamics of a population, age distribution may be the most significant, particularly when viewed in combination with average growth rates. The particular relevance of age distribution is that it tells us what to expect from a population in terms of growth over the next generation. If, for example, approximately 40–50 percent of a population is younger than the age of 15, that suggests that in the next generation about one-quarter of the total population will be women of childbearing age. When age distribution is combined with fertility rates (the average number of children born per woman in a population), an especially valid measurement

of future growth potential may be derived. A simple example: Nigeria, with a 2012 population of 170 million, has 40.9 percent of its population younger than the age of 15 and a fertility rate of 4.4; the United States, with a 2012 population of 313 million, has 20.1 percent of its population younger than the age of 15 and a fertility rate of 2.06. During the period in which those women presently younger than the age of 15 are in their childbearing years, Nigeria can be expected to add a total of approximately 150 million persons to its total population. Over the same period, the United States can be expected to add only 61 million.

Map 33 Average Life Expectancy at Birth

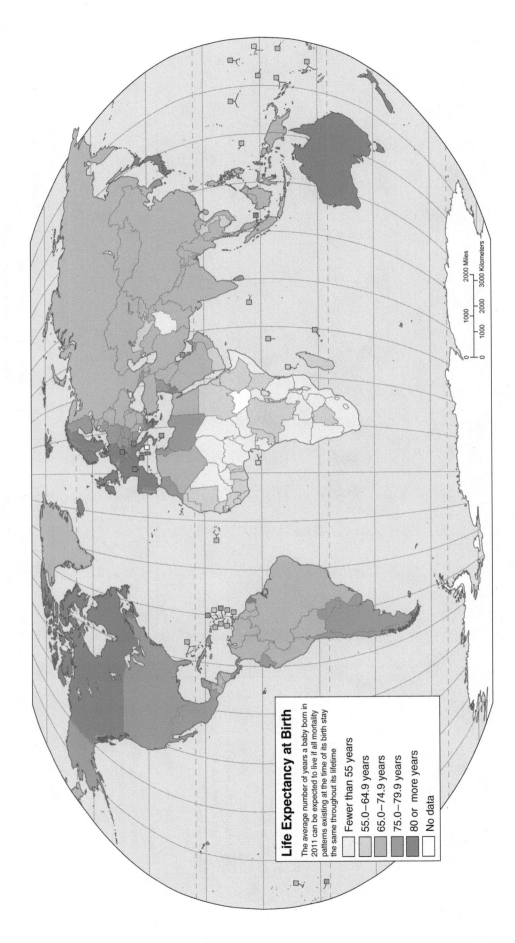

Life Expectancy at Birth

The average number of years a baby born in 2011 can be expected to live if all mortality patterns existing at the time of its birth stay the same throughout its lifetime

- Fewer than 55 years
- 55.0–64.9 years
- 65.0–74.9 years
- 75.0–79.9 years
- 80 or more years
- No data

Average life expectancy at birth is a measure of the average longevity of the population of a country. Like all average measures, it is distorted by extremes. For example, a country with a high mortality rate among children will have a low average life expectancy. Thus, an average life expectancy of 45 years does not mean that everyone can be expected to die at the age of 45. More normally, what the figure means is that a substantial number of children die between birth and 5 years of age, thus reducing the average life expectancy for the entire population. In spite of the dangers inherent in misinterpreting the data, average life expectancy (along with infant mortality and several other measures) is a valid way of judging the relative health of a population. It reflects the nature of the health care system, public sanitation and disease control, nutrition, and a number of other key human need indicators. As such, it is a measure of well-being that is significant in indicating economic development and predicting political stability.

Map 34

World Daily Per Capita Food Supply (Kilocalories)

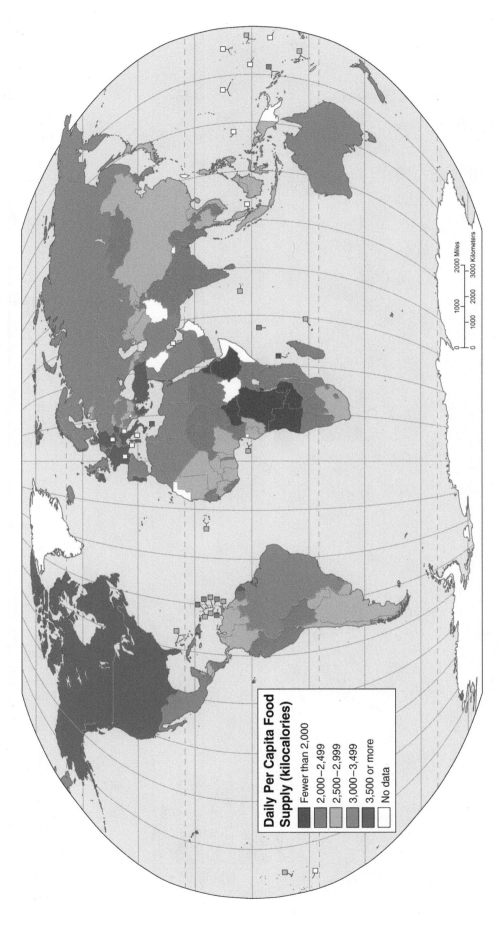

Daily Per Capita Food Supply (Kilocalories)

- Fewer than 2,000
- 2,000–2,499
- 2,500–2,999
- 3,000–3,499
- 3,500 or more
- No data

The data shown on this map, which indicate the presence or absence of critical food shortages, do not necessarily indicate the presence of starvation or famine. But they certainly do indicate potential problem areas for the next decade. The measurements are in calories from *all* food sources: domestic production, international trade, draw-down on stocks or food reserves, and direct foreign contributions or aid. The quantity of calories available is that amount, estimated by the UN's Food and Agriculture Organization (FAO),

that reaches consumers. The calories actually consumed may be lower than the figures shown, depending on how much is lost in a variety of ways: in home storage (to pests such as rats and mice), in preparation and cooking, through consumption by pets and domestic animals, and as discarded foods, for example. The estimate of need is not a global uniform value but is calculated for each country on the basis of the age and sex distribution of the population and the estimated level of activity of the population.

Map 35 The Earth's Hungry Millions

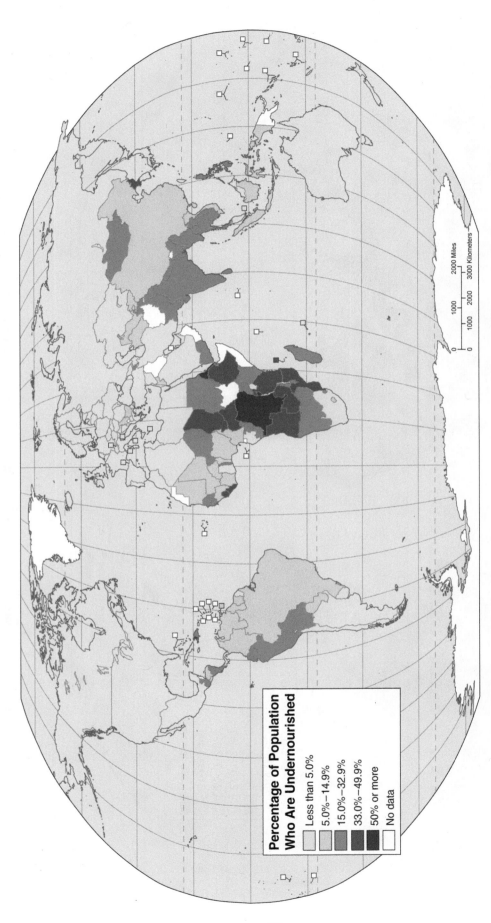

Percentage of Population Who Are Undernourished

- Less than 5.0%
- 5.0%–14.9%
- 15.0%–32.9%
- 33.0%–49.9%
- 50% or more
- No data

In order to maintain health and have sufficiently developed immune systems to fight off diseases, humans need to maintain a daily caloric intake of approximately 2,000 calories. Current estimates are that nearly 15% of the world's population have diets that do not constitute the minimum 2,000 calories for basic health maintenance. These countries are particularly susceptible to climate-induced famines which may lower already minimal food intake, or to food emergencies produced by warfare, civil unrest, economic instability, and poor governance. It is little surprise, when looking at the map, that the countries at greatest risk are in Africa and Asia where swelling populations have already approached (or, in some

instances, exceeded) carrying capacity or the ability of the environment to sustain a population at a given level. The causes of national malnutrition are as numerous as the countries that experience it but, again, the map is instructive. In Africa, where HIV/AIDS is prevalent among large segments of the working-age population, food production is limited simply by poor health (which, of course, begets low harvests which contributes to poorer health). In South Asia, the problem tends to be magnified by huge populations that have outstripped the ability of the environment to support them. Bangladesh, for example, with an area about the size of West Virginia, is one of the world's ten most populous countries.

-50-

Map 36   Child Malnutrition

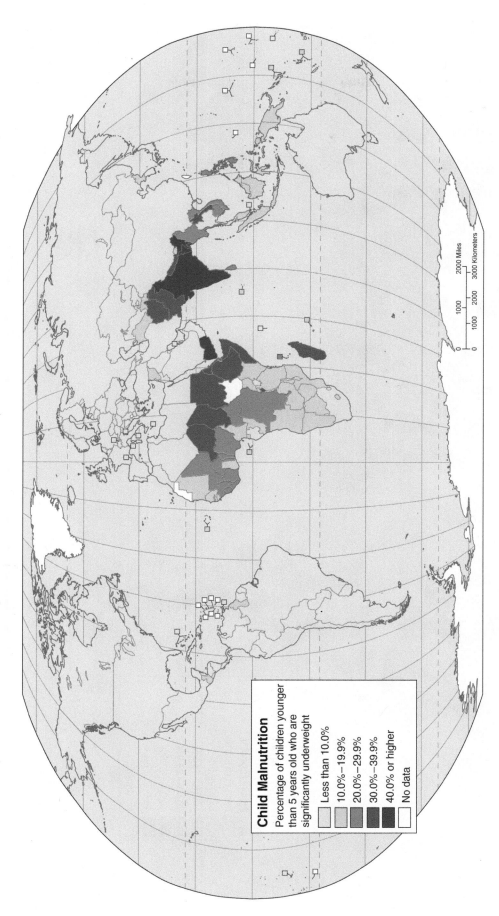

Child Malnutrition

Percentage of children younger than 5 years old who are significantly underweight

Less than 10.0%
10.0% – 19.9%
20.0% – 29.9%
30.0% – 39.9%
40.0% or higher
No data

The weight of poverty is not evenly spread among the members of a population, falling disproportionately upon the weakest and most disadvantaged members of society. In most societies, these individuals are children, particularly female children. Children simply do not compete as successfully as adults for their (meager) share of the daily food supply. Where food shortages prevail, children tend to have the quality of their future lives severely compromised by poor nutrition, which, in a downward spiral, robs them of the energy necessary to compete more effectively for food. Children who are inadequately fed are less likely to do well in school, are more prone to debilitating disease, and will more often become a drain on scarce societal resources than well-fed children. Recently, health care officials in the more developed world have become concerned over the trend to "overnutrition," leading to obesity and related health problems in the world's economically developed countries. Nevertheless, child malnourishment remains one of the primary distinguishing factors between the "haves" and "have-nots."

Map 37 Proportion of Daily Food Intake from Animal Products

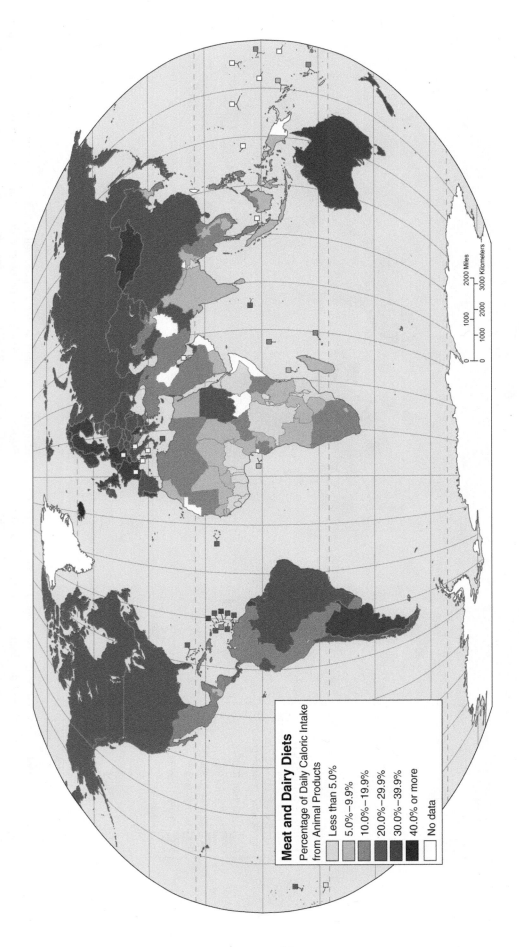

Meat and Dairy Diets

Percentage of Daily Caloric Intake
from Animal Products

- Less than 5.0%
- 5.0%—9.9%
- 10.0%—19.9%
- 20.0%—29.9%
- 30.0%—39.9%
- 40.0% or more
- No data

0   1000   2000 Miles
0   1000   2000   3000 Kilometers

As the world grows proportionally wealthier, the percentage of the diet from meat and fish grows accordingly. Although there is precious little scientific evidence to indicate that a vegetarian diet is, by definition, healthier than one richer in animal products, it is indisputable that more reliance on animal products produces a greater strain on environmental systems. Growing plants to feed animals to feed humans simply consumes a great deal more energy (in all forms) than growing plants to feed a human population directly. The long-term implications of altering land-use patterns to fit with greater wealth and, therefore, more meat consumption, are not generally favorable. And we are already beginning to notice the impact on fish population in the world's oceans as the result of an increasing demand for this food resource.

Map 38 Global Scourges: Major Infectious Diseases

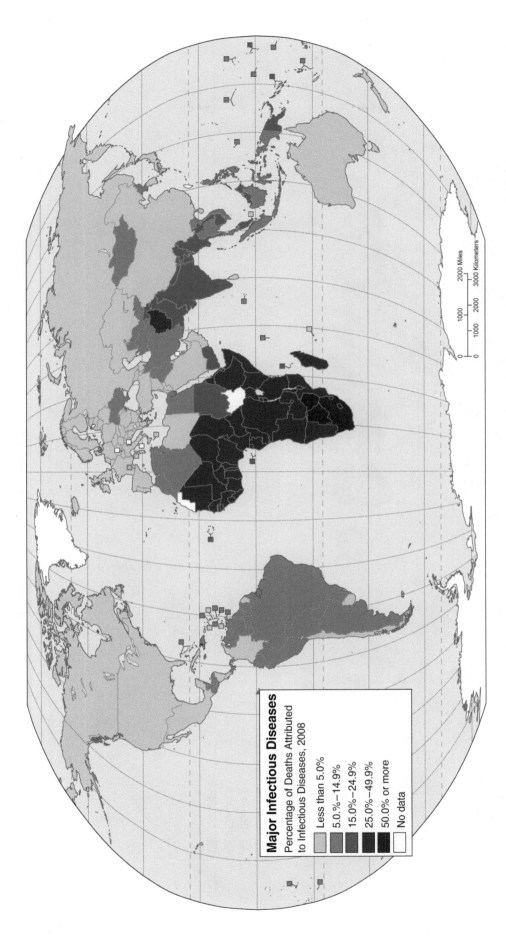

Major Infectious Diseases

Percentage of Deaths Attributed
to Infectious Diseases, 2008

- Less than 5.0%
- 5.0.%–14.9%
- 15.0%–24.9%
- 25.0%–49.9%
- 50.0% or more
- No data

Infectious diseases are the world's leading cause of premature death and at least half of the world's population is, at any time, at risk of contracting an infectious disease. Although we often think of infectious diseases as being restricted to the tropical world (malaria, dengue fever), many if not most of them have attained global proportions. A major case in point is HIV/AIDS, which quite probably originated in Africa but has, over the last two decades, spread throughout the entire world. Major diseases of the nineteenth century, such as cholera and tuberculosis, are making a major comeback in many parts of the world, in spite of being preventable or treatable. Part of the problem with infectious diseases is that they tend to be associated with poverty (poor nutrition, poor sanitation, substandard housing, and so on) and, therefore, are seen as a problem of undeveloped countries, with the consequent lack of funding for prevention and treatment. Infectious diseases are also tending to increase because lifesaving drugs, such as antibiotics and others used in the fight against diseases, are losing their effectiveness as bacteria develop genetic resistance to them. The problem of global warming is also associated with a spread of infectious diseases as many disease vectors (certain species of mosquito, for example) are spreading into higher latitudes with increasingly warm temperatures and are spreading disease into areas where populations have no resistance to them. Infectious diseases have become something greater than simply a health issue of poor countries. They are now major social problems with potentially enormous consequences for the entire world.

-53-

Map 39a   Major Infectious Diseases: HIV/AIDS

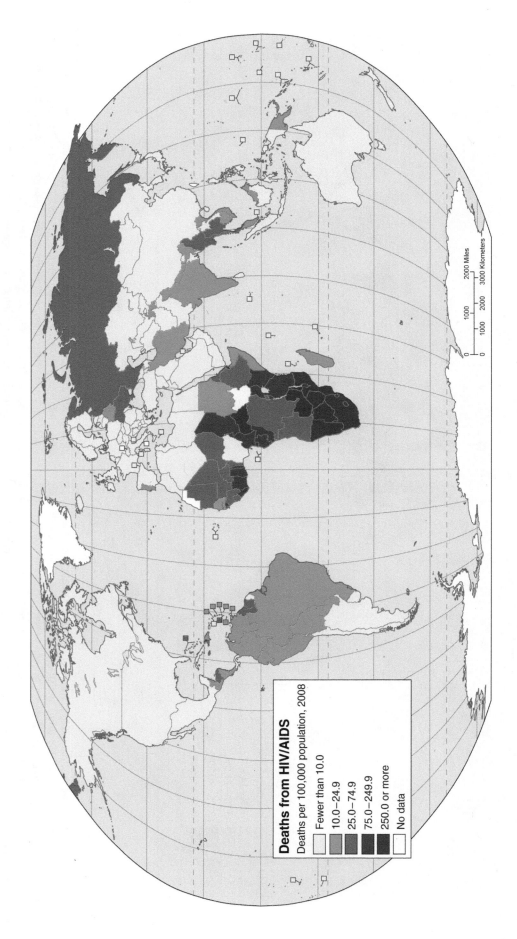

Deaths from HIV/AIDS

Deaths per 100,000 population, 2008

- Fewer than 10.0
- 10.0–24.9
- 25.0–74.9
- 75.0–249.9
- 250.0 or more
- No data

Of all the infectious diseases, the one that poses the greatest risks to public health worldwide is human immune deficiency which, if untreated, progresses to the deadly autoimmune deficiency, or AIDS. The highest incidence of AIDS among adult populations occurs in Sub-Saharan Africa, where public health systems are too poorly funded and developed to treat HIV cases, and the medications needed to preserve health and reasonable longevity are too expensive. Because the rate of adult cases is high, large numbers of children are also infected, having been born HIV-positive. Indeed, among all the world's children living with HIV, approximately 90 percent live in Sub-Saharan Africa. Other countries in the developing world also are high on the scale of HIV/AIDS incidence and the suspicion is that the official figures could go even higher if accurately reported. China, for example, may have rates that are 3 to 4 times greater than the official figures. Worldwide, HIV/AIDS has a devastating impact on family structures (many children are orphaned by both parents dying of AIDS) and on the economic growth of the developing countries.

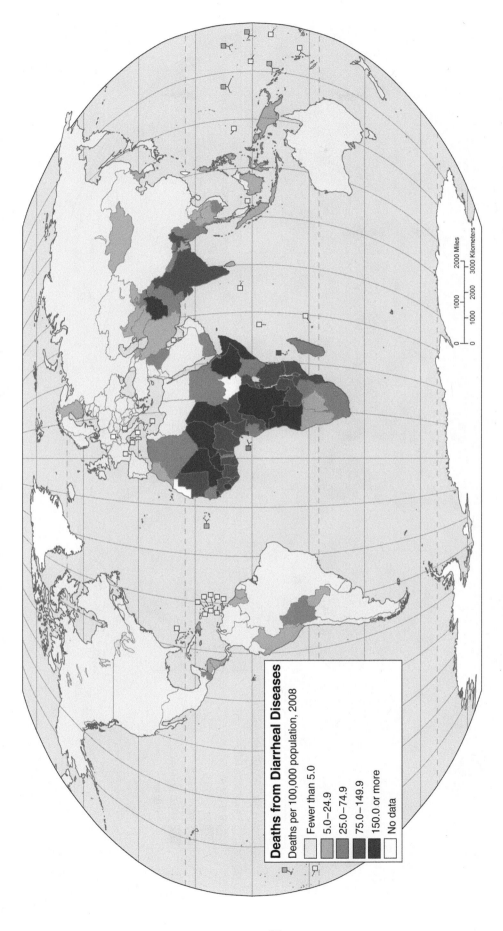

Deaths from Diarrheal Diseases
Deaths per 100,000 population, 2008

Fewer than 5.0
5.0–24.9
25.0–74.9
75.0–149.9
150.0 or more
No data

2000 Miles

3000 Kilometers

1000   2000

0   1000   2000

0   1000

Diarrhea, the passage of three or more loose stools per day, is caused by a wide variety of bacterial and parasitic organisms. Both preventable and treatable, the World Health Organization estimates diarrheal diseases are the second leading cause of death of children younger than the age of five. Approximately 1.5 million children die each year, and there are as many as two billion cases annually.

Map 39c Major Infectious Diseases: Malaria

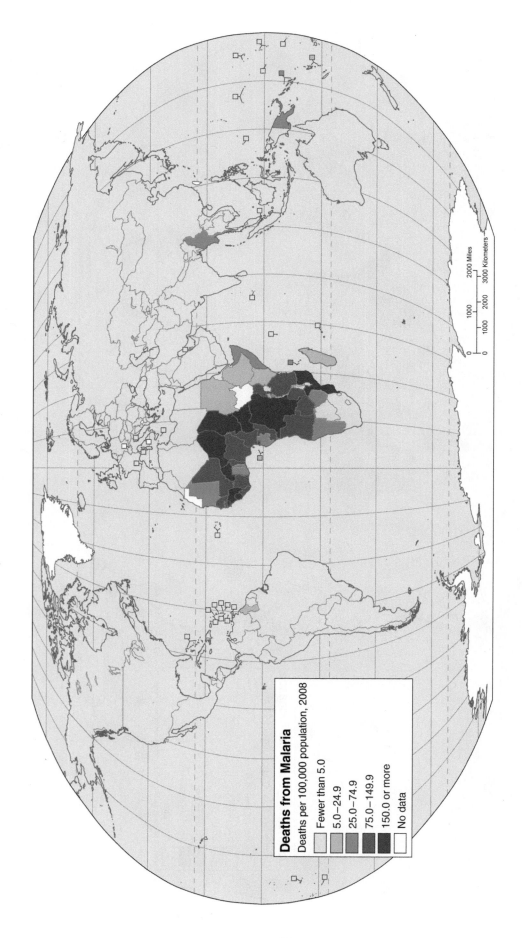

Deaths from Malaria

Deaths per 100,000 population, 2008

- Fewer than 5.0
- 5.0–24.9
- 25.0–74.9
- 75.0–149.9
- 150.0 or more
- No data

Endemic to the tropics, the World Health Organization estimates that half of the world's population is at risk of contracting malaria. Transmitted by mosquitoes, the disease induces fever and vomiting and can become life-threatening if not treated. Although effective treatments for the disease exist, many African countries still see high death rates—those in excess of 150 deaths for every 100,000 persons. Outside of Africa, death rates have dropped substantially since 2000. Various organizations have established the goal of eradication of malaria from Africa, where most of the deaths occur. Time will tell if those efforts are successful.

Map 39d Major Infectious Diseases: Tuberculosis

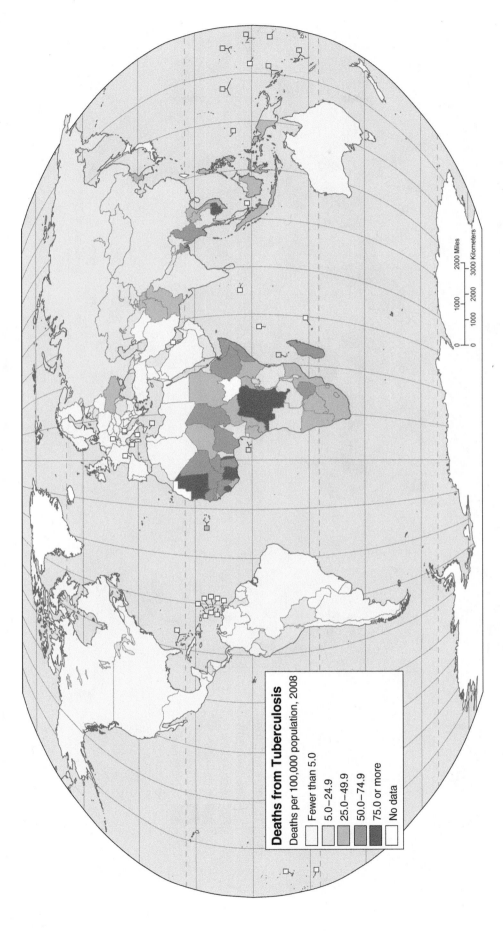

Deaths from Tuberculosis

Deaths per 100,000 population, 2008

- Fewer than 5.0
- 5.0–24.9
- 25.0–49.9
- 50.0–74.9
- 75.0 or more
- No data

Behind HIV/AIDS, tuberculosis, or TB, is the second leading killer globally among infectious diseases. In 2010, nearly nine million people were infected with TB with as many as 1.4 million dying from the disease. Like other diseases, most TB deaths occur in impoverished countries. Like malaria, tuberculosis is treatable, but an estimated one-third of the world's population has been infected with the disease and drug-resistance is now found throughout the world. The disease affects people of all ages but is more likely to be fatal among the elderly and those who are immuno-compromised—it is the leading killer of people infected with HIV.

Map 40 Illiteracy Rates

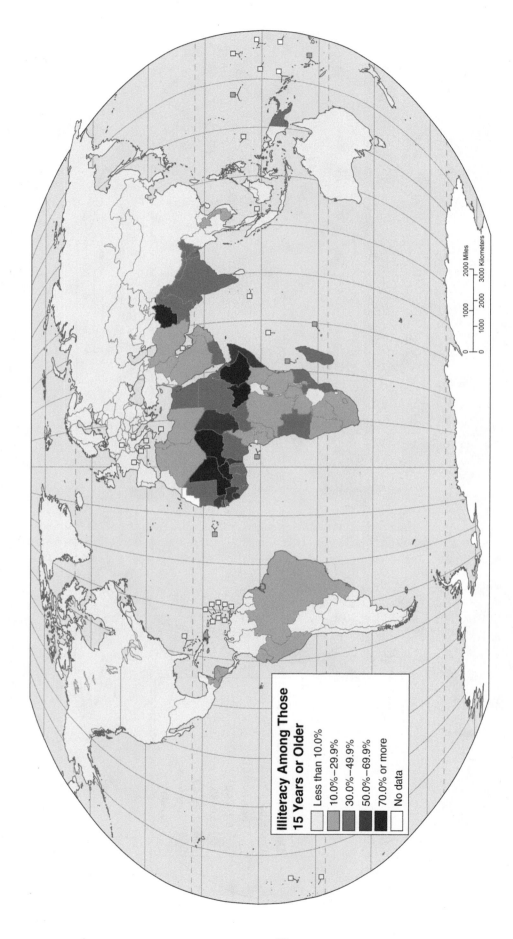

Illiteracy Among Those 15 Years or Older
- Less than 10.0%
- 10.0%—29.9%
- 30.0%—49.9%
- 50.0%—69.9%
- 70.0% or more
- No data

2000 Miles

3000 Kilometers

0   1000   2000

0   1000   2000

Illiteracy rates are based on the percentages of people age 15 or older (classed as adults in most countries) who are not able to write and read, with understanding, a brief, simple statement about everyday life written in their home or official language. As might be expected, illiteracy rates tend to be higher in the less-developed states, where educational systems are a low government priority. Rates of literacy or illiteracy also tend to be gender-differentiated, with women in many countries experiencing educational neglect or discrimination that makes it more likely they will be illiterate. In many developing countries, between five and ten times as many women will be illiterate as men, and the illiteracy rate for women may even exceed 90 percent. Both male and female illiteracy severely compromises economic development.

-58-

Map 41 Primary School Enrollment

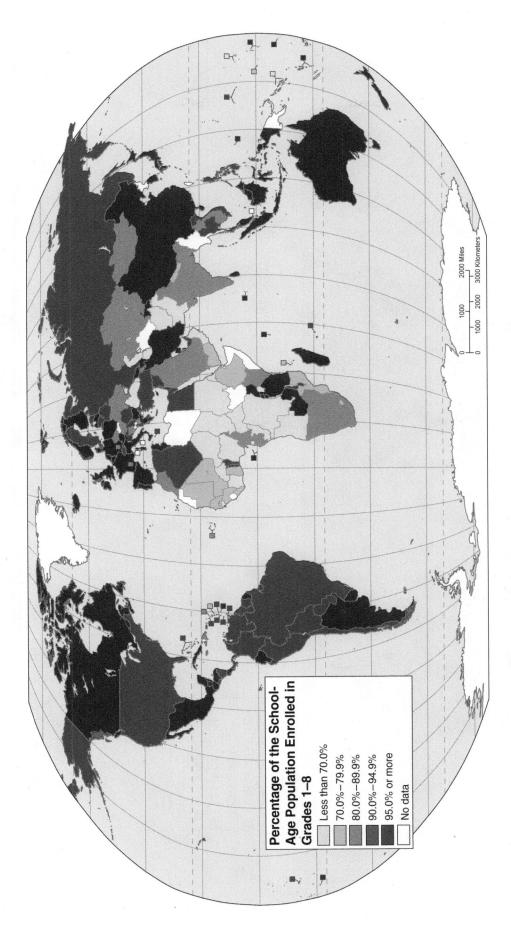

Percentage of the School-Age Population Enrolled in Grades 1–8

Less than 70.0%
70.0%–79.9%
80.0%–89.9%
90.0%–94.9%
95.0% or more
No data

Like many of the other measures illustrated in this atlas, primary school enrollment is a clear reflection of the division of the world into "have" and "have-not" countries. It is also a measure that has changed more rapidly over the last decade than demographic and other indicators of development, as countries of even very modest means have made concerted attempts to attain relatively high percentages of primary school enrollment. That they have been able to do so is good evidence of the fact that reasonably respectable levels of human development are feasible at even modest income levels. High primary school enrollment is also a reflection of the worldwide opinion that a major element in economic development is a well-educated, literate population. The links between human progress, as typified by higher levels of education, and economic growth are not automatic, however, and those countries without programs for maintaining the headway gained by improved education may be on the road to failure in terms of economic development.

Map 42 The Index of Human Development

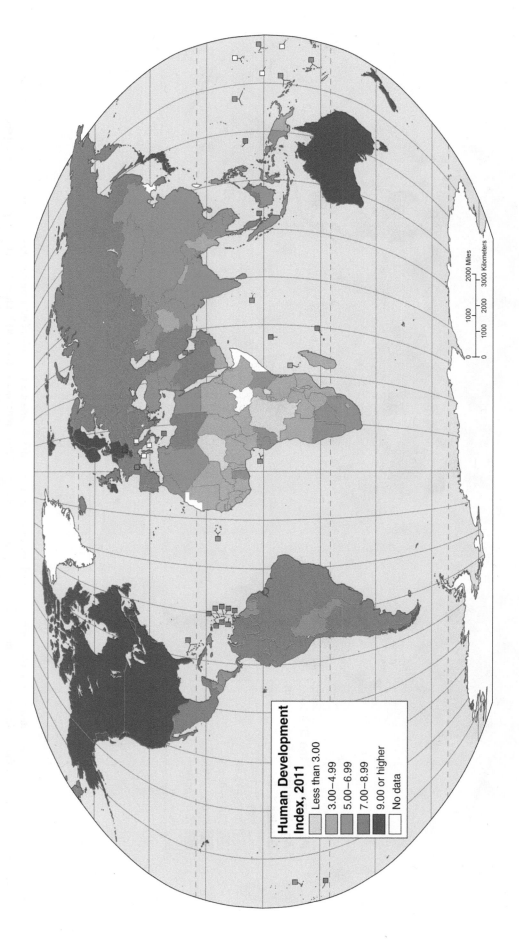

Human Development Index, 2011

- Less than 3.00
- 3.00–4.99
- 5.00–6.99
- 7.00–8.99
- 9.00 or higher
- No data

The development index upon which this map is based takes into account a wide variety of demographic, health, and educational data, including population growth, per capita gross domestic income, longevity, literacy, and years of schooling. The map reveals significant improvement in the quality of life in Middle and South America, although it is questionable whether the gains made in those regions can be maintained in the face of the dramatic population increases expected over the next 30 years. More clearly than anything else, the map illustrates the near-desperate situation in Africa and South Asia. In those regions, the unparalleled growth in population threatens to overwhelm all efforts to improve the quality of life. In Africa, for example, the population is increasing by 20 million persons per year. With nearly 45 percent of the continent's population aged 15 years or younger, this growth rate will accelerate as the women reach child-bearing age. Africa, along with South Asia, faces the very difficult challenge of providing basic access to health care, education, and jobs for a rapidly increasing population. The map also illustrates the striking difference in quality of life between those who inhabit the world's equatorial and tropical regions and those fortunate enough to live in the temperate zones, where the quality of life is significantly higher.

Map 43 Demographic Stress: The Youth Bulge

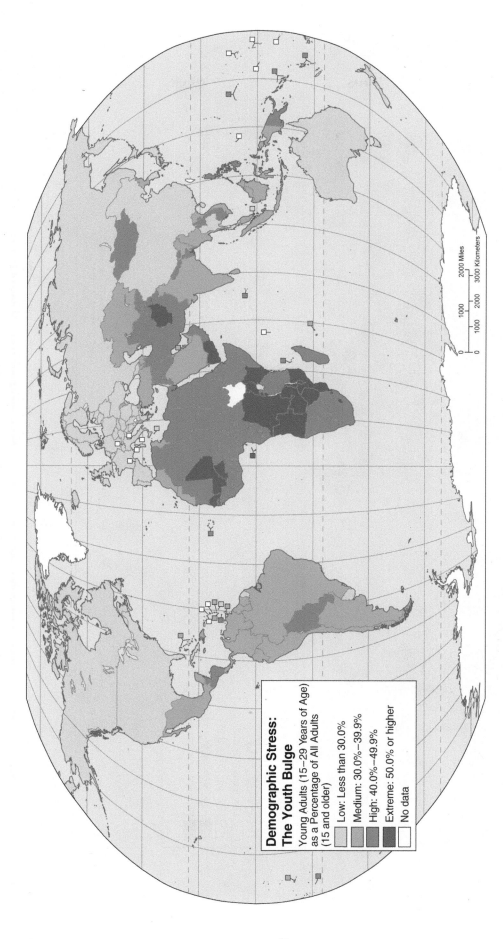

**Demographic Stress:
The Youth Bulge**

Young Adults (15–29 Years of Age)
as a Percentage of All Adults
(15 and older)

- Low: Less than 30.0%
- Medium: 30.0%–39.9%
- High: 40.0%–49.9%
- Extreme: 50.0% or higher
- No data

0   1000   2000   3000 Kilometers

0   1000   2000 Miles

One of the greatest stresses of the demographic transition is when the death rate drops as the result of better public health and sanitation and the birth rate remains high for the same reasons it has always been high in traditional, agricultural societies: (1) the need for enough children to supply labor, which is one of the few ways to increase agricultural production in a non-mechanized agricultural system, and (2) the need for enough children to offset high infant and child mortality rates so that some children will survive to take care of parents in their old age. With declining death rates and steady birth rates, population growth skyrockets—particularly among the youngest cohorts of a population (between the ages of birth and 15). Although this is, on the one hand, a demographic benefit because it increases the size of the labor force in the next generation and thereby helps to accelerate economic growth, it also means more people of child-bearing age in the next generation and, hence, greater numbers of births, which continue to swell the population in the youngest, most vulnerable, and most dependent portion of the population. The literature on population and conflict suggests that the larger the percentage of a population younger than the age of 25, the greater the chance for political violence and warfare.

-61-

Map 44   Demographic Stress: Rapid Urban Growth

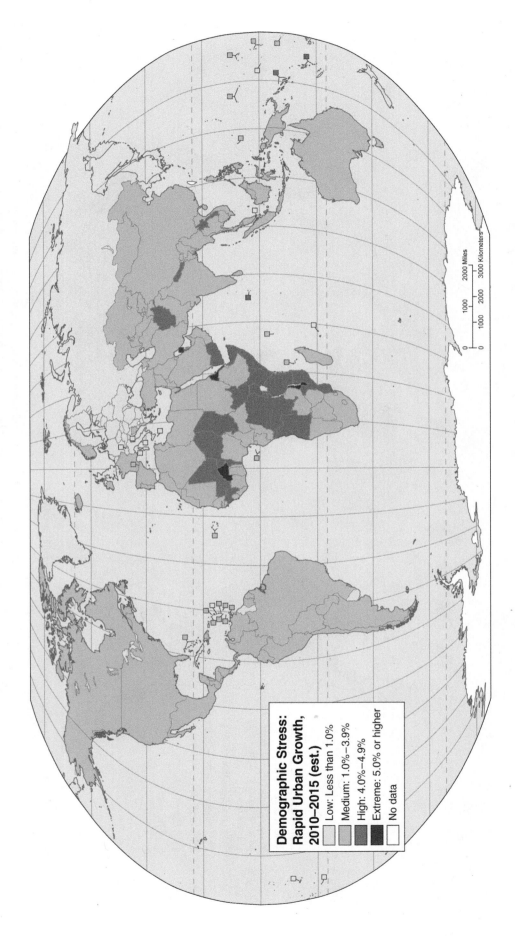

Demographic Stress: Rapid Urban Growth, 2010–2015 (est.)

- Low: Less than 1.0%
- Medium: 1.0%—3.9%
- High: 4.0%—4.9%
- Extreme: 5.0% or higher
- No data

The trend toward urbanization—an increasing percentage of a country's population living in a city—is normally a healthy, modern trend. But in many of the world's developing countries, increasing urbanization is the consequence of high physiologic population densities (too many people for the available agricultural land) and the flight of poorly educated, untrained rural poor to the cities where they hope (often in vain) to find employment. High urbanization exists in many of the world's poorest countries where the urban poor live in conditions that make the worst living conditions in the inner cities of developed countries look positively luxurious. In the urban slums of South America, Africa, and South Asia, millions of people live in temporary housing of cardboard and flattened aluminum cans, with no public services such as water, sewage, or electricity. In Africa, where only 40% of the population is urbanized (in comparison with more than 90% in North America and Europe), the population of urban poor is greater than the total urban population of the United States and the European Union. This represents an increasingly destabilizing element of modern urban societies.

Map 45 Demographic Stress: Competition for Cropland

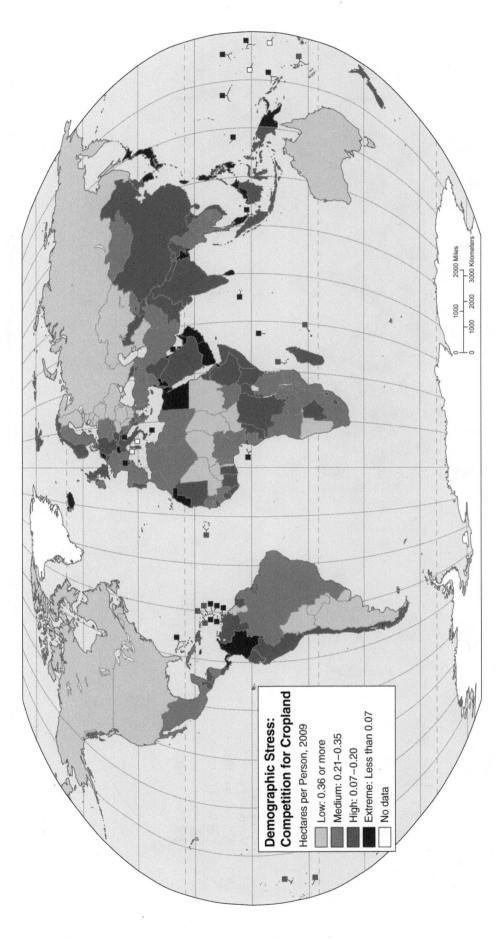

Demographic Stress: Competition for Cropland

Hectares per Person, 2009

- Low: 0.36 or more
- Medium: 0.21–0.35
- High: 0.07–0.20
- Extreme: Less than 0.07
- No data

Many of the world's countries have reached the point where available acres or hectares in cropland are less than the numbers of people wishing to occupy them. For developed countries, who pay for agricultural imports with industrial exports, this is not an alarming trend and therefore countries such as Germany or Italy (where the ratio between cropland and farmers is very low) have little to worry about. But in countries in South America, Africa, and South and East Asia, the trend toward more farmers and less available land *is* an alarming trend. In these developing regions, farmers depend on their crops for a relatively meager subsistence diet and—if they are lucky—a few

bushels of rice or corn to take to the local market to sell or exchange for the small surpluses of other farmers. Despite the broad global trends toward urbanization and more productive agriculture (and this generally means mechanized agriculture), farm occupation and subsistence cropping remain mainstays of the economy in Africa south of the Sahara, in much of western South America, and in much of South and East Asia. Here, the increasingly small margin between the numbers of farmers and the amount of available farmland can lead to conflict among tribal communities or even among members of a single family.

Map 46   Demographic Stress: Competition for Freshwater

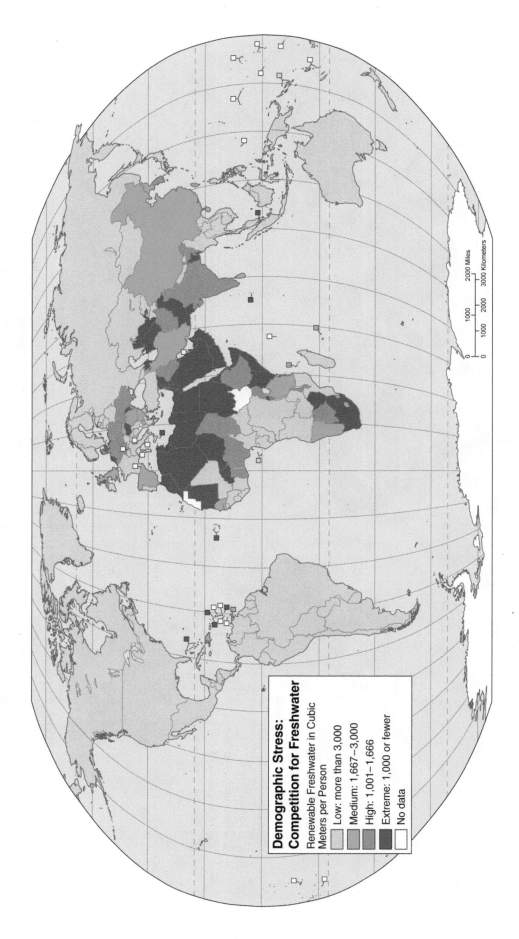

**Demographic Stress:
Competition for Freshwater**

Renewable Freshwater in Cubic
Meters per Person

Low: more than 3,000
Medium: 1,667–3,000
High: 1,001–1,666
Extreme: 1,000 or fewer
No data

As rates of urbanization have increased, the competition between city dwellers and farmers for water has also increased. And as the population of farmers relative to the available surface or ground water supply has accelerated, so rural competition for access to freshwater for irrigation has increased. Obviously, this trend is most obvious in the world's drier regions. On this map, the greatest stress factors relating to water availability are in North and East Africa, the Middle East, and Central and East Asia. The current conflict between Israelis and Palestinians in the West Bank goes far beyond religion or politics: Some is the result of the more affluent Israeli farmers being able to drill wells to tap ground water, which lowers the water table and causes previously accessible surface wells in Palestinian villages to go dry. Given that freshwater is, for all practical purposes, a nonrenewable resource, more conflicts over access to water are going to erupt in the future.

Map 47 Demographic Stress: Interactions of Demographic Stress Factors

Demographic Stress: Interactions of Demographic Stress Factors

- No apparent demographic stress
- One stress factor (high or extreme)
- Two stress factors (high or extreme)
- Three stress factors (high or extreme)
- Four stress factors (high or extreme)

0   1000   2000   2000 Miles
0   1000   2000   3000 Kilometers

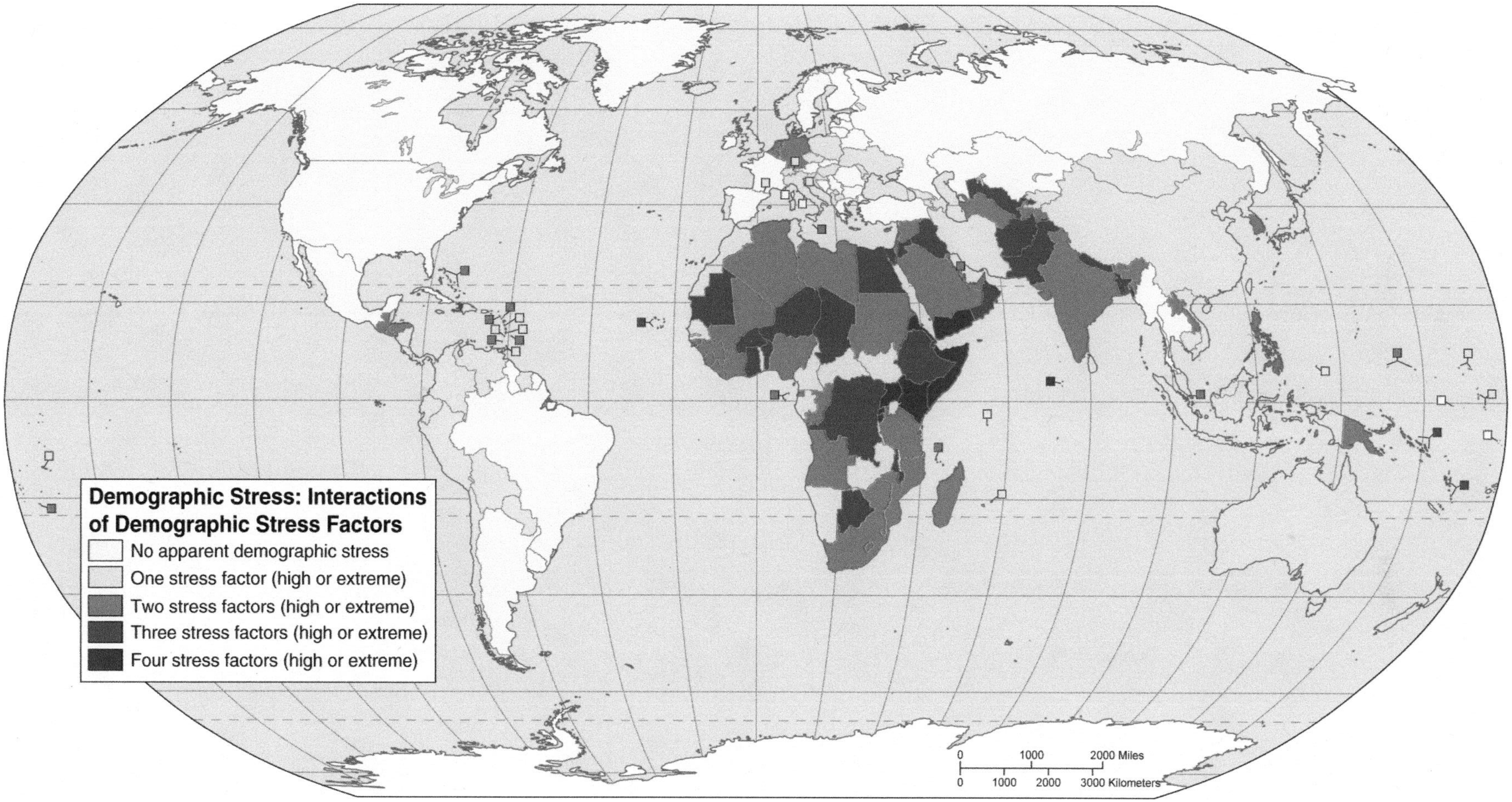

We have persisted in viewing internal and international conflicts as conflicts produced by religious differences, the acceptance or rejection of "freedom" or "democracy," or even, as argued by a prominent historian, a "Clash of Civilizations." Although religious, political, or historical differences cannot be ignored in interpreting the causes of conflict, neither can simple demographic stresses produced by too many new urban dwellers without jobs or hopes of jobs, by too many farmers and too little farmland, or by too little available freshwater. What we have repeatedly seen is countries suffering from multiple demographic risk factors are, were, and will be much more prone to civil conflict than countries with only one or two demographic risk factors. A look at this map provides a potential predictor of future civil unrest, violence, and even civil war.

Unit IV

Global Economic Patterns

Map 48 Gross National Income

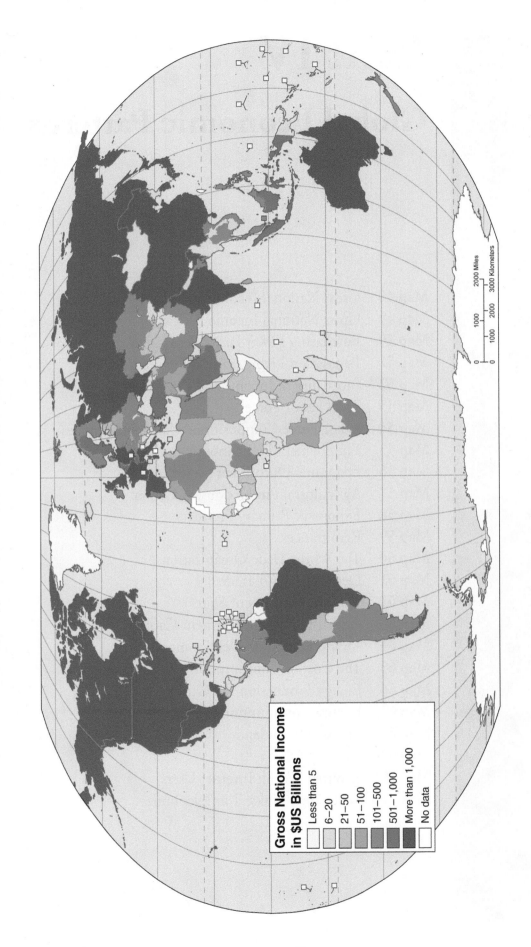

**Gross National Income
in $US Billions**

Less than 5
6–20
21–50
51–100
101–500
501–1,000
More than 1,000
No data

0   1000   2000 Miles

0   1000   2000   3000 Kilometers

Gross National Income (GNI) is the broadest measure of national income and measures the total claims of a country's residents to all income from domestic and foreign products during a year. Although GNI is often misleading and commonly incomplete, it is often used by economists, geographers, political scientists, policy makers, development experts, and others not only as a measure of relative well-being but also as an instrument of assessing the effectiveness of economic and political policies. What is wrong with GNI? First of all, it does not take into account a number of real economic factors, such as environmental deterioration, the accumulation or degradation of human and social capital, or the value of household work. Yet in spite of these deficiencies, GNI is still a reasonable way to assess the relative wealth of nations: the vast differences in wealth that separate the poorest countries from the richest. One of the more striking features of the map is the evidence it presents that such a small number of countries possess so many of the world's riches (keeping in mind that GNI provides no measure of the distribution of wealth within a country).

Map 49 Gross National Income Per Capita

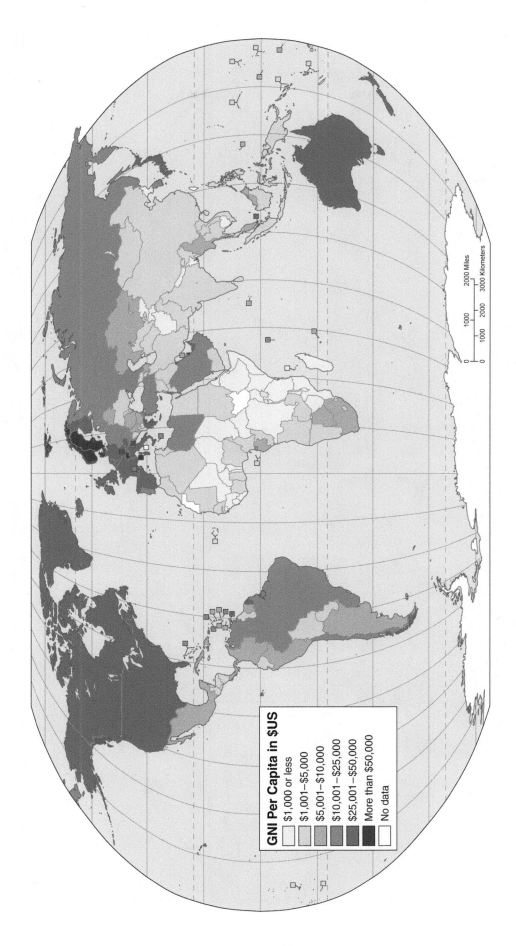

GNI Per Capita in $US

- $1,000 or less
- $1,001–$5,000
- $5,001–$10,000
- $10,001–$25,000
- $25,001–$50,000
- More than $50,000
- No data

0 1000 2000 3000 Kilometers
0 1000 2000 Miles

Gross National Income (GNI) in either absolute or per capita form should be used cautiously as a yardstick of economic strength because it does not measure the distribution of wealth among a population. There are countries (most notably, the oil-rich countries of the Middle East) where per capita GNI is high but where the bulk of the wealth is concentrated in the hands of a few individuals, leaving the remainder in poverty. Even within countries in which wealth is more evenly distributed (such as those in North America or Western Europe), there is a tendency for dollars or pounds sterling or euros to concentrate in the bank accounts of a relatively small percentage of the population.

Yet the maldistribution of wealth tends to be greatest in the less developed countries, where the per capita GNI is far lower than in North America and Western Europe, and poverty is widespread. In fact, a map of GNI per capita offers a reasonably good picture of comparative economic well-being. It should be noted that a low per capita GNI does not automatically condemn a country to low levels of basic human needs and services. There are a few countries, such as Costa Rica and Sri Lanka, that have relatively low per capita GNI figures but rank comparatively high in other measures of human well-being, such as average life expectancy, access to medical care, and literacy.

Map 50 Purchasing Power Parity Per Capita

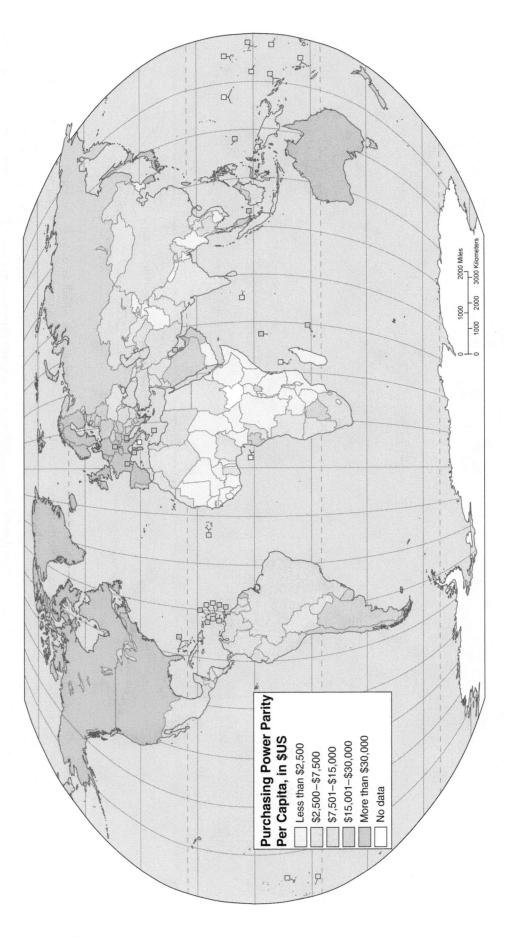

Purchasing Power Parity Per Capita, in $US

- Less than $2,500
- $2,500–$7,500
- $7,501–$15,000
- $15,001–$30,000
- More than $30,000
- No data

0 1000 2000 Miles
0 1000 2000 3000 Kilometers

Of all the economic measures that separate the "haves" from the "have-nots," perhaps per capita Purchasing Power Parity (PPP) is the most meaningful. Although per capita figures can mask significant uneven distributions within a country, they are generally useful for demonstrating important differences between countries. Per capita GNP and GDP (Gross Domestic Product) figures, and even per capita income, have the limitation of seldom reflecting the true purchasing power of a country's currency at home. In order to get around this limitation, international economists seeking to compare national currencies developed the PPP measure, which shows the level of goods and services that holders of a country's money can acquire locally. By converting all currencies to the "international dollar," the World Bank and other organizations using PPP can now show more truly

comparative values, because the new currency value shows the number of units of a country's currency required to buy the same quantity of goods and services in the local market as one U.S. dollar would buy in an average country. The use of PPP currency values can alter the perceptions about a country's true comparative position in the world economy. More than per capita income figures, PPP provides a valid measurement of the ability of a country's population to provide basic necessities. A glance at the map shows a clear-cut demarcation between temperate and tropical zones, with most of the countries with a PPP above $7,500 in the midlatitude zones and most of those with lower PPPs in the tropical and equatorial regions. Where exceptions to this pattern occur, they usually stem from a tremendous maldistribution of wealth among a country's population.

-70-

Map 51 Population Living on Less Than $2 Per Day

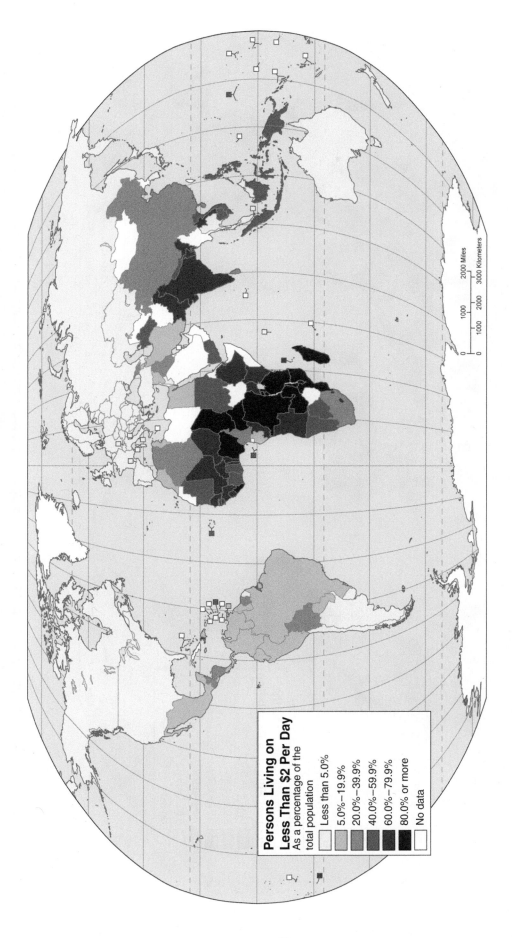

**Persons Living on
Less Than $2 Per Day**
As a percentage of the
total population

- Less than 5.0%
- 5.0%–19.9%
- 20.0%–39.9%
- 40.0%–59.9%
- 60.0%–79.9%
- 80.0% or more
- No data

1000   2000 Miles

0   1000   2000   3000 Kilometers

Map 50 demonstrates the disparity that exists between wealthy and poor countries, but it does not adequately portray the depth of poverty that exists in many countries. To do that, it is necessary to isolate the population living at the bottom of the economic scale. Nearly half of the world's population lives on less than $2.00 a day. In much of Africa, at least 60 percent of the population lives at this level and in a dozen countries, more than 80 percent of the population lives at this level. The extent of poverty at this scale

not only reflects absolute wealth, it also reflects a population's ability to access those basic necessities that those in the developed world take for granted: adequate food, shelter, clothing, education, and access to medical care. Additionally, it is important to note that $2.00 a day is an average, not a reflection of actual income earned each day. In many countries, people have irregular incomes meaning that there may be many days where no income is received.

Map 52   Inequality of Income and Consumption

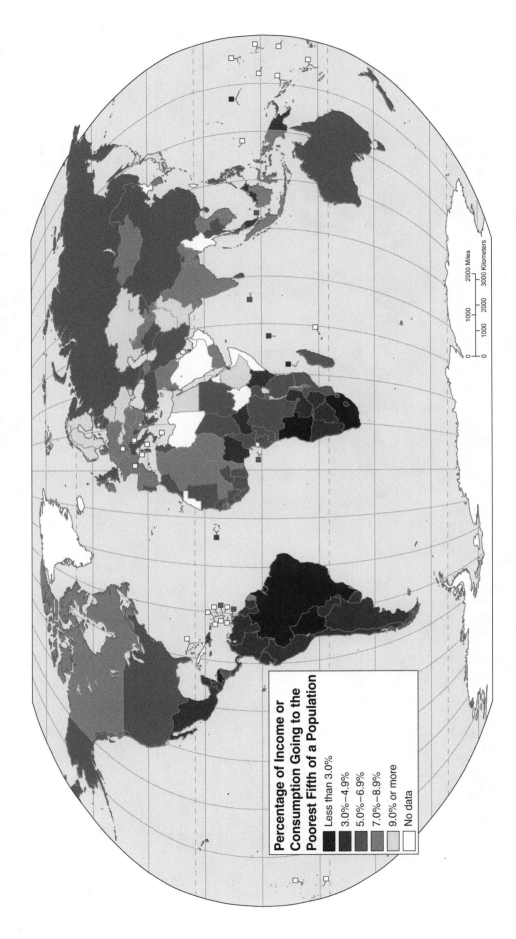

Percentage of Income or Consumption Going to the Poorest Fifth of a Population

- Less than 3.0%
- 3.0%–4.9%
- 5.0%–6.9%
- 7.0%–8.9%
- 9.0% or more
- No data

Although it is more than arguably true that the poor get poorer and the rich get richer (and that, by the way, is just as true in some highly developed areas such as the United States and the United Kingdom as it is in the lesser-developed regions), what is often ignored is the breadth of the inequality in distribution of incomes or in the levels of consumption of basic goods. In many of the world's developing countries—despite the rapid increases in economic growth of China and India over the last three decades—the poorest 20% of the population receives less than 7% of the income or consumption share. Although school participation rates have risen worldwide, in countries with the greatest inequalities of income, access to education remains low. If translated into ratios, the inequality ratio of many of the world's countries (including, as noted above, some in the highly developed world) is 8 or higher, meaning that the top 20% of the population spends and consumes at least 8 times as much per person as the bottom 20% of the population.

-72-

Map 53 Total Labor Force

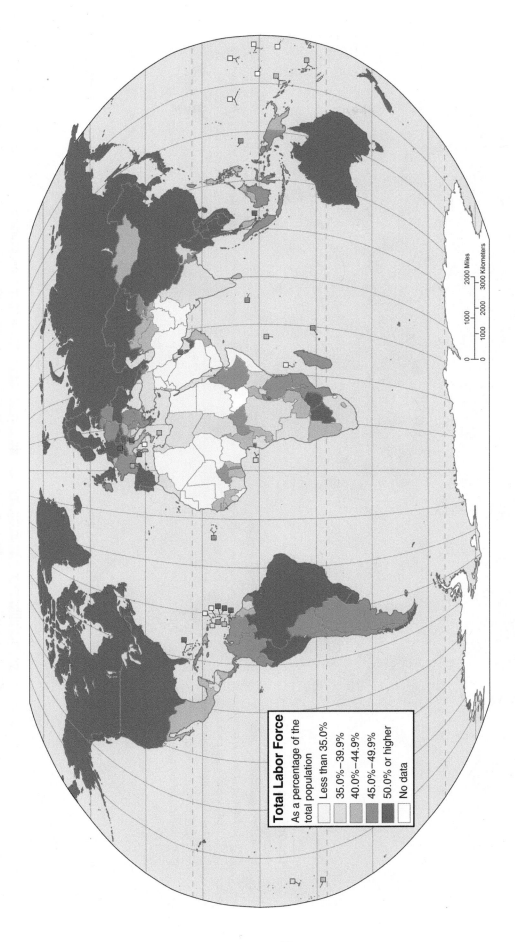

Total Labor Force

As a percentage of the total population

- Less than 35.0%
- 35.0%–39.9%
- 40.0%–44.9%
- 45.0%–49.9%
- 50.0% or higher
- No data

The term *labor force* refers to the economically active portion of a population, that is, all people who work or are without work but are available for and are seeking work to produce economic goods and services. The total labor force thus includes both the employed and the unemployed (as long as they are actively seeking employment). Labor force is considered a better indicator of economic potential than employment/ unemployment figures, because unemployment figures will include experienced workers with considerable potential who are temporarily out of work. Unemployment figures will also incorporate persons seeking employment for the first time (many recent college graduates, for example). Generally, countries with higher percentages of total population within the labor force will be countries with higher levels of economic development. This is partly a function of levels of education and training and partly a function of the age distribution of populations. In developing countries, substantial percentages of the total population are too young to be part of the labor force.

-73-

Map 54 Labor Force Composition: Males and Females

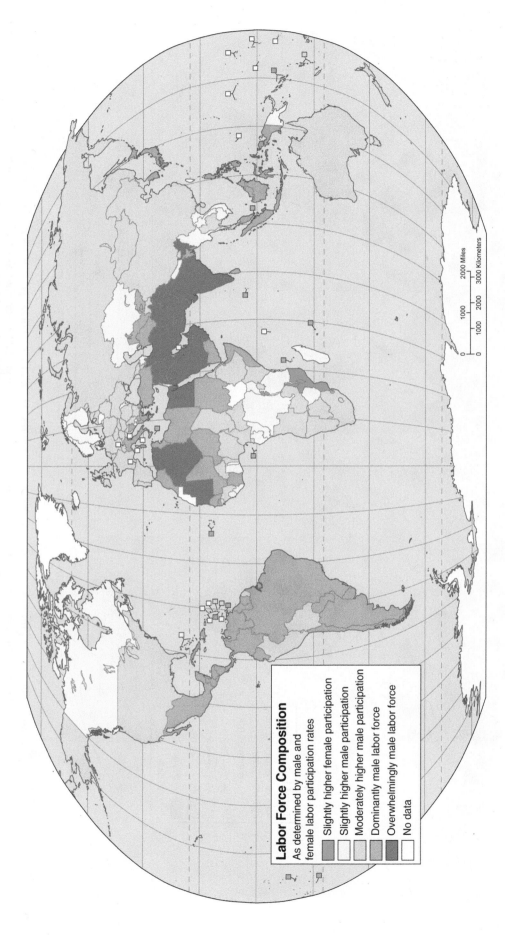

Labor Force Composition
As determined by male and female labor participation rates

Slightly higher female participation
Slightly higher male participation
Moderately higher male participation
Dominantly male labor force
Overwhelmingly male labor force
No data

1000   2000 Miles
0   1000   2000   3000 Kilometers

When examining the composition of the labor force, there are noticeable differences among countries as to the level of participation of women. Unlike other maps in this unit, the pattern of female participation in the labor force does not mimic the wealthy vs. poor country patterns. Although wealthy countries generally have much higher participation of females in the labor force, so do some of the world's poorest countries. The lack of a "wealthy vs. poor" pattern is influenced by two factors. First, the extent to which women participate in the labor force is strongly shaped by cultural forces. In many countries with low female participation, strong religious and historical forces heavily influence the roles of women. Secondly, in some countries women are working but their work is not adequately represented in labor force data because they are engaged in "informal" economic activities. In developing countries in particular, a significant percentage of the population consists of women engaged in household activities or subsistence cultivation, which are not reported to those collecting employment data. Thus, these people seldom appear on lists of either employed or unemployed seeking employment.

-74-

Map 55 Employment by Economic Activity

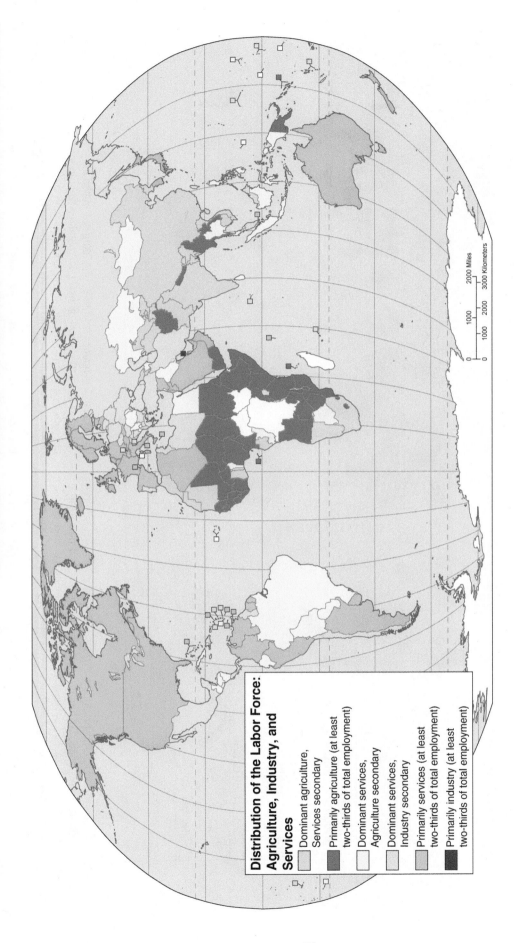

Distribution of the Labor Force: Agriculture, Industry, and Services

- Dominant agriculture, Services secondary
- Primarily agriculture (at least two-thirds of total employment)
- Dominant services, Agriculture secondary
- Dominant services, Industry secondary
- Primarily services (at least two-thirds of total employment)
- Primarily industry (at least two-thirds of total employment)

The employment structure of a country's population is one of the best indicators of the country's position on the scale of economic development. At one end of the scale are those countries with more than 40 percent of their labor force employed in agriculture. These are almost invariably the least developed, with high population growth rates, poor human services, significant environmental problems, and so on. In the middle of the scale are two types of countries: those with more than 20 percent of their labor force employed in industry and those with a fairly even balance among agricultural, industrial, and service employment but with at least 40 percent of their labor force employed in service activities. Generally, these countries have undergone the industrial revolution fairly recently

and are still developing an industrial base while building up their service activities. This category also includes countries with a disproportionate share of their economies in service activities primarily related to resource extraction. On the other end of the scale from the agricultural economies are countries with more than 20 percent of their labor force employed in industry and more than 50 percent in service activities. These countries are, for the most part, those with a highly automated industrial base and a highly mechanized agricultural system (the "postindustrial," developed countries). They also include, particularly in Middle and South America and Africa, industrializing countries that are also heavily engaged in resource extraction as a service activity.

-75-

Map 56 Economic Output Per Sector

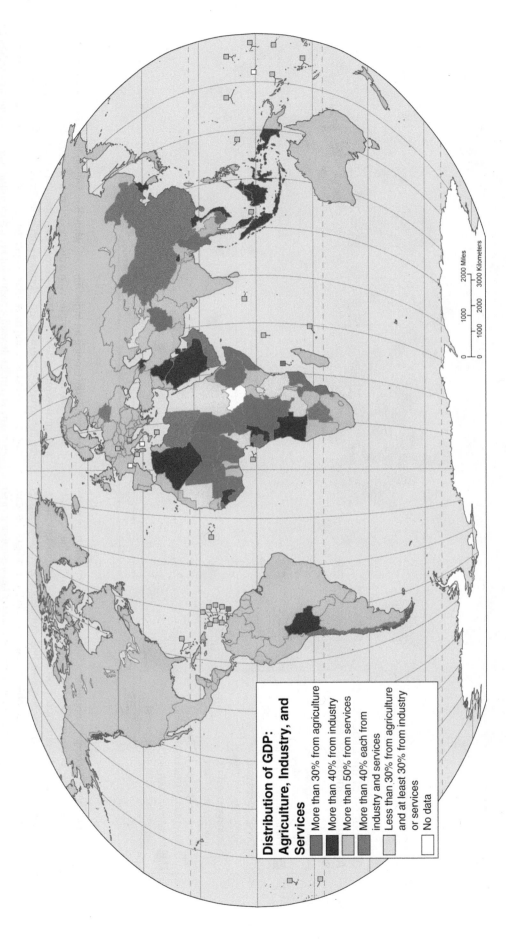

Distribution of GDP: Agriculture, Industry, and Services

- More than 30% from agriculture
- More than 40% from industry
- More than 50% from services
- More than 40% each from industry and services
- Less than 30% from agriculture and at least 30% from industry or services
- No data

0 1000 2000 3000 Kilometers
0 1000 2000 Miles

The percentage of the gross domestic product (the final output of goods and services produced by the domestic economy, including net exports of goods and nonfactor—nonlabor, noncapital—services) that is devoted to agricultural, industrial, and service activities is considered a good measure of the level of economic development. In general, countries with more than 40 percent of their GDP derived from agriculture are still in a *colonial dependency economy*—that is, raising agricultural goods primarily for the export market and dependent upon that market (usually the richer countries). Similarly, countries with more than 40 percent of GDP devoted to both agriculture and services often emphasize resource extractive (primarily mining and forestry) activities. These also tend to be *colonial dependency* countries, providing raw materials for foreign markets. Countries with more than 40 percent of their GDP obtained from industry are normally well along the path to economic development. Countries with more than half of their GDP based on service activities fall into two ends of the development spectrum. On the one hand are countries heavily dependent upon both extractive activities and tourism and other low-level service functions. On the other hand are countries that can properly be termed *postindustrial*: they have already passed through the industrial stage of their economic development and now rely less on the manufacture of products than on finance, research, communications, education, and other service-oriented activities.

Map 57 Agricultural Production Per Capita

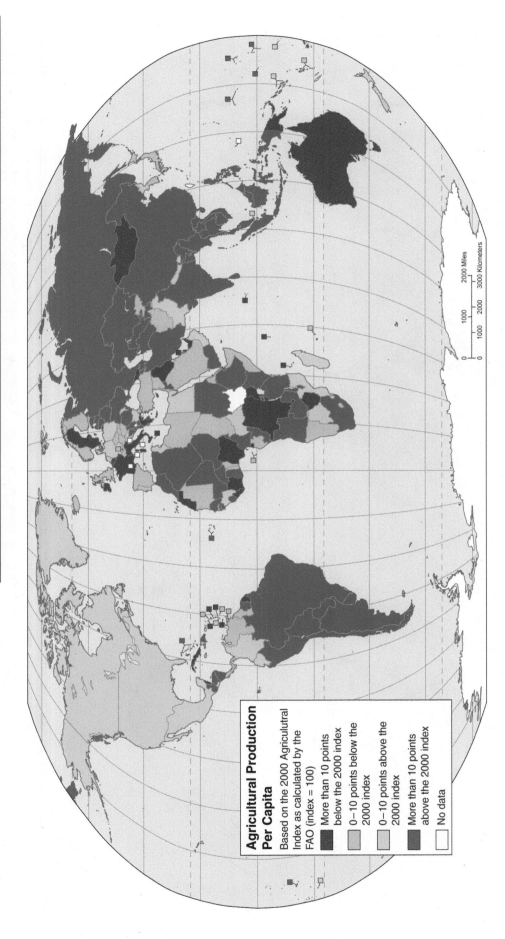

Agricultural Production Per Capita

Based on the 2000 Agriculultral Index as calculated by the FAO (index = 100)

- More than 10 points below the 2000 index
- 0–10 points below the 2000 index
- 0–10 points above the 2000 index
- More than 10 points above the 2000 index
- No data

0   1000   2000   3000 Kilometers
0   1000   2000 Miles

Agricultural production includes the value of all crops and livestock products originating within a country for the base year of 2010. The index value portrays the disposable output (after deductions for livestock feed and seed for planting) of a country's agriculture in comparison with that in 2000. Thus, the production values show not only the relative ability of countries to produce food but also show whether or not that ability has increased or decreased. In general, global food production has kept up with or very slightly exceeded population growth. However, there are significant regional variations in the trend of food production keeping up with or surpassing population growth. For example, agricultural production in Africa and in Middle America has fallen, while production in South America, Asia, and Europe has risen. In the case of Africa, the drop in production reflects a

population growing more rapidly than agricultural productivity. Where rapid increases in food production per capita exist (as in certain countries in South America, Asia, and Europe), most often the reason is the development of new agricultural technologies that have allowed food production to grow faster than population. In much of Asia, for example, the so-called Green Revolution of new, highly productive strains of wheat and rice made positive index values possible. Also in Asia, the cessation of major warfare allowed some countries (Cambodia, Laos, and Vietnam) to show substantial increases over the 2000 index. In some cases, a drop in production per capita reflects government decisions to limit production in order to maintain higher prices for agricultural products. Several countries in Western Europe fall into this category.

Map 58 Exports of Primary Products

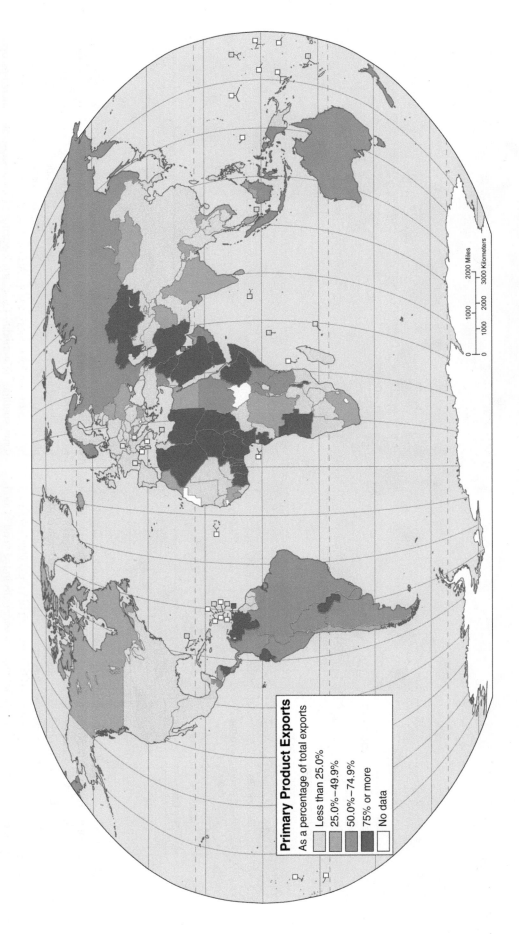

Primary Product Exports

As a percentage of total exports

- Less than 25.0%
- 25.0%–49.9%
- 50.0%–74.9%
- 75% or more
- No data

Primary products are those that require additional processing before they enter the consumer market: metallic ores that must be converted into metals and then into metal products such as automobiles or refrigerators; forest products such as timber that must be converted to lumber before they become suitable for construction purposes; and agricultural products that require further processing before being ready for human consumption. It is an axiom in international economics that the more a country relies on primary products for its export commodities, the more vulnerable its economy is to market fluctuations. Those countries with only primary products to export are hampered in their economic growth. A country dependent on only one or two products for export revenues is unprotected from economic shifts, particularly a changing market demand for its products. Imagine what would happen to the thriving economic status of the oil-exporting states of the Persian Gulf, for example, if an alternate source of cheap energy were found. A glance at this map shows that those countries with the lowest levels of economic development tend to be concentrated on primary products and, therefore, have economies that are especially vulnerable to economic instability.

Map **59** Remittances

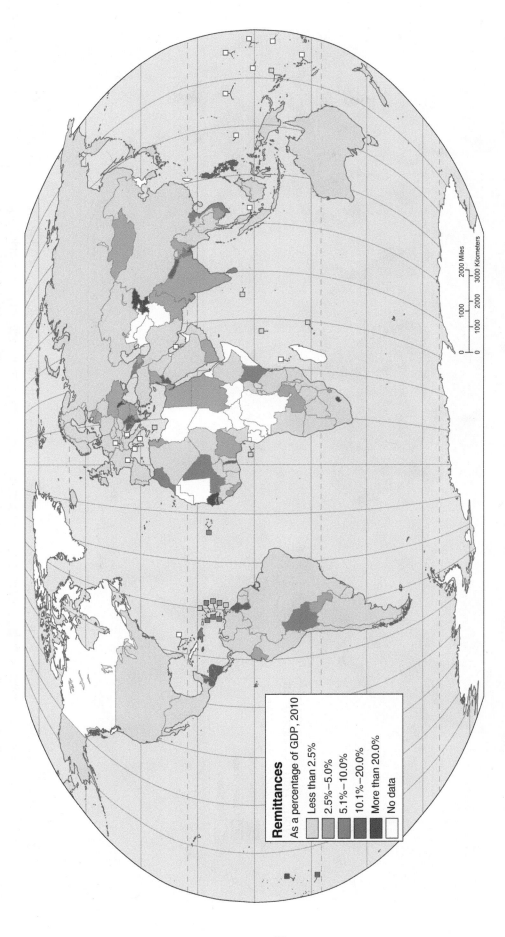

Remittances
As a percentage of GDP, 2010

- Less than 2.5%
- 2.5%–5.0%
- 5.1%–10.0%
- 10.1%–20.0%
- More than 20.0%
- No data

2000 Miles

0   1000   2000   3000 Kilometers

As seen in Map 27, the wealthy countries of the world also have sizeable immigrant populations. In countries where economic pull factors are particularly strong, there may also be a large population of foreign workers who ultimately may return to the home country. Whether the move to a new country is temporary or permanent, the ties to the home country oftentimes remain strong. Economic ties can be particularly strong, as demonstrated by remittances. A remittance simply is the transfer of money from an individual to his or her home country. During the mid-2000s remittances exceeded $250 billion. In some smaller countries such as Honduras and Tajikistan, remittances comprise more than a quarter of GDP. As would be expected, the bulk of this money is coming from workers employed in North America, Western Europe, and Japan.

Map 60 The World Trade Organization

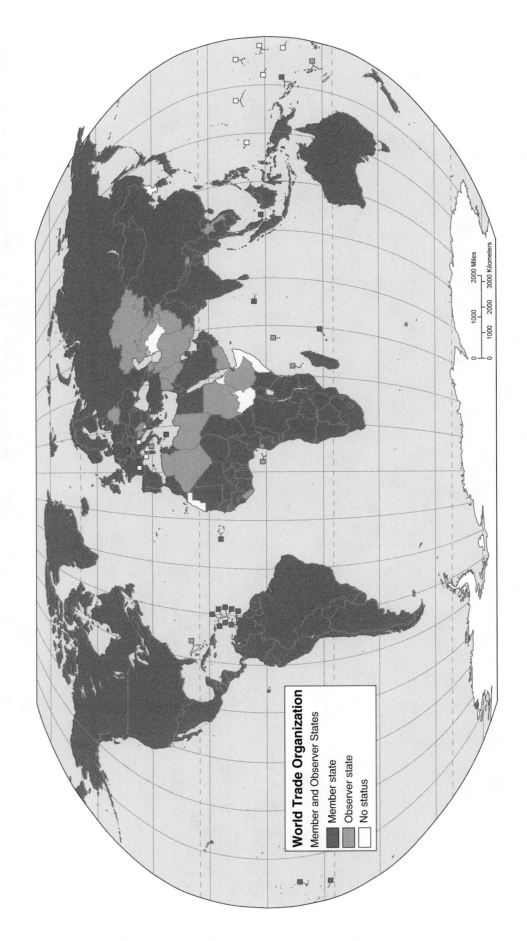

World Trade Organization
Member and Observer States

- Member state
- Observer state
- No status

0 1000 2000 Miles
0 1000 2000 3000 Kilometers

After World War II, the General Agreement on Tariffs and Trade (GATT) sponsored several rounds of negotiations, especially related to lower tariffs but also considering issues such as dumping and other nontariff questions. The last round of negotiations under GATT took place in Uruguay in 1986–1994 and set the stage for the World Trade Organization (WTO), which was formally established in 1995. Today the WTO has 155 members with more than 27 "observer" governments. With the exception of the Holy See (Vatican), observer governments are expected to begin negotiations for full membership within five years of becoming observers. The objective of the WTO is to help international trade flow smoothly and fairly and to assure more stable and secure supplies of goods to consumers. To this end, it administers trade agreements, acts as a forum for trade negotiations, settles trade disputes, reviews national trade policies, assists developing countries through technical assistance and training programs, and cooperates with other international organizations. Increased globalization of the world's economy makes the administrative role of the WTO of increasing importance in the twenty-first century. The headquarters of the WTO is in Geneva, Switzerland.

Map 61   Dependence on Trade

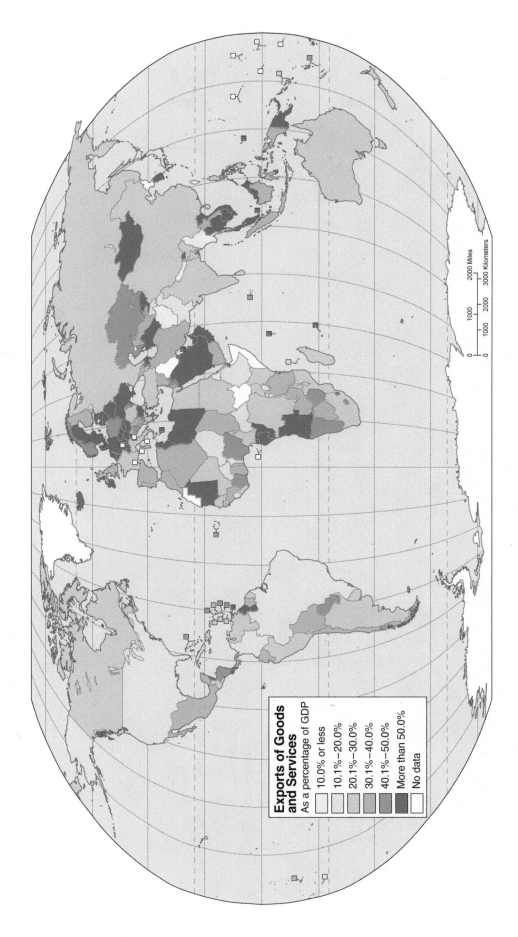

Exports of Goods and Services
As a percentage of GDP

- 10.0% or less
- 10.1%–20.0%
- 20.1%–30.0%
- 30.1%–40.0%
- 40.1%–50.0%
- More than 50.0%
- No data

As the global economy becomes more and more a reality, the economic strength of virtually all countries is increasingly dependent upon trade. For many developing nations, with relatively abundant resources and limited industrial capacity, exports provide the primary base upon which their economies rest. Even countries such as the United States, Japan, and Germany, with huge and diverse economies, depend on exports to generate a significant percentage of their employment and wealth. Without imports, many products that consumers want would be unavailable or more expensive; without exports, many jobs would be eliminated.

Map 62 The Indebtedness of States

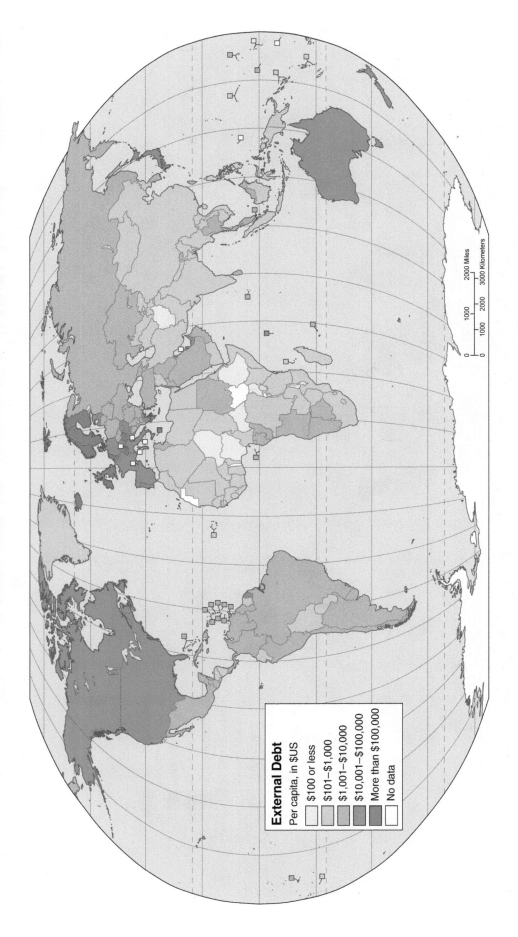

External Debt

Per capita, in $US

- $100 or less
- $101–$1,000
- $1,001–$10,000
- $10,001–$100,000
- More than $100,000
- No data

External debt is money or credit owed to foreign lenders. It generally is comprised of bonds and treasury bills (in the case of the United States) that are sold to foreign lenders and money owed to banks, governments, and international financial institutions. External debt is highly fluid and many countries of the world have "sustainable debt"—where a country can meet its debt service obligations. Other countries, particularly those in the developing world, have levels of debt, typically to international financial institutions such as the IMF and World Bank, that are beyond the governments' ability to repay. These countries typically are in a constant "catch-up" situation in which they fall increasingly further into debt and cannot meet their obligations without receiving partial or total forgiveness of debt. The IMF and World Bank have jointly put together debt reduction packages for 35 of the 41 countries identified as Heavily Indebted Poor Countries (HIPC). Approximately one-third of the U.S. national debt is owed to foreign governments, with China and Japan accounting for more than 44 percent of those holdings.

Map 63 Global Flows of Investment Capital

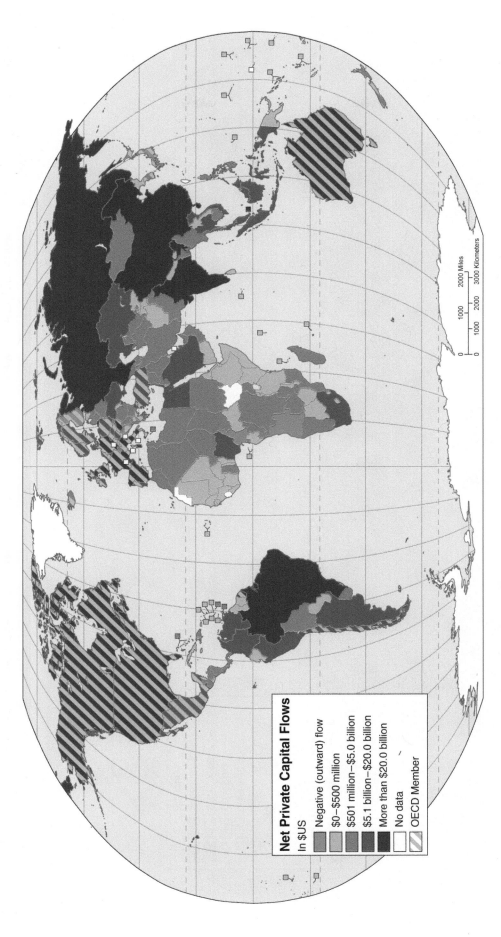

Net Private Capital Flows

In $US

- Negative (outward) flow
- $0–$500 million
- $501 million–$5.0 billion
- $5.1 billion–$20.0 billion
- More than $20.0 billion
- No data
- OECD Member

International capital flows include private debt and nondebt flows from one country to another, shown on the map as flows into a country. Nearly all of the capital comes from those countries that are members of the Organization for Economic Cooperation and Development (OECD), shown in black on the map. Capital flows include commercial bank lending, bonds, other private credits, foreign direct investment, and portfolio investment. Most of these flows are indicators of the increasing influence developed countries exert over the developing economies. Foreign direct investment or FDI, for example, is a measure of the net inflow of investment monies used to acquire long-term management interest in businesses located somewhere other than in the economy of the investor. Usually this means the acquisition of at least 10 percent of the stock of a company by a foreign investor and is, then, a measure of what might be termed "economic colonialism": control of a region's economy by foreign investors that could, in the world of the future, be as significant as colonial political control was in the past. International capital flows have increased greatly in the last decade as the result of the increasing liberalization of developing countries, the strong economic growth exhibited by many developing countries, and the falling costs and increased efficiency of communication and transportation services.

-83-

Map 64 Aiding Economic Development

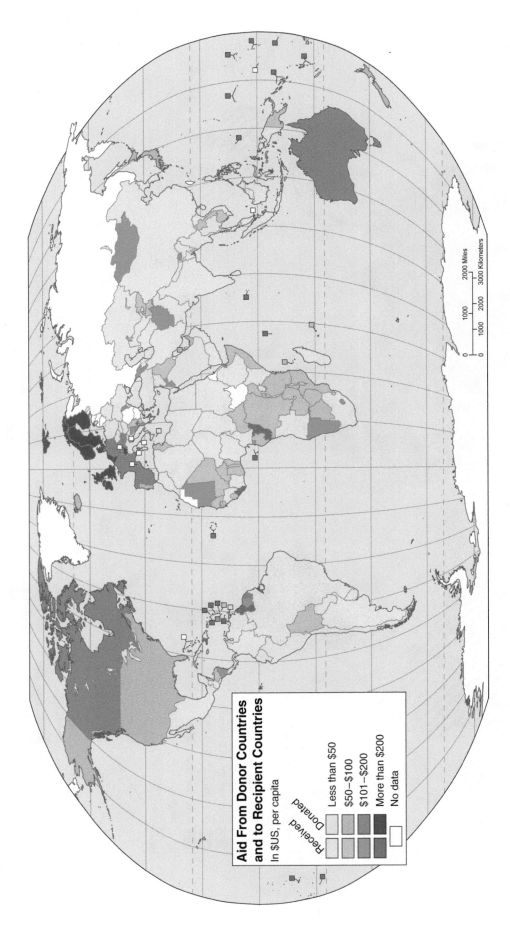

**Aid From Donor Countries
and to Recipient Countries**

In $US, per capita

Received	Donated	
		Less than $50
		$50–$100
		$101–$200
		More than $200
		No data

Over the last few years, official development assistance to developing countries from the member countries of the Organization for Economic Cooperation and Development has risen dramatically, with the United States as the world's number one donor country, giving more than 25 percent of the total of development assistance. Development assistance—or "foreign aid," as it is sometimes called—is widely recognized as benefiting both the donor and the recipient. Developing countries that increase their levels of per capita income through economic development have more money to spend on products from more highly developed countries. Increased development increases the capacity to foster not just economic but political change. In some parts of the world, such as Sub-Saharan Africa, foreign aid is the largest single source of external finance, far exceeding foreign investments.

Map 65   The Cost of Consumption

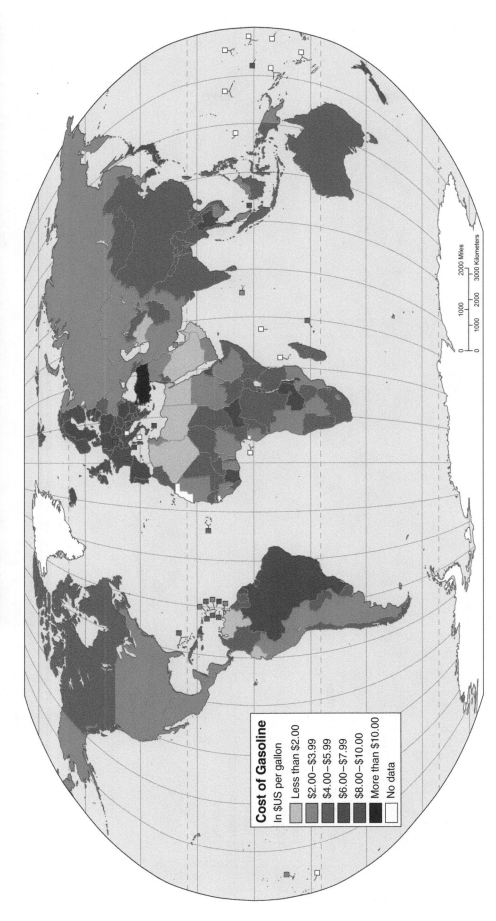

Cost of Gasoline
In $US per gallon

- Less than $2.00
- $2.00–$3.99
- $4.00–$5.99
- $6.00–$7.99
- $8.00–$10.00
- More than $10.00
- No data

The year 2008 brought massive increases in the price of gasoline at the pump, creating considerable consternation among the American driving population, in particular. It is one thing to point out that gasoline prices are and historically have been considerably higher in Europe than in North America. But it is another to note that European spatial patterns of places of work and places of residence are significantly different than in North America. The North American metropolitan area evolved its spatial patterns in conjunction with the rise of privately owned automobiles; European cities, on the other hand, had spatial patterns well established centuries or even millennia before the automobile emerged as a common mode of transportation. As a consequence, such things as the journey to work are very different for many Europeans who can walk from where they live to where they work than it is for Americans and Canadians who often

live considerable distances from their places of employment. A daily commute of 75 miles one way would not be considered unusual in America. In Europe it is virtually unheard of. There is also the component of the scale of organization of human activities: because of the very large country in which Americans live, their spatial movements are customarily more extensive than those of Europeans who live in countries the size of American states. So, yes, gasoline is much more expensive in Europe than in North America. But in North America, the increase in the cost of gasoline to and above $4.00 per gallon produces significantly more economic impact than proportionally similar increases in Europe. Similarly, the reduction in gasoline prices by the end of 2011 had a greater impact in North America—although its impact was overshadowed by larger forces.

-85-

Map 66   Energy Production Per Capita

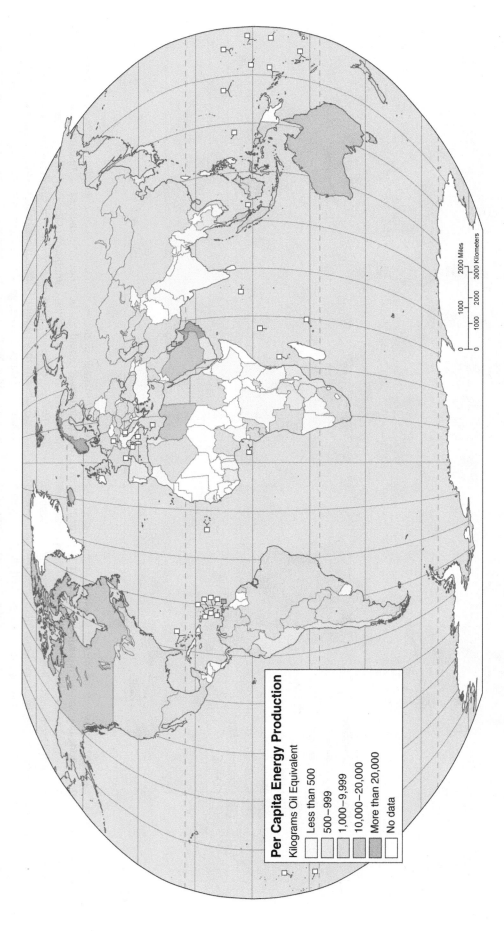

Per Capita Energy Production

Kilograms Oil Equivalent

- Less than 500
- 500—999
- 1,000—9,999
- 10,000—20,000
- More than 20,000
- No data

Energy production per capita is a measure of the availability of mechanical energy to assist people in their work. This map shows the amount of all kinds of energy—solid fuel (primarily coal), liquid fuel (primarily petroleum), natural gas, geothermal, wind, solar, hydroelectric, nuclear, waste recycling, and indigenous heat pumps—produced per person in each country. With some exceptions, wealthier countries produce more energy per capita than poor ones. Countries such as Japan and many European states rank among the world's wealthiest, but are energy-poor and produce relatively little of their own energy.

They have the ability, however, to pay for imports. On the other hand, countries such as those of the Persian Gulf or the oil-producing states of Central and South America may rank relatively low on the scale of economic development but rank high as producers of energy. In many poor countries, especially in Central and South America, Africa, South Asia, and East Asia, large proportions of energy come from traditional fuels such as firewood and animal dung. Indeed for many in the developing world, the real energy crisis is a shortage of wood for cooking and heating.

Map **67** Energy Consumption Per Capita

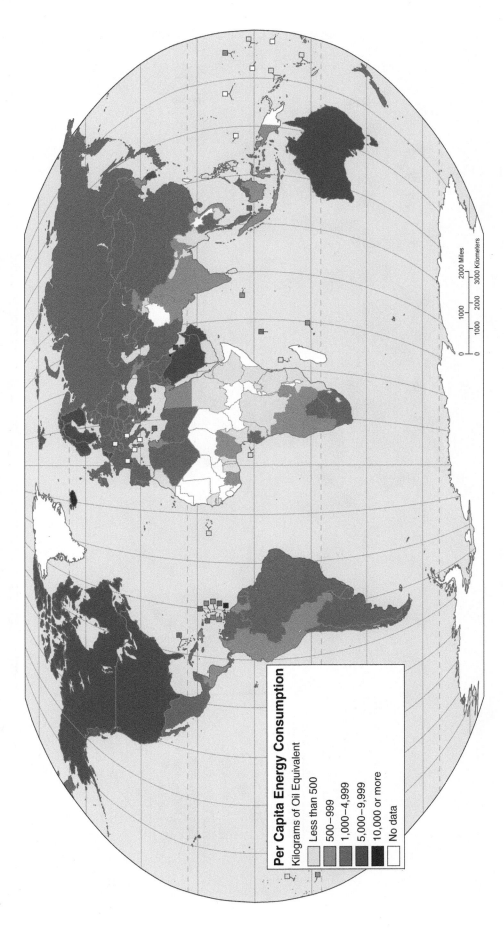

Per Capita Energy Consumption

Kilograms of Oil Equivalent

- Less than 500
- 500–999
- 1,000–4,999
- 5,000–9,999
- 10,000 or more
- No data

Of all the quantitative measures of economic well-being, energy consumption per capita may be the most expressive. All of the countries defined by the World Bank as having high incomes consume at least 100 gigajoules of commercial energy (the equivalent of about 3.5 metric tons of coal) per person per year, with some, such as the United States and Canada, having consumption rates in the 300 gigajoule range (the equivalent of more than 10 metric tons of coal per person per year). With the exception of the oil-rich Persian Gulf states, where consumption figures include the costly "burning off" of excess energy in the form of natural gas flares at wellheads, most of the highest-consuming countries are in the Northern Hemisphere, concentrated in North America and Western Europe. At the other end of the scale are low-income countries, whose consumption rates are often less than one percent of those of the United States and other high consumers. These figures do not, of course, include the consumption of noncommercial energy—the traditional fuels of firewood, animal dung, and other organic matter—widely used in the less developed parts of the world.

Map **68**   Energy Dependency

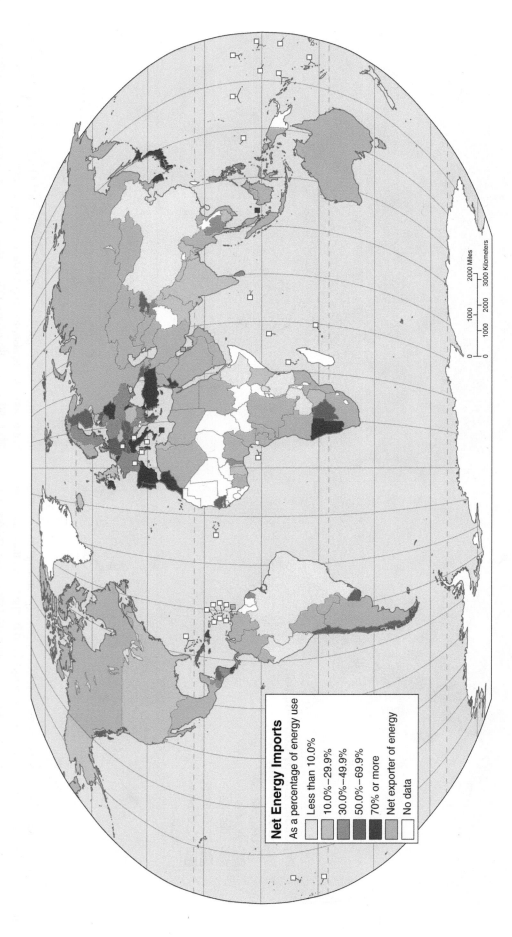

Net Energy Imports
As a percentage of energy use

Less than 10.0%
10.0%–29.9%
30.0%–49.9%
50.0%–69.9%
70% or more
Net exporter of energy
No data

0   1000   2000 Miles
0   1000   2000   3000 Kilometers

The patterns on the map show dependence on commercial energy before transformation to other end-use fuels such as electricity or refined petroleum products; energy from traditional sources such as fuelwood or dried animal dung is not included. Energy dependency is the difference between domestic consumption and domestic production of commercial energy and is most often expressed as a net energy import or export. A few of the world's countries are net exporters of energy: most are importers. The growth in global commercial energy use over the last decade indicates growth in the modern sectors of the economy—industry, transportation, and urbanization—particularly in the lesser developed countries. Still, the primary consumers of energy—and those having the greatest dependence on foreign sources of energy—are the more highly developed countries of Europe, North America, and Japan.

Map 69   Flows of Oil

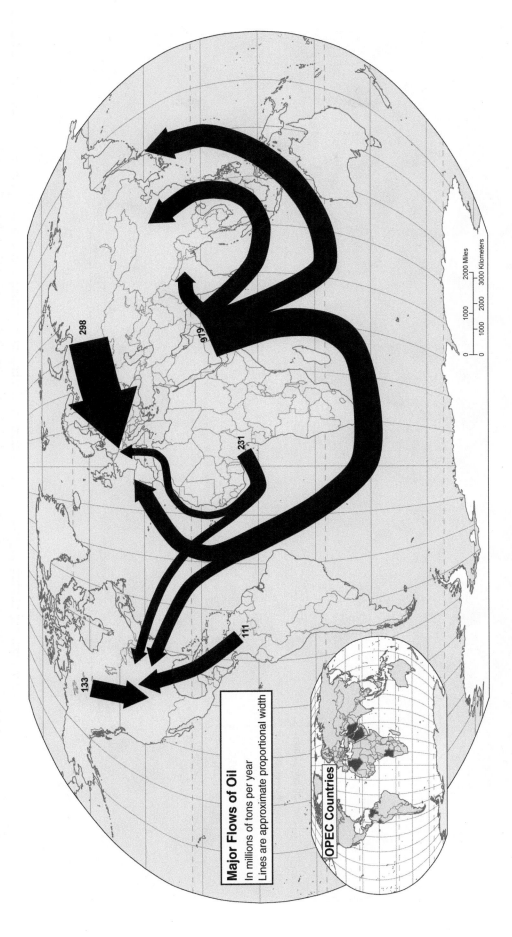

Major Flows of Oil
In millions of tons per year
Lines are approximate proportional width

979

298

231

133

111

OPEC Countries

0 1000 2000 Miles
0 1000 2000 3000 Kilometers

The pattern of oil movements from producing region to consuming region is one of the dominant facts of contemporary international maritime trade. Supertankers carry a million tons of crude oil and charge rates in excess of $0.10 per ton per mile, making the transportation of oil not only a necessity for the world's energy-hungry countries, but also an enormously profitable proposition. One of the major negatives of these massive oil flows is the damage done to the oceanic ecosystems—not just from well-publicized and dramatic events such as the 2010 *Deepwater Horizon* oil spill but from the incalculable amounts of oil from leakage, scrubbings, purgings, and so on, which are a part of the oil transport technology. As seen above, much of the supply of the world's oil comes from countries belonging to the Organization of the Petroleum Exporting Countries (OPEC). In 2009, OPEC members controlled nearly two-thirds of the world's known oil reserves and more than one-third of the world's production. It is clear from the map that the primary recipients of these oil flows are the world's most highly developed economies.

Map 70 A Wired World: Internet Users

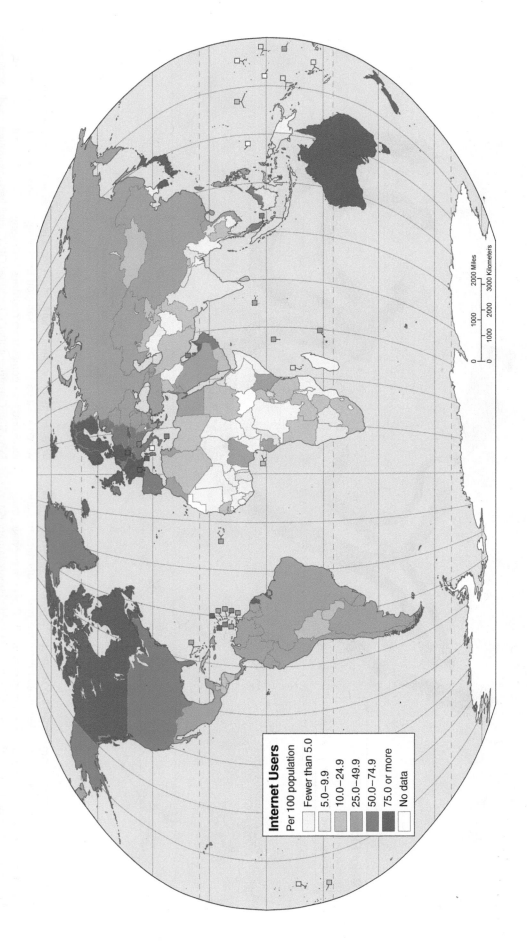

Internet Users
Per 100 population

- Fewer than 5.0
- 5.0–9.9
- 10.0–24.9
- 25.0–49.9
- 50.0–74.9
- 75.0 or more
- No data

0   1000   2000 Miles
0   1000   2000   3000 Kilometers

It is interesting to contemplate that a short quarter of a century ago, such a map could not have been created. The emergence of immediate, long-distance connectivity via the Internet has been one of the most important components of globalization. We now live, as author Thomas Friedman has noted, in a "flat world" where lines of connection are more important than distance and where virtually instantaneous connections have altered—perhaps forever—the way that we do business, exchange information, and transform our cultures. Some of the recent transformations we have seen in the emergence of countries such as China and India as major players in the international economy are, in part, the consequence of access to the Internet. Originally conceived as a quick way for academics to exchange information, the Internet has become a cultural and social phenomenon that far exceeds its original purpose. Whether or not this will result in an eventual benefit to human well-being remains to be seen. Does the benefit of quick communication result in a cost to the complexity and richness of human cultures worldwide?

Map 71   Traditional Links: The Telephone

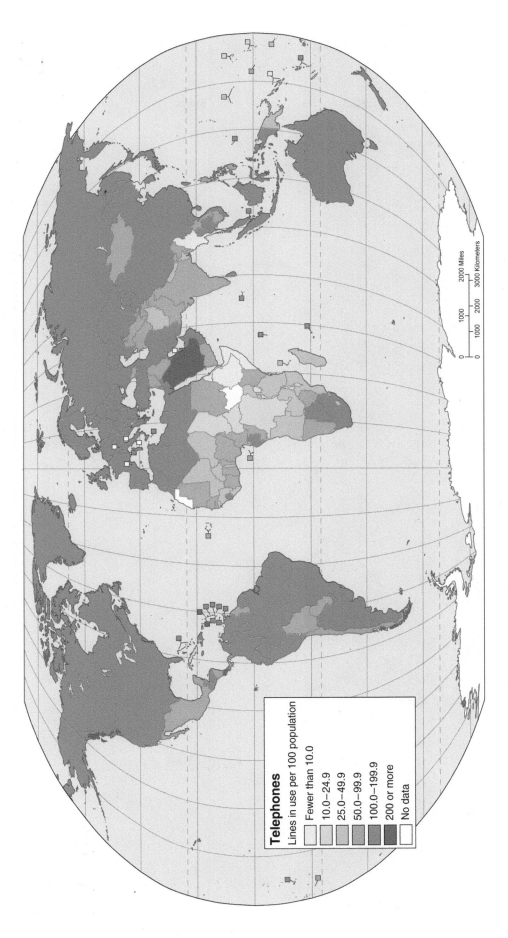

Telephones

Lines in use per 100 population

- Fewer than 10.0
- 10.0–24.9
- 25.0–49.9
- 50.0–99.9
- 100.0–199.9
- 200 or more
- No data

Not all of the world's communications take place via computers and the Internet. A lot of person-to-person connection is still carried out via the telephone, and access to telephone connections is perhaps as good an indication of economic development or, more important, the potential for economic development as anything else. The map clearly shows the prevalence of telephone connectivity in the developed world. But it also shows an increasingly high degree of access to telephones in major countries in the developing world, such as India and China. If these countries are to continue to develop their economies at the pace of the last decade, then their degree of communication—including access to telephones—will also have to increase. The data shown on this map include users of both land lines and cellular phones. By the end of 2008, for the first time, the number of cellular phone users worldwide exceeded the number of those using the traditional land lines. As cellular phone complexity increases to include e-mail and other computer functions, the gap between cellular users and traditional phone users can be expected to widen.

Unit V

Global Patterns of Environmental Disturbance

Map 72 Global Air Pollution: Sources and Wind Currents

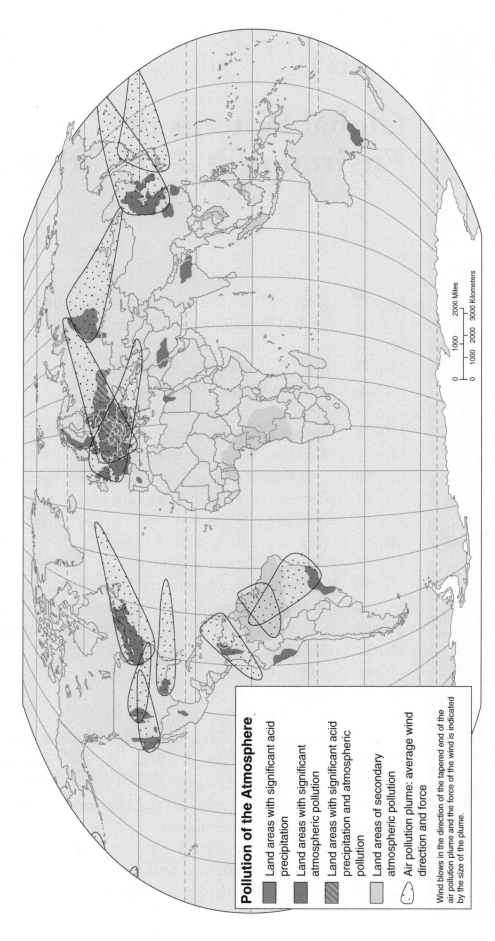

Pollution of the Atmosphere

- Land areas with significant acid precipitation
- Land areas with significant atmospheric pollution
- Land areas with significant acid precipitation and atmospheric pollution
- Land areas of secondary atmospheric pollution
- Air pollution plume: average wind direction and force

Wind blows in the direction of the tapered end of the air pollution plume and the force of the wind is indicated by the size of the plume.

0   1000   2000 Miles
0   1000   2000   3000 Kilometers

Almost all processes of physical geography begin and end with the flows of energy and matter among land, sea, and air. Because of the primacy of the atmosphere in this exchange system, air pollution is potentially one of the most dangerous human modifications in environmental systems. Pollutants such as various oxides of nitrogen or sulfur cause the development of acid precipitation, which damages soil, vegetation, and wildlife and fish. Air pollution in the form of smog is often dangerous for human health. And most atmospheric scientists believe that the efficiency of the atmosphere in retaining heat—the so-called greenhouse effect—is being enhanced by increased carbon dioxide, methane, and other gases produced by agricultural and industrial activities. The result, they fear, will be a period of global warming that will dramatically alter climates in all parts of the world.

Map 73 The Acid Deposition Problem: Air, Water, Soil

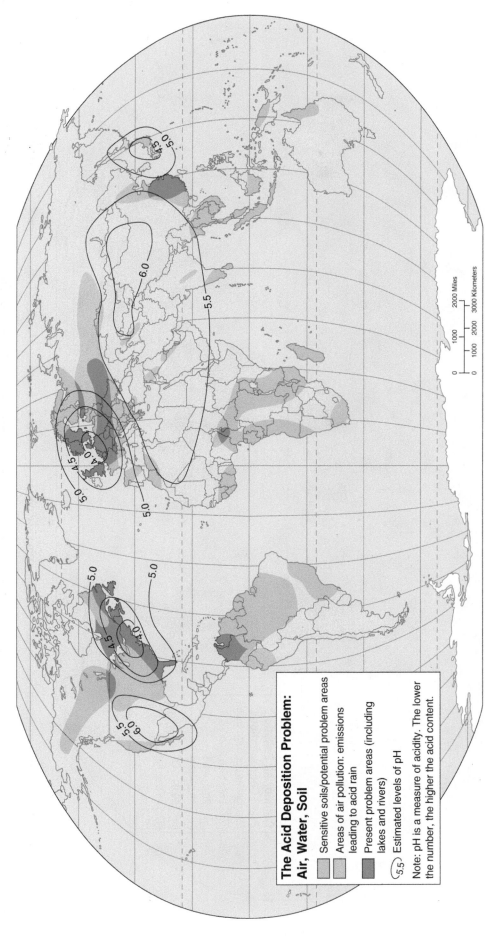

The Acid Deposition Problem: Air, Water, Soil

Sensitive soils/potential problem areas

Areas of air pollution: emissions leading to acid rain

Present problem areas (including lakes and rivers)

5.5 Estimated levels of pH

Note: pH is a measure of acidity. The lower the number, the higher the acid content.

The term "acid precipitation" refers to increasing levels of acidity in snowfall and rainfall caused by atmospheric pollution. Oxides of nitrogen and sulfur resulting from incomplete combustion of fossil fuels (coal, oil, and natural gas) combine with water vapor in the atmosphere to produce weak acids that then "precipitate" or fall along with water or ice crystals. Some atmospheric acids formed by this process are known as "dry-acid" precipitates and they too will fall to earth, although not necessarily along with rain or snow. In some areas of the world, the increased acidity of streams and lakes stemming from high levels of acid precipitation or dry acid fallout has damaged or destroyed aquatic life. Acid precipitation and dry acid fallout also harm soil systems and vegetation, producing a characteristic burned appearance in forests that lends the same quality to landscapes that forest fires would. The region most dramatically impacted by acid precipitation is Central Europe where decades of destructive environmental practices, including the burning of high sulfur coal for commercial, industrial, and residential purposes, has produced the destruction of hundreds of thousands of acres of woodlands—a phenomenon described by the German foresters who began their study of the area following the lifting of the Iron Curtain as *Waldsterben*: "Forest Death".

Map 74 Outdoor Air Pollution

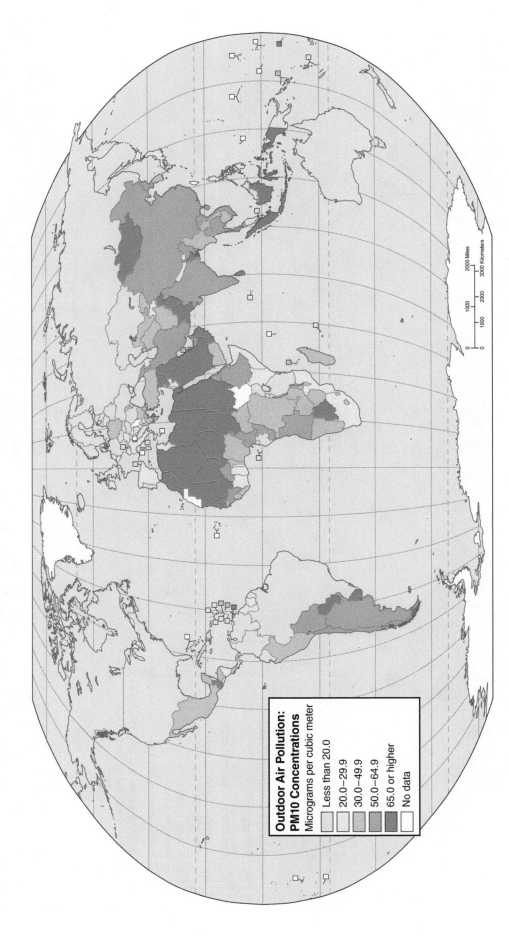

**Outdoor Air Pollution:
PM10 Concentrations**
Micrograms per cubic meter

- Less than 20.0
- 20.0–29.9
- 30.0–49.9
- 50.0–64.9
- 65.0 or higher
- No data

Air pollution is a "catch all" term referring to a wide range of pollutants comprised primarily of particulate matter and cases such as ozone, sulfur dioxide, and nitrogen dioxide. Of all pollutants, particulate matter—pieces of solid or liquid material that are suspended in the air—most adversely affect people. Small particles, those with diameters of less than 10 microns (PM10), are of particular concern as they can be found in large quantities, and have significant health effects. Particulate matter has both natural and anthropogenic sources. Examples of naturally produced particulates include dust, smoke from wildfires, and ash from volcanic eruptions. Anthropogenic sources include agricultural operations,

diesel trucks, power plants, and other industrial processes. Prolonged exposure to particulate matter increases the risk of cardiovascular and respiratory diseases. Typically cities have much higher PM10 levels than surrounding rural areas, evidenced by haze and smog that plague many large urban areas, and mortality rates in highly polluted cities can be 10 to 20 percent higher than in less-polluted cities. Although the data are aggregated at the country level in the map above, the levels actually reflect PM10 levels in the cities of the countries.

Map 75 Global Carbon Dioxide Emissions

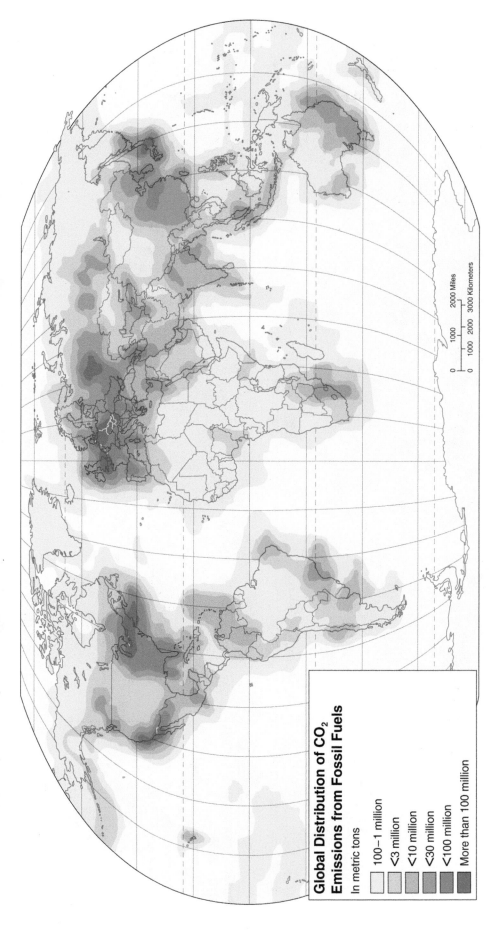

Global Distribution of CO₂ Emissions from Fossil Fuels

In metric tons

- 100–1 million
- <3 million
- <10 million
- <30 million
- <100 million
- More than 100 million

One of the most important components of the atmosphere is the gas carbon dioxide (CO₂), the byproduct of animal respiration, decomposition, and combustion. During the past 200 years, atmospheric CO₂ has risen dramatically, largely as the result of the tremendous increase in fossil fuel combustion brought on by the industrialization of the world's economy and the burning and clearing of forests by the expansion of farming. Although CO₂ by itself is relatively harmless, it is an important "greenhouse gas." The gases in the atmosphere act like the panes of glass in a greenhouse roof, allowing light in but preventing heat from escaping. The greenhouse capacity of the atmosphere is crucial for organic life and is a purely natural component of the global energy cycle. But too much CO₂ and other greenhouse gases such as methane could cause the earth's atmosphere to warm up too much, producing the global warming that atmospheric scientists are concerned about. Researchers estimate that if greenhouse gases such as CO₂ continue to increase at their present rates, the earth's mean temperature could rise between 1.5 and 4.5 degrees Celsius by the middle of the present century. Such a rise in global temperatures would produce massive alterations in the world's climate patterns.

Map 76 The Earth Warms, 1976–2006

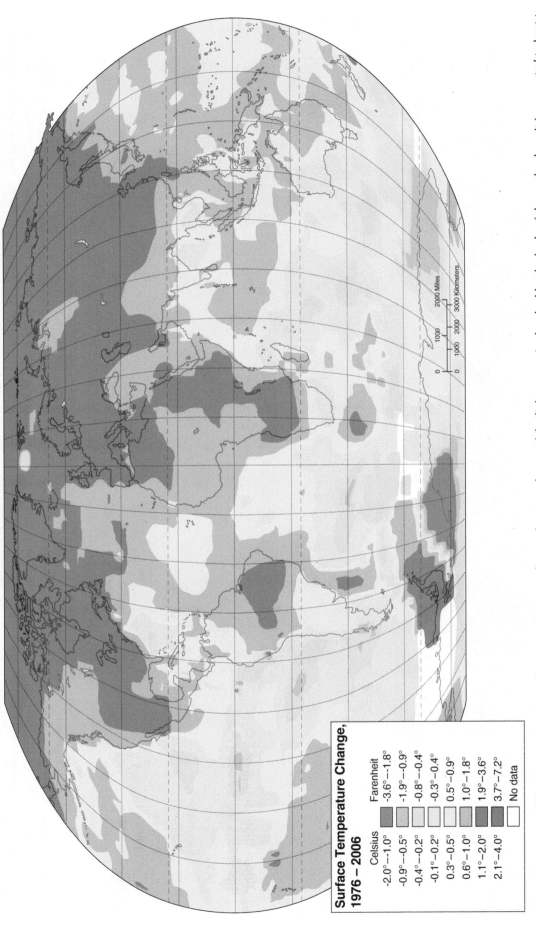

Surface Temperature Change, 1976 – 2006

Celsius	Farenheit
-2.0°--1.0°	-3.6°--1.8°
-0.9°--0.5°	-1.9°--0.9°
-0.4°--0.2°	-0.8°--0.4°
-0.1°--0.2°	-0.3°--0.4°
0.3°--0.5°	0.5°--0.9°
0.6°--1.0°	1.0°--1.8°
1.1°--2.0°	1.9°--3.6°
2.1°--4.0°	3.7°--7.2°
No data	

Although there may still be disagreement about the nature of present climate change and its link to human activities, that disagreement is largely political and economic, rather than scientific. The evidence from a generation of scientific studies of the atmosphere is compelling: the world is warming more rapidly than at any time in the past and the bulk of this warming trend is the consequence of human alterations in the atmospheric proportion of the "greenhouse gases," such as carbon dioxide and methane, that help to trap heat radiated from the earth and release it back into space more slowly. The atmosphere is a massive heat engine: short wave (light) energy from the sun passes through the atmosphere where some of it is absorbed and converted to heat; a larger percentage of the light energy is absorbed by the body of the earth where it is converted to heat to warm the atmosphere. The "greenhouse effect," as this process is known, is a good thing. Were it not for the greenhouse effect, life on earth as we know it would be impossible. The temperature changes shown on this map are the result of an accelerating greenhouse effect and, in turn, cause changes in precipitation patterns (see next map), storms of greater intensity and frequency, and rising sea levels. The earth has experienced these changes before. What makes the current changes different is that they are occurring over decades rather than over millennia, making it very difficult (if not impossible) for natural and human systems to adapt to them.

-98-

Map 77 Global Precipitation Changes, 1976–2006

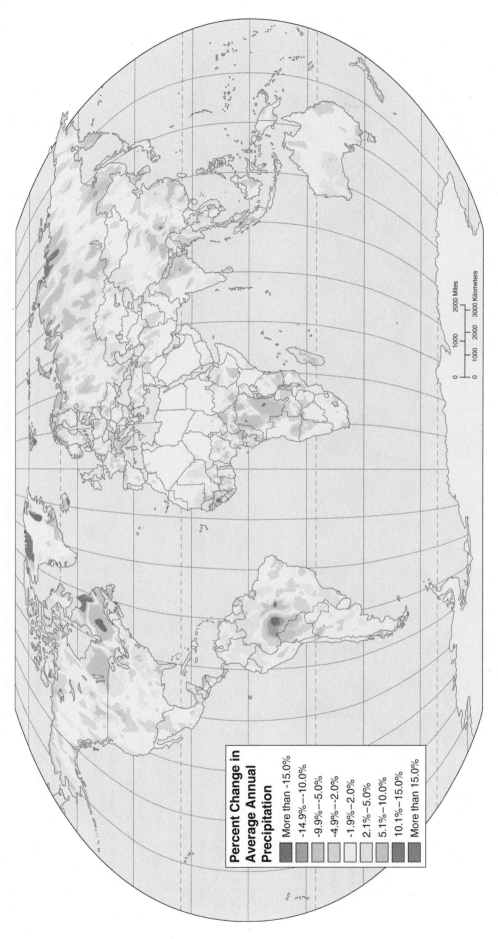

Percent Change in Average Annual Precipitation

- More than -15.0%
- -14.9%–-10.0%
- -9.9%–-5.0%
- -4.9%–-2.0%
- -1.9%–-2.0%
- 2.1%–5.0%
- 5.1%–10.0%
- 10.1%–15.0%
- More than 15.0%

0   1000   2000 Miles
0   1000   2000   3000 Kilometers

The links between global temperature change and alterations in the geographical distribution of precipitation are so complex as to be only marginally understood. What can be said with certainty is that increasing global temperatures will produce altered precipitation patterns. We have already begun to experience some of this in the form of prolonged droughts in the interior regions of North America; increased precipitation in the previously semi-arid region of the Sahel in Africa; and severe and long-term drought in the southern portions of the Amazon Basin, adjacent areas in South America, and in those sections of Africa south of the Sahel. There is an approximate tendency for dry regions to become drier and for wet regions to become wetter with increasing temperatures, although this is not universally the case. The primary problem is that persistent, long-term changes in precipitation patterns will cause significant disruptions in established agricultural patterns such as those in the United States, China, India, and Brazil—four of the world's most productive agricultural countries.

-99-

Map **78**   Potential Global Temperature Change

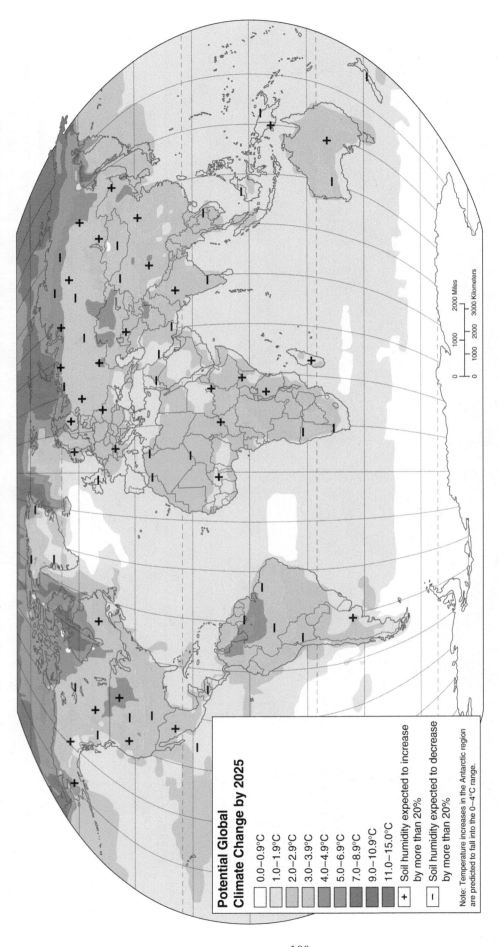

**Potential Global
Climate Change by 2025**

☐	0.0–0.9°C
☐	1.0–1.9°C
☐	2.0–2.9°C
☐	3.0–3.9°C
☐	4.0–4.9°C
☐	5.0–6.9°C
☐	7.0–8.9°C
☐	9.0–10.9°C
☐	11.0–15.0°C

+   Soil humidity expected to increase
    by more than 20%

−   Soil humidity expected to decrease
    by more than 20%

Note: Temperature increases in the Antarctic region
are predicted to fall into the 0–4°C range.

0   1000   2000 Miles

0   1000   2000   3000 Kilometers

By the end of the first quarter of the twenty-first century, the world's population will probably be facing climatic conditions quite different from those of the last quarter of the twentieth century: significantly increased temperatures, particularly in the higher latitudes (closer to the poles), and significantly altered precipitation patterns with some of the world's great agricultural regions experiencing long-term and persistent drought. The increase of temperatures in the higher latitudes means greater melting of the Arctic polar ice sheet (ice floating on water), greater melting of the Greenland ice sheet (ice of incredible thickness now resting on land), and greater melting of the Antarctic ice sheet—similar in structure but larger than that of Greenland. There are some benefits of polar ice melting

(as long as you are not a polar bear), most notably the opening of sea lanes across the Arctic Ocean—the final establishment of the Northwest Passage sought for by European explorers for several centuries. But there are many more hazards in the form of even small increases in sea level—a meter or two would be enough to produce a major increase in damage by even moderate tropical storms, let alone hurricanes. And many low-lying coastal and island communities run the risk of complete inundation by rising sea waters. Is all this inevitable just because a computer model such as the one that generated the above map predicts it? No. But ignoring the evidence gathered and analyzed so carefully by objective scientists would be both foolhardy and arrogant.

Map 79 Water Resources: Availability of Renewable Water Per Capita

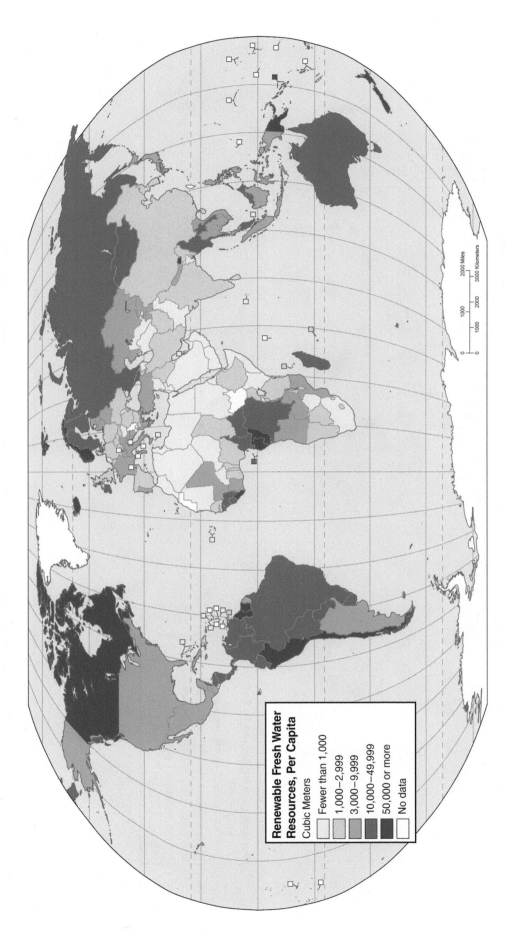

Renewable Fresh Water Resources, Per Capita

Cubic Meters

- Fewer than 1,000
- 1,000–2,999
- 3,000–9,999
- 10,000–49,999
- 50,000 or more
- No data

Renewable water resources are usually defined as the total water available from streams and rivers (including flows from other countries), ponds and lakes, and groundwater storage or aquifers. Not included in the total of renewable water would be water that comes from such nonrenewable sources as desalinization plants or melted icebergs. Although the concept of renewable or flow resources is a traditional one in resource management, in fact, few resources, including water, are truly renewable when their use is excessive.

The water resources shown here are indications of that principle. A country such as the United States possesses truly enormous quantities of water. But the United States also uses enormous quantities of water. The result is that, largely because of excessive use, the availability of renewable water is much less than in many other parts of the world where the total supply of water is significantly less.

Map 80 Water Resources: Access to Safe Drinking Water

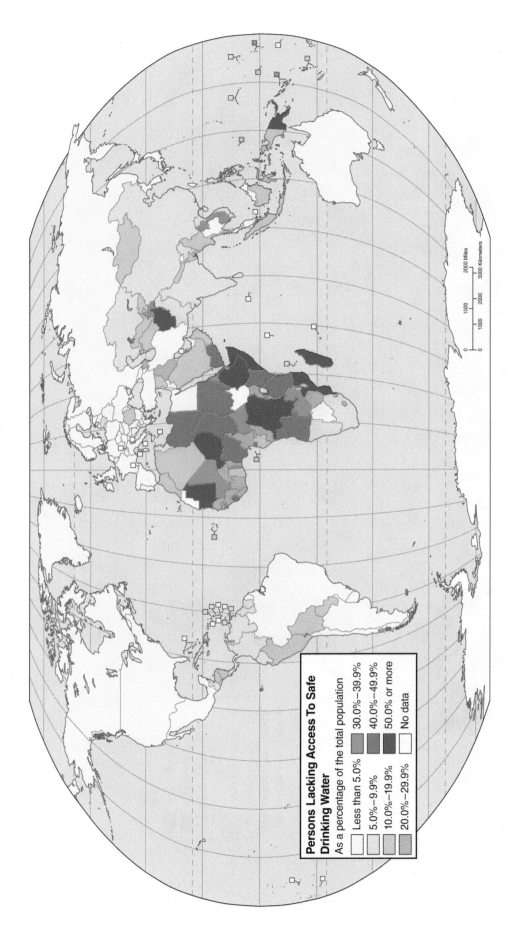

Persons Lacking Access To Safe Drinking Water
As a percentage of the total population

- Less than 5.0%
- 5.0%–9.9%
- 10.0%–19.9%
- 20.0%–29.9%
- 30.0%–39.9%
- 40.0%–49.9%
- 50.0% or more
- No data

0   1000   2000 Miles
0   1000   2000   3000 Kilometers

Drinking water is water that is safe enough for human consumption, and not everyone in the world has reasonable access to water from an improved source—a source that is protected from outside contamination. In fact, the World Health Organization/UNICEF Joint Monitoring Program for Water Supply and Sanitation estimates that more than 750 million people do not use an improved source for drinking water. Contaminated drinking water is a contributor to diarrheal diseases (see Map 39b), which disproportionally affects children younger than the age of five. As part of the Millennium Development Goals established in 2000, the United Nations targeted a reduction of the proportion of people without sustainable access to safe drinking water and basic sanitation by one-half. Since 1990, more than two billion people have gained access to improved sources of drinking water.

Map 81   Water Resources: Annual Withdrawal Per Capita

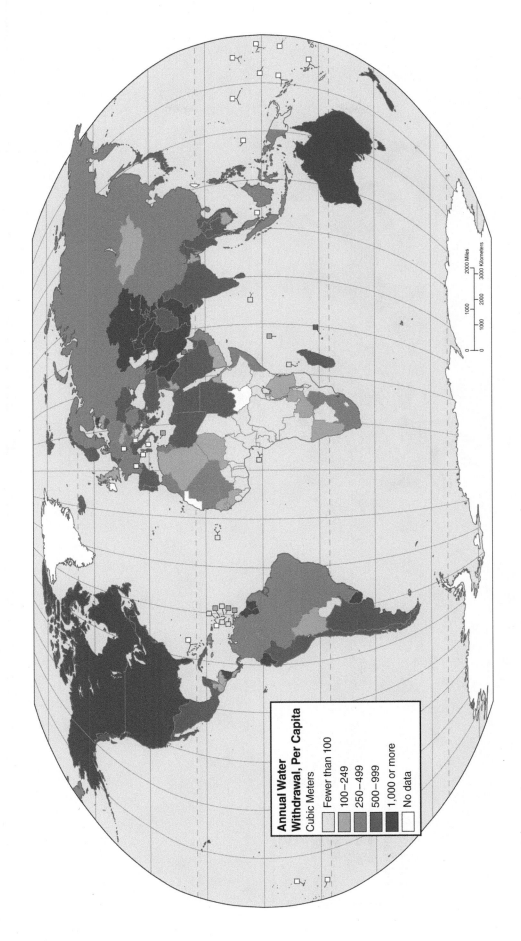

Annual Water Withdrawal, Per Capita
Cubic Meters

- Fewer than 100
- 100–249
- 250–499
- 500–999
- 1,000 or more
- No data

Water resources must be viewed like a bank account in which deposits and withdrawals are made. As long as the deposits are greater than the withdrawals, a positive balance remains. But when the withdrawals begin to exceed the deposits, sooner or later (depending on the relative sizes of the deposits and withdrawals) the account becomes overdrawn. For many of the world's countries, annual availability of water is insufficient to cover the demand. In these countries, reserves stored in groundwater are being tapped, resulting in depletion of the water supply (think of this as shifting money from a savings account to a checking account). The water supply can maintain its status as a renewable resource only if deposits continue to be greater than withdrawals, and that seldom happens. In general, countries with high levels of economic development and countries that rely on irrigation agriculture are the most spendthrift when it comes to their water supplies.

Map 82 Water Stress: Shortage, Abundance, and Population Density

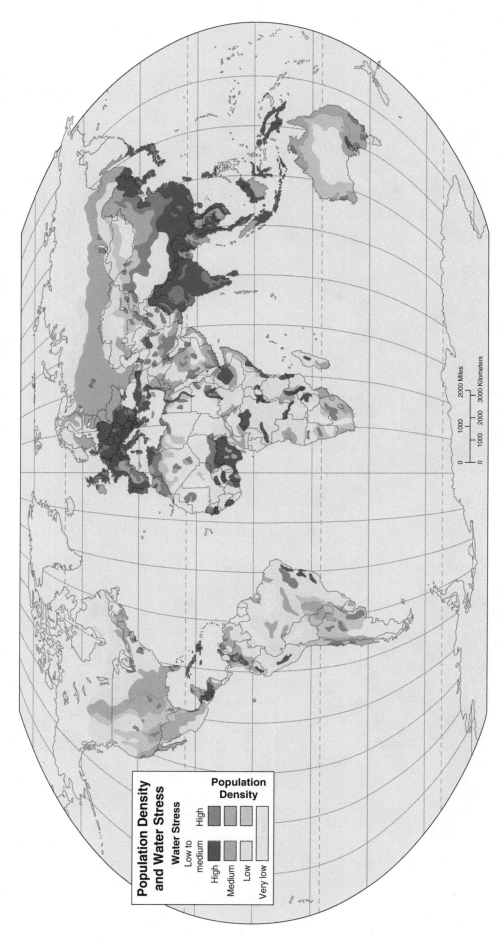

Population Density and Water Stress

Water Stress

- Low to medium
- High

Population Density
- High
- High
- Medium
- Low
- Very low

0   1000   2000 Miles

0 1000 2000 3000 Kilometers

Maps such as the previous two, based on national-level data for water consumption and availability, should be used only to obtain national-level understanding. Information on water withdrawal and availability are regionally and locally based geographic phenomena and are linked not just with water supplies but with the density of human populations. Even areas (such as New England in the United States) in which water availability is high and withdrawal rates are relatively low show areas of stress in regions of high population density (cities such as Boston). This map, originally produced by scientists at the University of New Hampshire, attempts to show those areas of the world where populations will tend to be at high, medium, and low risk of stress because of water availability. It is important to note that many of the world's prime agricultural regions, such as the Great Plains of the United States or the Argentine Pampas, show the potential for high risk of water stress in the immediate future. Why is this important? Because the greatest single use of water on the planet is for irrigation (nearly 70 percent of the world's water use) and it is the continued expansion of irrigation systems that allows the increase in agricultural production that feeds the earth's more than 7 billion persons.

-104-

Map 83   Water Stress: The Ogallala Aquifer

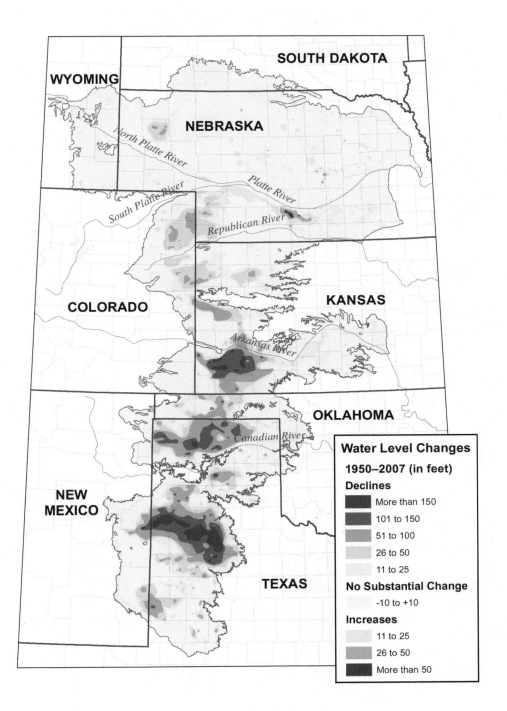

An aquifer is an underground water source, usually consisting of a porous medium of gravel, sand, and bedrock, filled with water trapped by a layer of impermeable stone. Aquifers may feed rivers and lakes through springs. Water is also withdrawn from aquifers for human uses, particularly irrigation. An aquifer is recharged by precipitation and stream run-off. Most aquifers represent thousands of years of accumulation of stored water. The Ogallala aquifer underlies 174,000 square miles of the Great Plains and has total water storage about equal to that of Lake Huron. Although it is a major source of water for agriculture, industry, and human consumption, it is irrigation agriculture that is the dominant use. When aquifers such as the Ogallala are tapped for irrigation, however, the rates of withdrawal far exceed the rates of recharge. Since the beginning of major use of the Ogallala aquifer in the 1950s, water storage has dropped by nearly 10 percent and there have been water level declines of more than 150 feet in the Texas panhandle and southwestern Kansas. Although resource managers have often thought of groundwater as a renewable resource, in fact it is just as nonrenewable in the human timeframe as an oil field or iron mine. When the Ogallala aquifer is depleted, it will take tens of thousands of years to restore it.

Map 84 Pollution of the Oceans

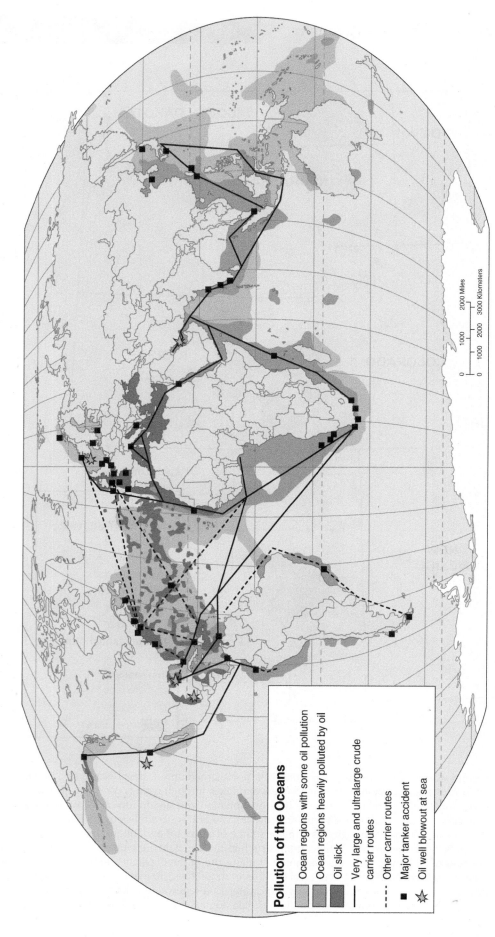

Pollution of the Oceans

- Ocean regions with some oil pollution
- Ocean regions heavily polluted by oil
- Oil slick
- Very large and ultralarge crude carrier routes
- ---- Other carrier routes
- ■ Major tanker accident
- ✦ Oil well blowout at sea

0   1000   2000 Miles

0 1000 2000 3000 Kilometers

The pollution of the world's oceans has long been a matter of concern to physical geographers, oceanographers, and other environmental scientists. The great circulation systems of the ocean are one of the controlling factors of the earth's natural environment, and modifications to those systems have unknown consequences. This map is based on what we can measure: (1) areas of oceans where oil pollution has been proven to have inflicted significant damage to ocean ecosystems and life forms (including phytoplankton, the oceans' primary food producers, equivalent to land-based vegetation) and (2) areas of oceans where unusually high concentrations of hydrocarbons from oil spills may have inflicted some damage to the oceans' biota. A glance at the map shows that there are few areas of the world's oceans where some form of pollution is not a part of the environmental system. What the map does not show in detail, because of the scale, are the dramatic consequences of large individual pollution events: the devastation produced by the 1991 Gulf War in the Persian Gulf, or the 2010 *Deepwater Horizon* oil spill in the Gulf of Mexico.

Map 85 Food Supply from Marine and Freshwater Systems

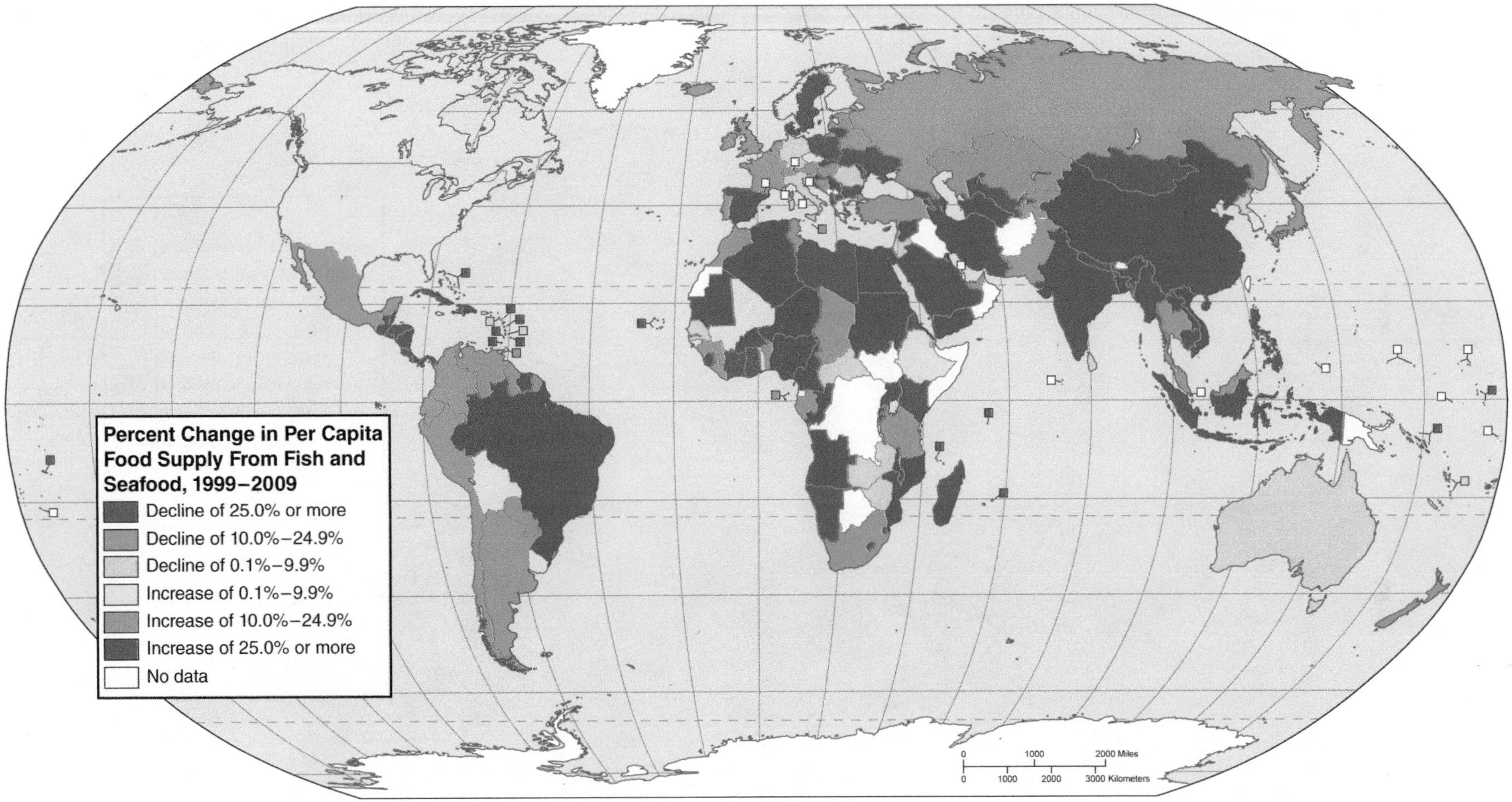

Percent Change in Per Capita Food Supply From Fish and Seafood, 1999–2009

- Decline of 25.0% or more
- Decline of 10.0%–24.9%
- Decline of 0.1%–9.9%
- Increase of 0.1%–9.9%
- Increase of 10.0%–24.9%
- Increase of 25.0% or more
- No data

Not that many years ago, food supply experts were confidently predicting that the "starving millions" of the world of the future could be fed from the unending bounty of the world's oceans. Although the annual catch from the sea helped to keep hunger at bay for a time, by the late 1980s it had become apparent that without serious human intervention in the form of aquaculture, the supply of fish would not be sufficient to offset the population/food imbalance that was beginning to affect so many of the world's regions. The development of factory-fishing with advanced equipment to locate fish and process them before they went to market increased the supply of food from the ocean, but in that increase was sown the seeds of future problems. The factory-fishing system, efficient in

terms of economics, was costly in terms of fish populations. In some well-fished areas, the stock of fish that was viewed as near infinite just a few decades ago has dwindled nearly to the point of disappearance. This map shows both increases and decreases in the amount of individual countries' food supplies from the ocean. The increases are often the result of more technologically advanced fishing operations. The decreases are usually the result of the same thing: increased technology has brought increased harvests, which has reduced the supply of fish and shellfish and that, in turn, has increased prices. Most of the countries that have experienced sharp decreases in their supply of food from the world's oceans are simply no longer able to pay for an increasingly scarce commodity.

Map 86 Changes in Cropland

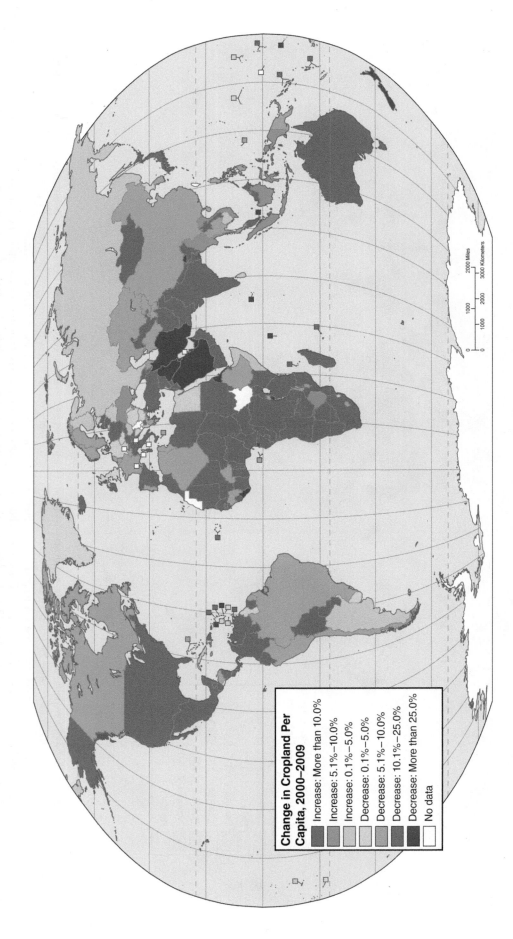

Change in Cropland Per Capita, 2000–2009

- Increase: More than 10.0%
- Increase: 5.1%–10.0%
- Increase: 0.1%–5.0%
- Decrease: 0.1%–5.0%
- Decrease: 5.1%–10.0%
- Decrease: 10.1%–25.0%
- Decrease: More than 25.0%
- No data

As population has increased rapidly throughout the world, the area of cultivated land has increased at the same time; in fact, the amount of farmland per person has gone up slightly. Unfortunately, the figures that show this also tell us that because most of the best (or even good) agricultural land in 1985 was already under cultivation, most of the agricultural area added since the 1990s involves land that would have been viewed as marginal by the fathers and grandfathers of present farmers—marginal in that it was too dry, too wet, too steep to cultivate, too far from a market, and so on. The continued expansion of agricultural area is one reason that serious famine and starvation have struck only a few regions of the globe. But land, more than any other resource we deal with, is finite, and the expansion cannot continue indefinitely. Future gains in agricultural production are most probably going to come through more intensive use of existing cropland, heavier applications of fertilizers and other agricultural chemicals, and genetically engineered crops requiring heavier applications of energy and water, than from an increase in the amount of the world's cropland.

Map 87 Food Staples Under Stress: The Fight Against Genetic Simplicity

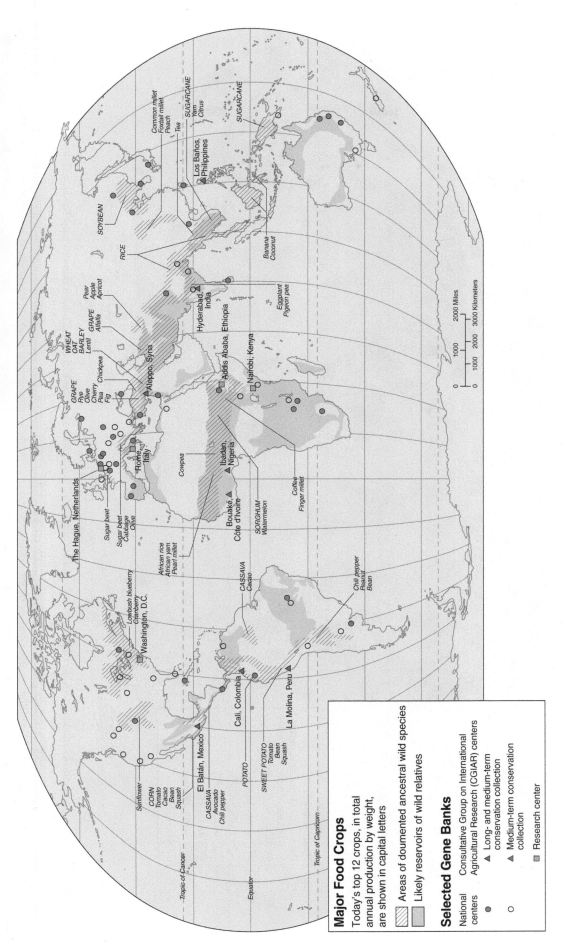

Major Food Crops

Today's top 12 crops, in total annual production by weight, are shown in capital letters

▨ Areas of documented ancestral wild species

▨ Likely reservoirs of wild relatives

Selected Gene Banks

National centers
● Long- and medium-term conservation collection
○ Medium-term conservation collection

Consultative Group on International Agricultural Research (CGIAR) centers
▲ Long- and medium-term conservation collection
▲ Medium-term conservation collection
▪ Research center

The point has been made elsewhere in this atlas that the loss of biological diversity is one of the greatest threats to ecosystem stability. That is certainly true when we speak of the threatened loss of diversity among those plants that provide the earth's population with the bulk of its food supply. As farming becomes more specialized and certain strains of wheat, rice, and corn, achieve dominance because of their increased productivity levels, there is great danger in "putting all our eggs in one basket." Plant geneticists are attempting to protect the world against the loss of genetic diversity and the potential starvation that could result from crop failure of the favored strains by preserving as much of the genetic material of both domesticated relatives and wild ancestors of the most-used grains and other crops. Protecting the world's genetic reserves of plants is more than just an academic exercise. It is insurance against the biological degradation that future population growth is going to create and, therefore, provides some measure of assurance that future food production will be able to meet future demand.

Map 88 World Pastureland, 2009

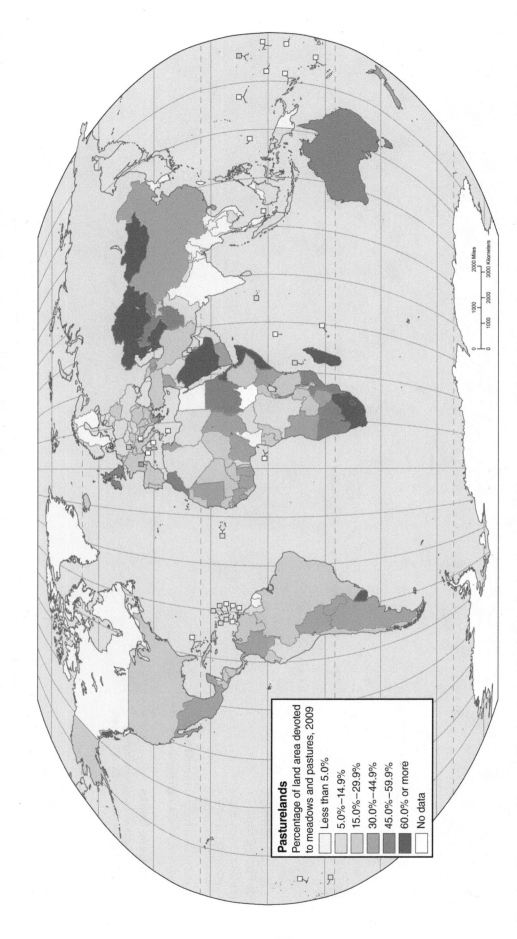

Pasturelands

Percentage of land area devoted to meadows and pastures, 2009

- Less than 5.0%
- 5.0%–14.9%
- 15.0%–29.9%
- 30.0%–44.9%
- 45.0%–59.9%
- 60.0% or more
- No data

0   1000   2000   3000 Kilometers

0   1000   2000 Miles

More than 25 percent of the world's surface is considered pastureland—either native wild grasses or human made pastures created by clearing forests and planting grass and other herbaceous forage plants for livestock feed. Two types of pasture predominate: the prairie and steppe grasses of the midlatitudes and the savanna grasses of the subtropics and tropics. The map depicts pastureland as a proportion of land area of individual countries. You will note that many of the countries with high levels of pastureland (Argentina and Australia, for example) are among the world's leading exporters of livestock products. Other countries with high levels of pastureland (Saudi Arabia and the Central Asian states)

consume the bulk of their products domestically. The significance of the distribution and use of pastureland is that—whether it is in the African Sahel, China, Brazil, or the United States—the world's pasturelands are deteriorating rapidly under increasing demands to produce more animal products than even a wealthier world can afford to pay for (see Map 37). Grassland degradation, particularly in areas where pastoral nomads use their animals as their chief source of food and income, creates woody scrublands where the carrying capacity for grazing animals is sharply diminished. In short, most of the world's pasturelands are overgrazed, and the world's supply of animal products is in jeopardy.

Map 89   Fertilizer Use

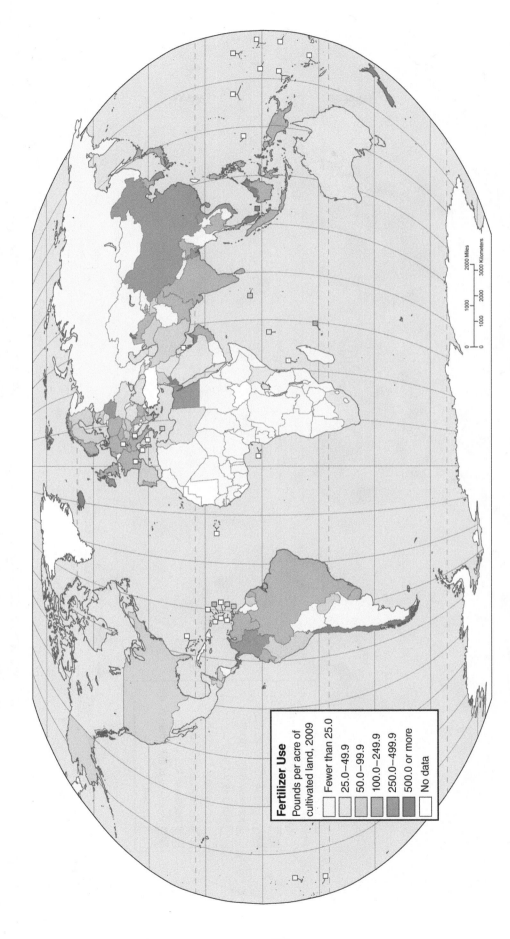

Fertilizer Use

Pounds per acre of
cultivated land, 2009

- Fewer than 25.0
- 25.0–49.9
- 50.0–99.9
- 100.0–249.9
- 250.0–499.9
- 500.0 or more
- No data

The use of fertilizer to maintain the productivity of agricultural lands is a wonderful agricultural invention—as long as the fertilizers used are natural rather than artificial. In most of the world's developed countries, such as those in Europe and North America, the use of animal manure to fertilize fields has decreased dramatically over the past century, in favor of artificial fertilizers that are cheaper and easier to use and—what is most important—increase crop yields more dramatically. The danger here is that artificial fertilizers normally have high concentrations of nitrates that tend to convert to nitrites in the soil, reducing the ability of soil bacteria to extract "free" nitrogen from the atmosphere.

As more artificial fertilizer is used, natural soil fertility is decreased, creating the demand for more artificial fertilizers. In some areas, overuse of artificial fertilizers has actually created soils that are too "hot" chemically to produce crops. Countries with high fertilizer use in the developing world—northern South America, Southwest, South, and East Asia—still tend to use more natural fertilizers. But as farmers in those areas gain more ability to buy and use artificial fertilizers, their soils will also begin to suffer from overfertilization. Global agriculture needs to come to grips with the need to maintain productivity but to do so in a manner that is sustainable.

Map 90   Changes in Forest Cover

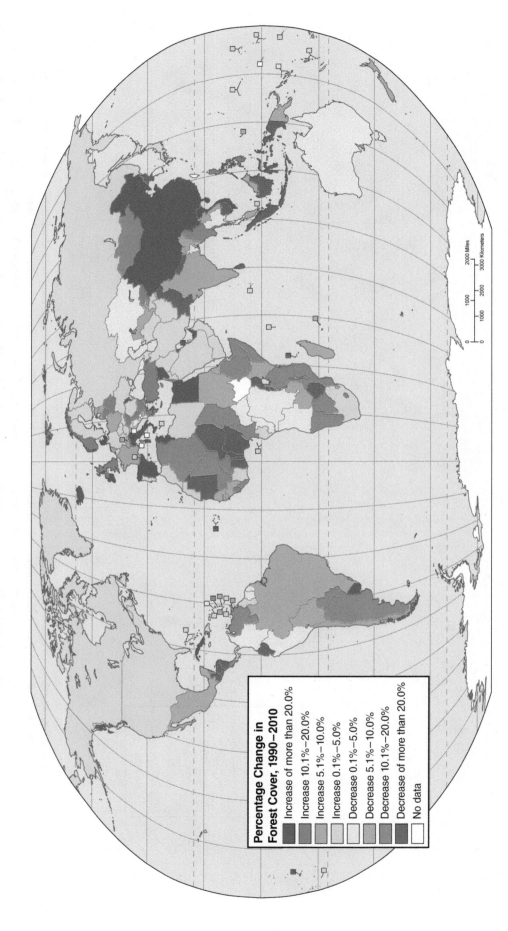

Percentage Change in Forest Cover, 1990–2010

- Increase of more than 20.0%
- Increase 10.1%–20.0%
- Increase 5.1%–10.0%
- Increase 0.1%–5.0%
- Decrease 0.1%–5.0%
- Decrease 5.1%–10.0%
- Decrease 10.1%–20.0%
- Decrease of more than 20.0%
- No data

One of the most discussed environmental problems is that of deforestation. For most people, deforestation means clearing of tropical rain forests for agricultural purposes. Yet nearly as much forest land per year—much of it in North America, Europe, and Russia— is impacted by commercial lumbering as is cleared by tropical farmers and ranchers. Even in the tropics, much of the forest clearance is undertaken by large corporations producing high-value tropical hardwoods for the global market in furniture, ornaments, and other fine wood products. Still, it is the agriculturally driven clearing of the great rain forests of the Amazon Basin, west and central Africa, Middle America, and Southeast Asia that draws public attention. Although much concern over forest clearance focuses on the relationship between forest clearance and the reduction in the capacity of the world's vegetation system to absorb carbon dioxide (and thus delay global warming), of just as great concern are issues having to do with the loss of biodiversity (large numbers of plants and animals), the near-total destruction of soil systems, and disruptions in water supply that accompany clearing.

Map 91   World Timber Production

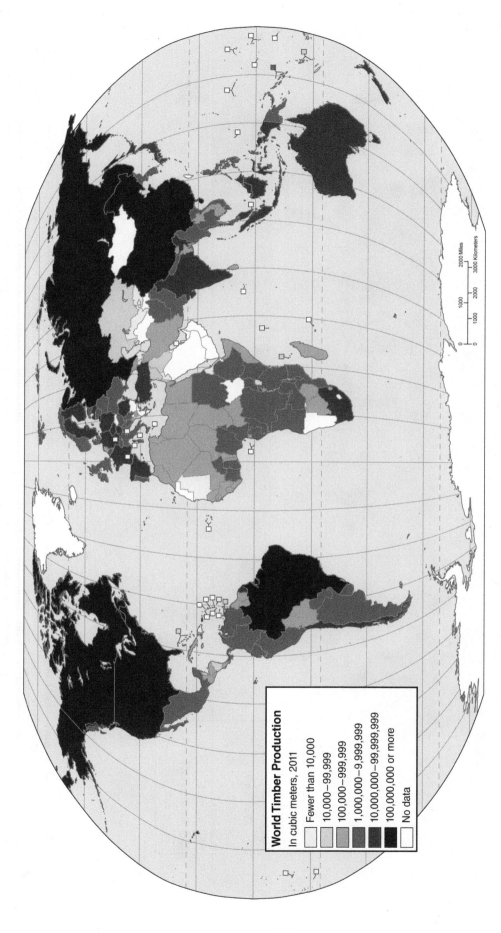

World Timber Production
In cubic meters, 2011

- Fewer than 10,000
- 10,000–99,999
- 100,000–999,999
- 1,000,000–9,999,999
- 10,000,000–99,999,999
- 100,000,000 or more
- No data

With the exception of water and arable land, forests have played a larger part in the transition from hunting-gathering to the emergence of modern industrial economies than any other natural resource. The chief uses of forests are for lumber, fuelwood, and pulp and paper products. The per capita use of wood does not tend to vary greatly between the developed and developing countries. In developed areas such as North America, Europe, and Russia, the bulk of forest products is used for lumber (much of it exported to forest-poor areas) and for pulp and paper products. In the developing countries in South America, Africa, and Asia (excluding Russia), more wood is used for fuel than for any other purpose.

In those parts of the world where forests are are used for lumber and pulp-paper products, more-or-less sustainable forest systems have been developed. These systems are often artificially maintained and are much simpler (and, hence, more fragile) ecosystems than the rich, complex natural ecosystems they have replaced. Where forests provide the chief source of fuel, the forest systems tend to remain healthier but are shrinking in size in proportion to the increase in human populations. Of the major forest ecosystems, those in greatest jeopardy are the tropical forests that are being cleared for agricultural land and the boreal or northern forests of Canada and Russia that are being simplified by overcutting.

-113-

Map 92   The Loss of Biodiversity: Globally Threatened Animal Species

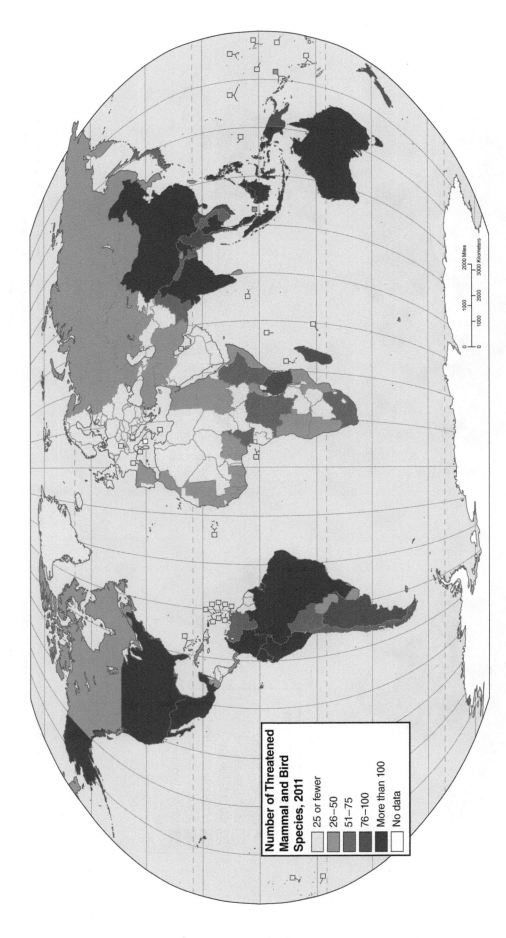

Number of Threatened Mammal and Bird Species, 2011
- 25 or fewer
- 26–50
- 51–75
- 76–100
- More than 100
- No data

Threatened species are those in grave danger of going extinct. Their populations are becoming restricted in range, and the size of the populations required for sustained breeding is nearing a critical minimum. *Endangered species* are in immediate danger of becoming extinct. Their range is already so reduced that the animals may no longer be able to move freely within an ecozone, and their populations are at the level where the species may no longer be able to sustain breeding. Most species become threatened first and then endangered as their range and numbers continue to decrease. When people think of animal extinction, they think of large herbivorous species such as the rhinoceros

or fierce carnivores such as lions, tigers, or grizzly bears. Certainly these animals make almost any list of endangered or threatened species. But there are literally hundreds of less conspicuous animals that are equally threatened. Extinction is normally nature's way of informing a species that it is inefficient. But conditions in the late twentieth century are controlled more by human activities than by natural evolutionary processes. Species that are endangered or threatened fall into that category because, somehow, they are competing with us or with our domesticated livestock for space and food. And in that competition the animals are always going to lose.

Map 93   Disappearing Giants: The Reduction in "Conservative" Species

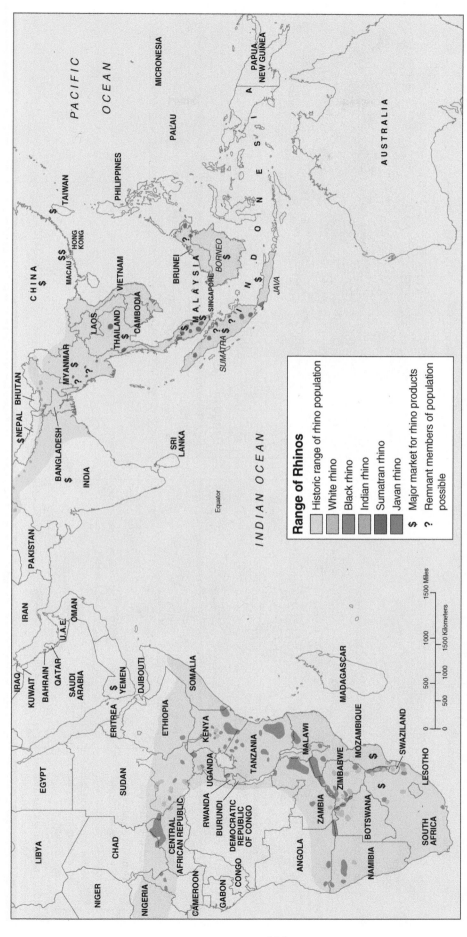

Range of Rhinos

- Historic range of rhino population
- White rhino
- Black rhino
- Indian rhino
- Sumatran rhino
- Javan rhino
- $ Major market for rhino products
- ? Remnant members of population possible

The species singled out for special attention have been given that attention not just because they represent major species in greatest danger of extinction but because they are signal species (indicating the health of ecosystems) as well as conservative or aristocratic species. These species often suffer most from human infringement on traditional land or from overhunting. Rhinos, elephants, and tigers fall into the conservative or aristocratic categories: they are large, they breed slowly, they do not produce large numbers of offspring in a single breeding season, and they may have long gestation periods. Conservative species are also often those that are both in great demand for their hides, horns, tusks, or fur (in other words, non-edible products that demand

extremely high prices in the luxury market), and that tend to compete most directly with humans for the same resources—even if that resource is simply space. Part of the rapid loss of such aristocratic species as rhinos, elephants, and tigers is the value of ivory, tiger hides, or rhinoceros horn in the global market; but part of the rapid loss is that these species occupy space that is also needed for cropland or grazing land. In other words, these species compete with humans for the same resources and, in any such competition, it is the humans who are going to win. And a win for humans represents a huge loss for those animals that may have formerly been the ecological dominants in their home ranges.

-115-

Map 93 Disappearing Giants: The Reduction in "Conservative" Species

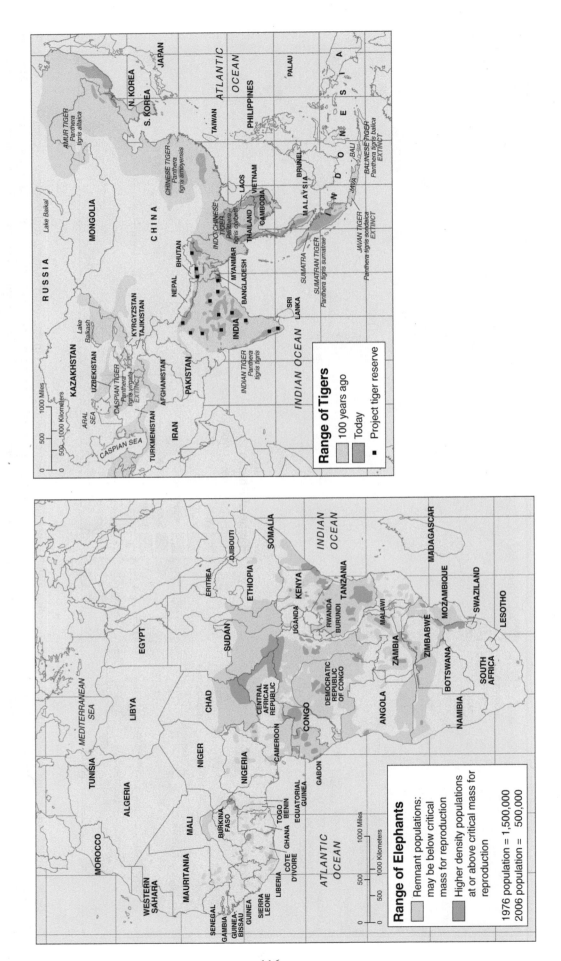

Map 94 The Loss of Biodiversity: Globally Threatened Plant Species

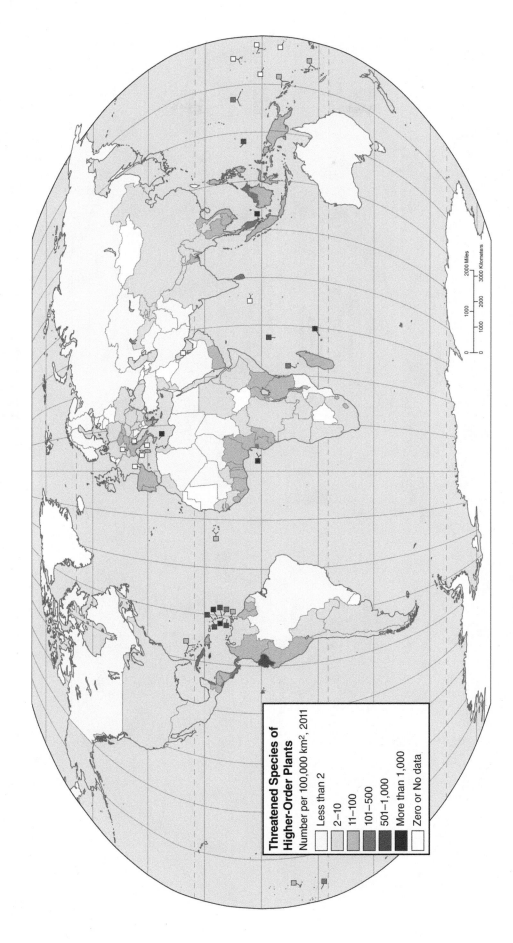

Threatened Species of Higher-Order Plants
Number per 100,000 km², 2011

- Less than 2
- 2–10
- 11–100
- 101–500
- 501–1,000
- More than 1,000
- Zero or No data

Although most people tend to be more concerned about the animals on threatened and endangered species lists, the fact is that many more plants are in jeopardy, and the loss of plant life is, in all ecological regions, a more critical occurrence than the loss of animal populations. Plants are the primary producers in the ecosystem; that is, plants produce the food upon which all other species in the food web, including human beings, depend for sustenance. It is plants from which many of our critical medicines come, and it is plants that maintain the delicate balance between soil and water in most of the world's regions. When environmental scientists speak of a loss of biodiversity, what they are most often describing is a loss of the richness and complexity of plant life that lends stability to ecosystems. Systems with more plant life tend to be more stable than those with less. For these and other reasons, the scientific concern over extinction is greater when applied to plants than to animals. It is difficult for people to become as emotional over a teak tree as they would over an elephant. But as great a tragedy as the loss of the elephant would be, the loss of the teak would be greater.

Map 95   The Areas of Greatest Loss and Hotspots of Biodiversity

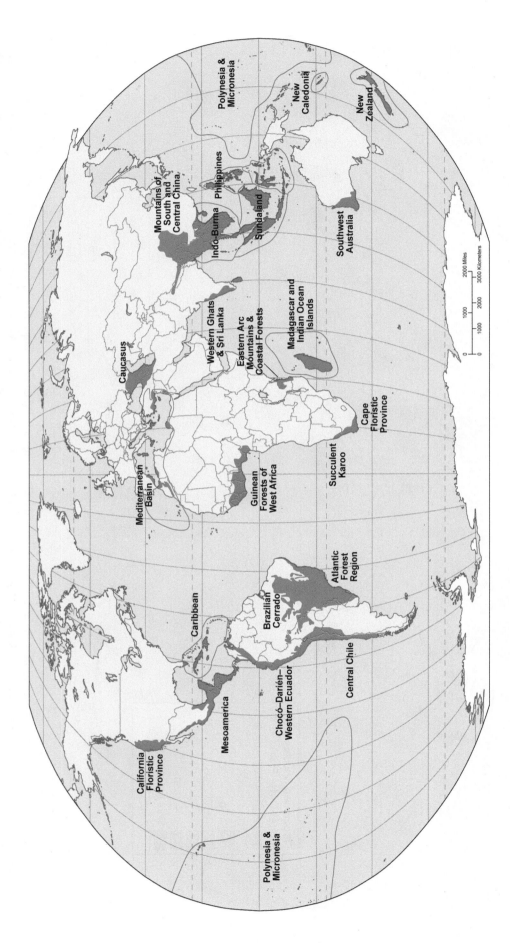

Polynesia & Micronesia

New Caledonia

New Zealand

Mountains of South and Central China

Philippines

Indo-Burma

Sundaland

Southwest Australia

Caucasus

Western Ghats & Sri Lanka

Eastern Arc Mountains & Coastal Forests

Madagascar and Indian Ocean Islands

Mediterranean Basin

Guinean Forests of West Africa

Cape Floristic Province

Succulent Karoo

Caribbean

Brazilian Cerrado

Atlantic Forest Region

Mesoamerica

Chocó–Darién–Western Ecuador

Central Chile

California Floristic Province

Polynesia & Micronesia

2000 Miles
3000 Kilometers

1000

2000

0

1000

0

Where we have normally thought of tropical forest basins such as Amazonia as the world's most biologically diverse ecosystems, recent research has discovered the surprising fact that a number of hotspots of biological diversity exist outside the major tropical forest regions. These hotspot regions contain slightly less than 2 percent of the world's total land area but may contain up to 60 percent of the total world's terrestrial species of plants and animals. Geographically, the hotspot areas are characterized by vertical zonation (that is, they tend to be hilly to mountainous regions), long known to be a factor in biological complexity. They are also in coastal locations or near large bodies of water, locations that stimulate climatic variability and, hence, biological complexity. Although some of the hotspots are sparsely populated, others, such as Sundaland, are occupied by some of the world's densest populations. Protection of the rich biodiversity of these hotspots is, most biologists feel, of crucial importance to the preservation of the world's biological heritage. The greatest actual or potential losses of bird and mammal species tend to occur in those areas of the world where there is the most active competition for living spaces between humans and animals. Thus highly populated areas tend to suffer the greatest loss of native species.

Map 96   The Risks of Desertification

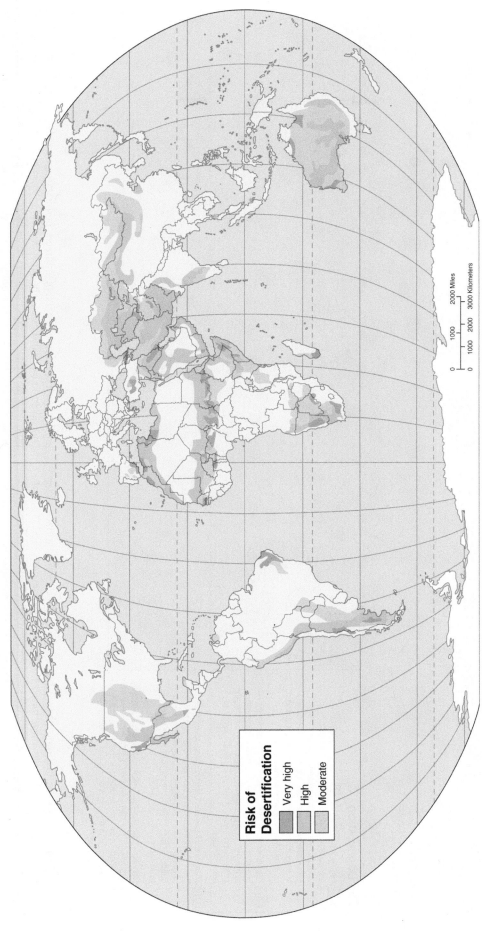

Risk of Desertification

Very high
High
Moderate

0   1000   2000 Miles
0   1000   2000   3000 Kilometers

The awkward-sounding term *desertification* refers to a reduction in the food-producing capacity of drylands through vegetation, soil, and water changes that culminate in either a drier climate or in soil and plant systems that are less efficient in their use of water. Most of the world's existing drylands—the shortgrass steppes, the tropical savannas, the bunchgrass regions of the desert fringe—are fairly intensively used for agriculture and are, therefore, subject to the kinds of pressures that culminate in desertification. Most desertification is a natural process that occurs near the margins of desert regions. It is caused by dehydration of the soil's surface layers during periods of drought and by high water loss through evaporation in an environment of high temperature and high winds. This natural process

is greatly enhanced by human agricultural activities that expose topsoil to wind and water erosion. Among the most important practices that cause desertification are (1) overgrazing of rangelands, resulting from too many livestock on too small an area of land; (2) improper management of soil and water resources in irrigation agriculture, leading to accelerated erosion and to salt buildup in the soil; (3) cultivation of marginal terrain with soils and slopes that are unsuitable for farming; (4) surface disturbances of vegetation (clearing of thorn scrub, mesquite, chaparral, and similar vegetation) without soil protection efforts being made or replanting being done; and (5) soil compaction by agricultural implements, domesticated livestock, and rain falling on an exposed surface.

Map 97 Global Soil Degradation

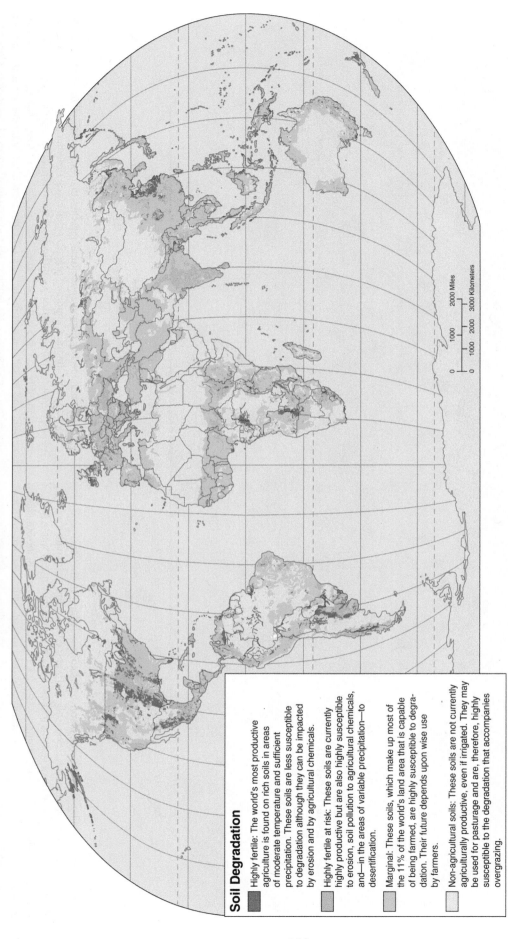

Soil Degradation

Highly fertile: The world's most productive agriculture is found on rich soils in areas of moderate temperature and sufficient precipitation. These soils are less susceptible to degradation although they can be impacted by erosion and by agricultural chemicals.

Highly fertile at risk: These soils are currently highly productive but are also highly susceptible to erosion, soil pollution to agricultural chemicals, and—in the areas of variable precipitation—to desertification.

Marginal: These soils, which make up most of the 11% of the world's land area that is capable of being farmed, are highly susceptible to degradation. Their future depends upon wise use by farmers.

Non-agricultural soils: These soils are not currently agriculturally productive, even if irrigated. They may be used for pasturage and are, therefore, highly susceptible to the degradation that accompanies overgrazing.

0 1000 2000 2000 Miles

0 1000 2000 3000 Kilometers

Recent research has shown that more than 3 billion acres of the world's surface suffer from serious soil degradation, with more than 22 million acres so severely eroded or poisoned with chemicals that they can no longer support productive crop agriculture. Most of this soil damage has been caused by poor farming practices, overgrazing of domestic livestock, and deforestation. These activities strip away the protective cover of natural vegetation forests and grasslands, allowing wind and water erosion to remove the topsoil that contains necessary nutrients and soil microbes for plant growth. But millions of acres of topsoil have been degraded by chemicals as well. In some instances these chemicals are the result of overapplication of fertilizers, herbicides, pesticides, and other agricultural chemicals. In other instances, chemical deposition from industrial and urban wastes and from acid precipitation has poisoned millions of acres of soil. As the map shows, soil erosion and pollution are not problems just in developing countries with high population densities and increasing use of marginal lands. They also afflict the more highly developed regions of mechanized, industrial agriculture. Although many methods for preventing or reducing soil degradation exist, they are seldom used because of ignorance, cost, or perceived economic inefficiency.

Map 98 The Degree of Human Disturbance

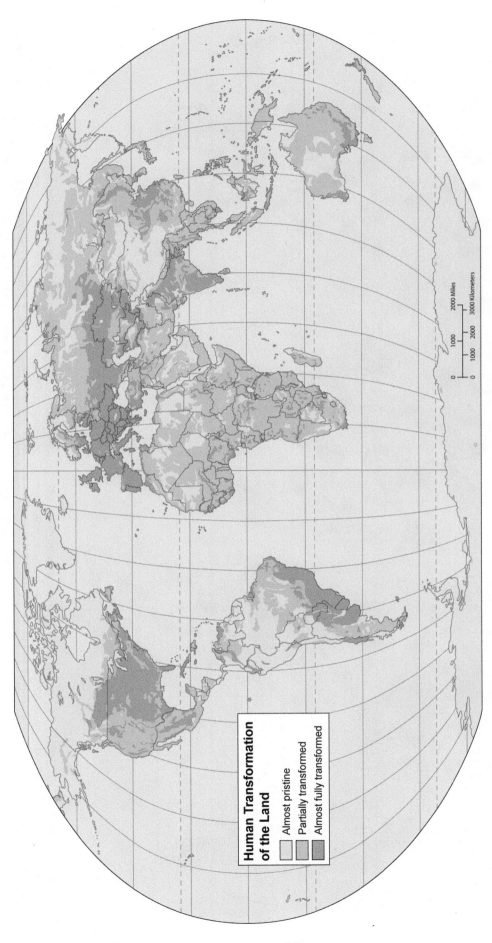

Human Transformation of the Land

Almost pristine
Partially transformed
Almost fully transformed

0 1000 2000 Miles
0 1000 2000 3000 Kilometers

The data on human disturbance have been gathered from a wide variety of sources, some of them conflicting and not all of them reliable. Nevertheless, at a global scale this map fairly depicts the state of the world in terms of the degree to which humans have modified its surface. The almost pristine areas, covered with natural vegetation, generally have population densities less than 10 persons per square mile. These areas are, for the most part, in the most inhospitable parts of the world: too high, too dry, too cold for permanent human habitation in large numbers. The partially transformed areas are normally agricultural areas, either subsistence (such as shifting cultivation) or extensive (such as livestock grazing). They often contain areas of secondary vegetation, regrown

after removal of original vegetation by humans. They are also sometimes marked by a density of livestock in excess of carrying capacity, leading to overgrazing, which further alters the condition of the vegetation. The almost fully transformed areas are those of permanent and intensive agriculture and urban settlement. The primary vegetation of these regions has been removed, with no evidence of regrowth or with current vegetation that is quite different from natural (potential) vegetation. Soils are in a state of depletion and degradation, and, in drier lands, desertification is a factor of human occupation. The disturbed areas match closely those areas of the world with the densest human populations.

Map 99   The Green and Not So Green World

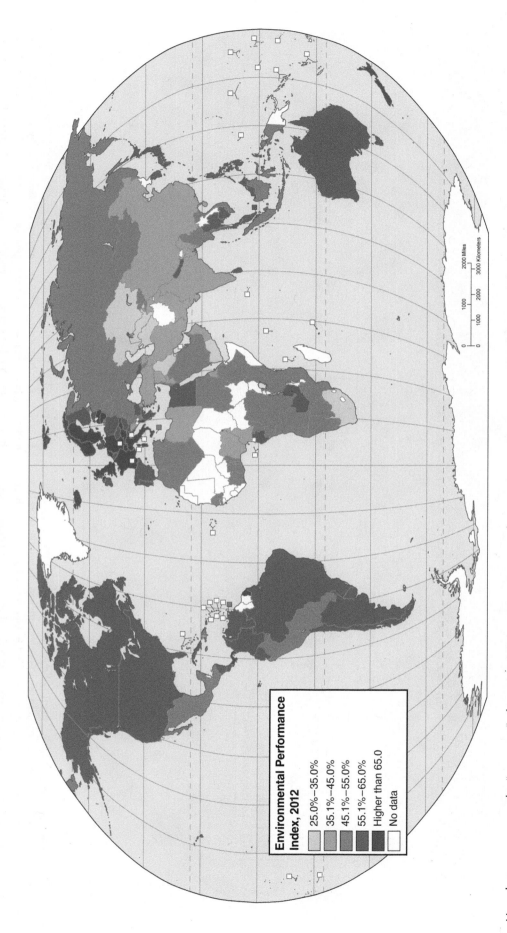

Environmental Performance Index, 2012

- 25.0%–35.0%
- 35.1%–45.0%
- 45.1%–55.0%
- 55.1%–65.0%
- Higher than 65.0
- No data

How does one measure the "greenness" of a country? Does one measure greenhouse gas emissions or water resources or land in parks and preserves? One measure, developed by Yale and Columbia Universities, is the Environmental Performance Index, which measures the extent to which countries perform relative to goals established in their various environmental policies. As might be expected, the "greenest" countries also tend to be relatively affluent and in cooler climates. There are obvious exceptions to this generalization (Costa Rica, for example) but, in general, we find countries such as Switzerland, Sweden, Norway, Finland, France, New Zealand, and Denmark ranking high on the index whereas poorer

countries in warmer climates (much of South and East Asia, much of Sub-Saharan Africa) generally tend to rank lower on the index. There is a pretty clear correlation between affluence and "greenness"—that is, wealthy countries not only recognize the importance of "being green" but have the ability to pay for it. Still, some of the world's more affluent countries, such as the United States and Australia, rank rather lower on the scale. The relatively low ranking (not quite in the top third of the countries in the index) attained by the United States—despite its excellent record in dealing with environmental health issues—results from its persistent reliance on fossil fuels and various impacts related to ecosystem vitality.

Unit VI

Global Political Patterns

Map 100   Political Systems

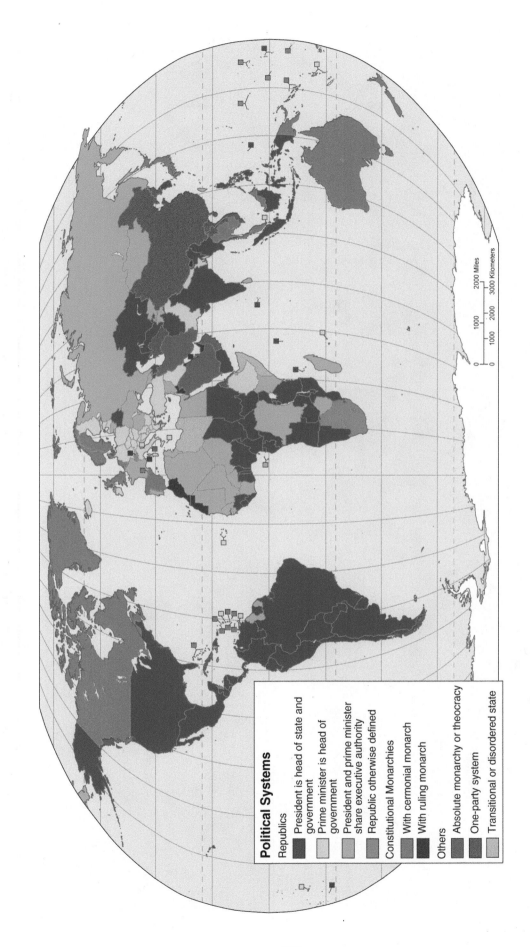

Political Systems

Republics
- President is head of state and government
- Prime minister is head of government
- President and prime minister share executive authority
- Republic otherwise defined

Constitutional Monarchies
- With cermonial monarch
- With ruling monarch

Others
- Absolute monarchy or theocracy
- One-party system
- Transitional or disordered state

World political systems have changed dramatically during the last two decades and may change even more in the future. The categories of political systems shown on the map are subject to some interpretation: most of the world's states are democracies, either republics or constitutional monarchies. In republics the head of state is either a president or prime minister, although in some instances a country may have both. One-party systems are states where single-party rule is constitutionally guaranteed or where a one-party regime is a fact

of political life. Monarchies are countries with heads of state who are members of a royal family. In a constitutional monarchy, such as the U.K. and the Netherlands, the monarchs are titular heads of state only. Theocracies are countries in which rule is within the hands of a priestly or clerical class; today, this means primarily fundamentalist Islamic countries such as Iran. Finally, transitional states have had ongoing or recently concluded civil war in which the government either is not functional or is in the process of being formed.

Map 101 · The Emergence of the State

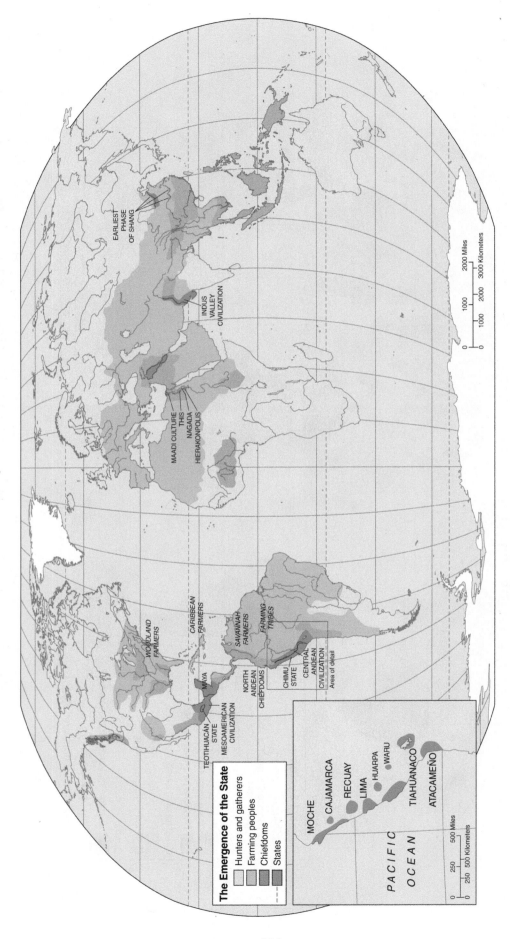

The Emergence of the State

- Hunters and gatherers
- Farming peoples
- Chiefdoms
- States

WOODLAND FARMERS

CARIBBEAN FARMERS

SAVANNAH FARMERS

FARMING TRIBES

MAYA

NORTH ANDEAN CHIEFDOMS

TEOTIHUACAN STATE

MESOAMERICAN CIVILIZATION

CHIMU STATE

CENTRAL ANDEAN CIVILIZATION

Area of detail

MAADI CULTURE
THIS
NAGADA
HIERAKONPOLIS

INDUS VALLEY CIVILIZATION

EARLIEST PHASE OF SHANG

PACIFIC OCEAN

MOCHE
CAJAMARCA
RECUAY
LIMA
HUARPA
WARU
TIAHUANACO
ATACAMEÑO

0   250   500 Miles
0   250   500 Kilometers

0   1000   2000 Miles
0   1000   2000   3000 Kilometers

Agriculture is the basis of the development of the state, a form of complex political organization. Social stratification based on wealth and power creates different classes or groups, some of whom no longer work the land. An agricultural surplus supports those who perform other functions for society, such as artisans and craftspeople, soldiers and police, priests and kings. Thus, over time, egalitarian hunters and gatherers shifted to state-level societies in some parts of the world. The first true states are shown in the rust-colored areas of this map.

Agriculture is the basis of the development of the state, a form of complex political organization. Geographers and anthropologists believe that agriculture allowed for larger concentrations of population. Farmers do not need to be as mobile as hunters and gatherers to make a living, and people living sedentarily can have larger families than those constantly on the move. Ideas about access to land and ownership also change as people develop in the social and political hierarchies that come with the transition from a hunting-gathering

-126-

Map 102   Organized States and Chiefdoms, A.D. 1500

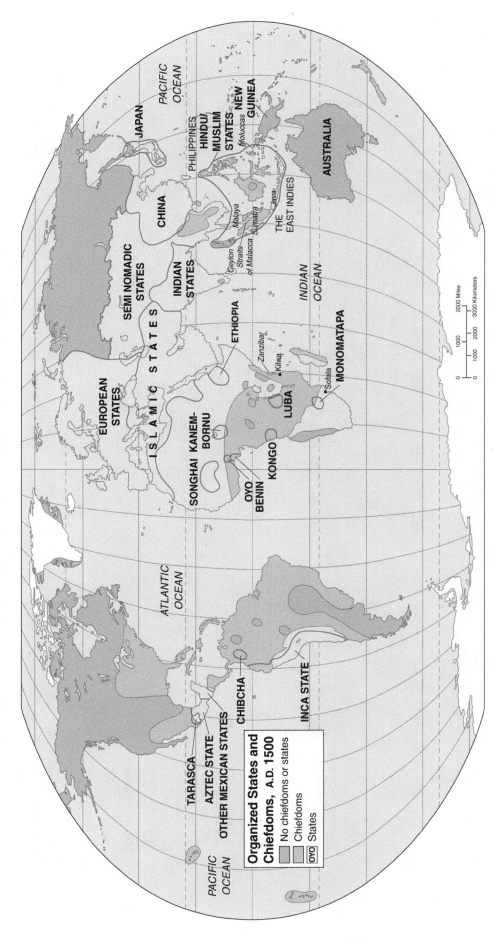

As Europeans began to expand outward through exploration and settlement from the late fifteenth to the eighteenth centuries, it was inevitable that they would find the complex political organizations of chiefdoms and states in many different parts of the world. Both chiefdoms and states are large-scale forms of political organization in which some people have privileged access to land, power, wealth, and prestige. Chiefdoms are kin-based societies in which wealth is distributed from upper to lower classes. States are organized on the basis of socioeconomic classes, headed by a centralized government that is led by an elite. States in non-European areas included, just as they did in Europe, full-time bureaucracies and specialized subsystems for such activities as military action, taxation, operation of state religions, and social control. In many of the colonial regions of the world after the fifteenth century, Europeans actually found it easier to gain control of organized states and chiefdoms because those populations were already accustomed to some form of institutionalized central control.

-127-

Map 103   European Colonialism, 1500–2000

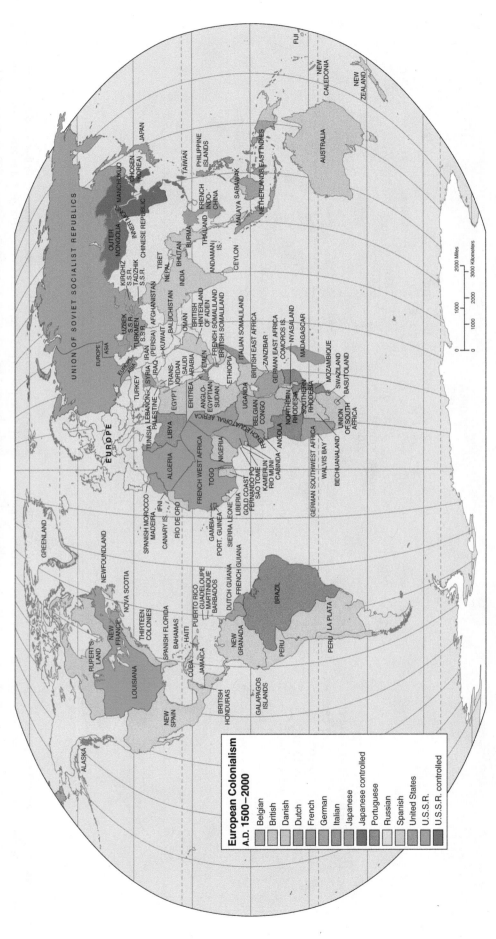

European powers have controlled many parts of the world during the last 500 years. The period of European expansion began when European explorers sailed the oceans in search of new trading routes and ended after World War II when many colonies in Africa and Asia gained independence. The process of colonization was very complex but normally involved the acquisition, extraction, or production of raw materials (including minerals, forest products, products from the sea, agricultural products, and animal furs/pelts) from the areas being controlled by the European colonial power in exchange for items of European manufacture. The concept of colonial dependency implied an economic structure in which the European country obtained raw materials from the colonial country in exchange for those manufactured items upon which populations in the colonial areas quickly came to depend. The colors on this map represent colonial control at its maximum extent and do not take into account shifting colonial control. In North America, for example, "New France" became British territory and "Louisiana" became Spanish territory after the Seven Years' (French and Indian) War. There are also two non-European countries—the United States and Japan—shown on the map as colonial powers. Both countries followed the classical European-style colonial model, albeit over relatively small areas and for relatively short periods of time.

Map 104   Sovereign States: Duration of Independence

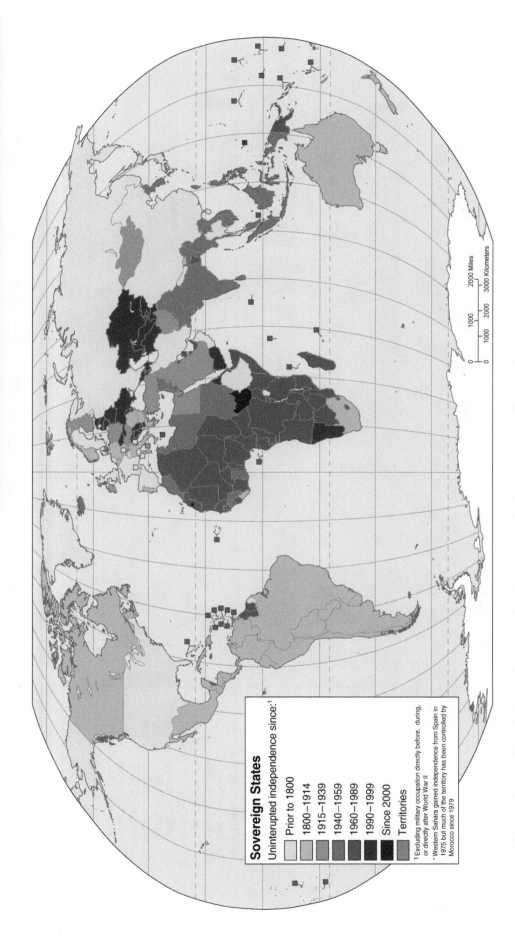

Sovereign States

Uninterrupted independence since:[1]

- Prior to 1800
- 1800–1914
- 1915–1939
- 1940–1959
- 1960–1989
- 1990–1999
- Since 2000
- Territories

[1] Excluding military occupation directly before, during, or directly after World War II

* Western Sahara gained independence from Spain in 1975 but much of the territory has been controlled by Morocco since 1979

0   1000   2000 Miles

0   1000   2000   3000 Kilometers

Most countries of the modern world, including such major states as Germany and Italy, became independent after the beginning of the nineteenth century. Of the world's current countries, only 27 were independent in 1800. (Ten of the 27 were in Europe; the others were Afghanistan, China, Colombia, Ethiopia, Haiti, Iran, Japan, Mexico, Nepal, Oman, Paraguay, Russia, Taiwan, Thailand, Turkey, the United States, and Venezuela). Following 1800, there have been five great periods of national independence. During the first of these (1800–1914), most of the mainland countries of the Americas achieved independence. During the second period (1915–1939), the countries of Eastern Europe emerged as independent entities. The third period (1940–1959) includes World War II and the years that followed, when independence for African and Asian nations that had

been under control of colonial powers first began to occur. During the fourth period (1960–1989), independence came to the remainder of the colonial African and Asian nations, as well as to former colonies in the Caribbean and the South Pacific. More than half of the world's countries came into being as independent political entities during this period. In the last decade of the twentieth century, the breakup of the existing states of the Soviet Union, Yugoslavia, and Czechoslovakia created 22 countries where only 3 had existed before. Since 2000, there have been four new additions to the world's states. Timor-Leste became independent in 2002, the breakup of the former Yugoslavia continued with the separation of Montenegro (2006) and Kosovo (2008), and South Sudan separated from Sudan in 2011.

Map 105　Post–Cold War International Alliances

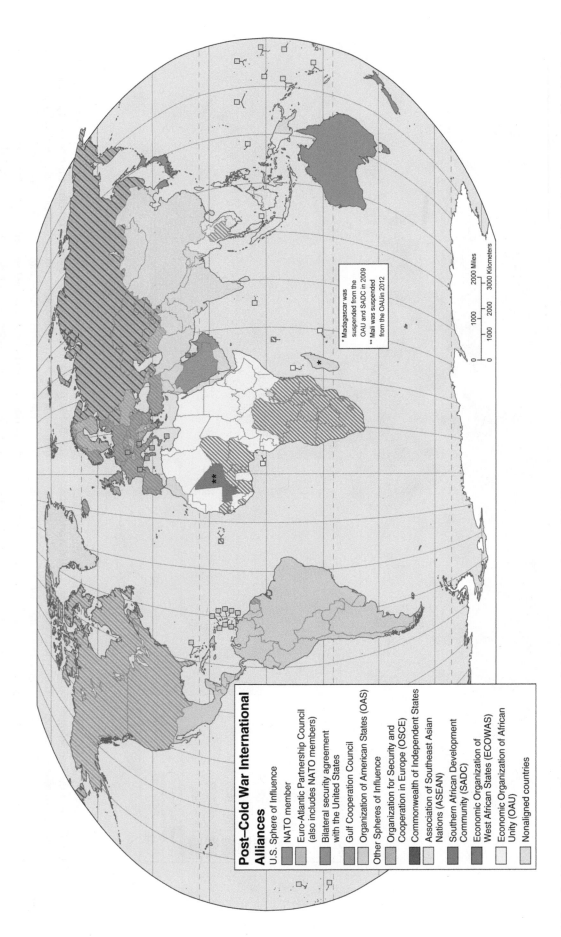

Post–Cold War International Alliances

U.S. Sphere of Influence
- NATO member
- Euro-Atlantic Partnership Council (also includes NATO members)
- Bilateral security agreement with the United States
- Gulf Cooperation Council
- Organization of American States (OAS)

Other Spheres of Influence
- Organization for Security and Cooperation in Europe (OSCE)
- Commonwealth of Independent States
- Association of Southeast Asian Nations (ASEAN)
- Southern African Development Community (SADC)
- Economic Organization of West African States (ECOWAS)
- Economic Organization of African Unity (OAU)
- Nonaligned countries

* Madagascar was suspended from the OAU and SADC in 2009
** Mali was suspended from the OAU in 2012

0　1000　2000 Miles
0　1000　2000　3000 Kilometers

When the Warsaw Pact dissolved in 1992, the North Atlantic Treaty Organization (NATO) was left as the only major military alliance in the world. Some former Warsaw Pact members (Czechia, Hungary, and Poland) have joined NATO, and others are petitioning for entry. The bipolar division of the world into two major military alliances is over, at least temporarily, leaving the United States alone as the world's dominant political and military power. But other international alliances, such as the Commonwealth of Independent States (including most of the former republics of the Soviet Union), will continue to be important. It may well be that during the first few decades of the twenty-first century economic alliances will begin to overshadow military ones in their relevance for the world's peoples.

-130-

Map 106   Political Realms: Regional Changes, 1945–2003

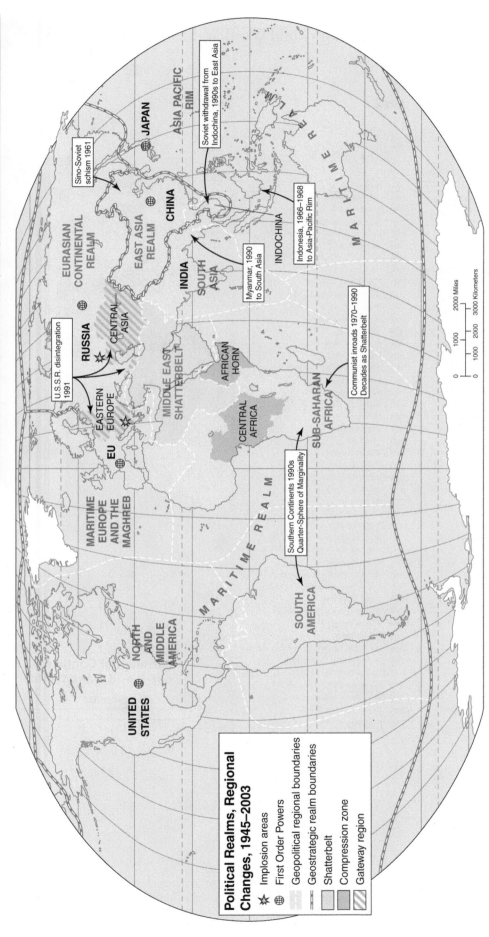

Political Realms, Regional Changes, 1945–2003

- ☆ Implosion areas
- ⊕ First Order Powers
- ▨ Geopolitical regional boundaries
- — Geostrategic realm boundaries
- Shatterbelt
- Compression zone
- Gateway region

U.S.S.R. disintegration 1991

Sino-Soviet schism 1961

Soviet withdrawal from Indochina, 1990s to East Asia

Indonesia, 1966–1968 to Asia-Pacific Rim

Myanmar, 1990 to South Asia

Communist inroads 1970–1990 Decades as Shatterbelt

Southern Continents 1990s Quarter-Sphere of Marginality

EURASIAN CONTINENTAL REALM

EAST ASIA REALM

CHINA

JAPAN

ASIA PACIFIC RIM

MARITIME REALM

RUSSIA

CENTRAL ASIA

EASTERN EUROPE

EU

MARITIME EUROPE AND THE MAGHREB

MIDDLE EAST SHATTERBELT

AFRICAN HORN

CENTRAL AFRICA

SUB-SAHARAN AFRICA

INDIA

SOUTH ASIA

INDOCHINA

NORTH AND MIDDLE AMERICA

SOUTH AMERICA

UNITED STATES

0   1000   2000   3000 Kilometers
0   1000   2000 Miles

The Cold War following World War II shaped the major outlines of today's geopolitical relations. The Cold War included three phases. In the first, from 1945–1956, the Maritime Realm established a ring around the Continental Eurasian Realm in order to prevent its expansion. This phase included the Korean War (1950–1953), the Berlin Blockade (1948), the Truman doctrine and Marshall Plan (1947), and the founding of NATO (1949) and the Warsaw Pact (1955). Most of the world fell within one of the two realms: the Maritime (dominated by the United States) or the Eurasian Continental Realm (dominated by the Soviet Union). The Soviet Union sought to establish a ring of satellite states to protect it from a repeat of the invasions of World War II. The United States and other Maritime Realm states, in turn, sought to establish a ring of allies around the Continental Realm to prevent its expansion. South Asia was politically independent, but under pressure from both realms. During the second phase (1957–1979), Communist forces from the Continental Eurasian Realm penetrated deeply into the Maritime Realm. The Berlin Wall went up in 1961, Soviet missiles in Cuba ignited a crisis in 1962, and the United States became increasingly involved in the war in Vietnam (late 1960s). The Soviet Union sought increased political and military presence along important waterways including those in the Middle East, Southeast Asia, and the Caribbean. These regions became especially dangerous shatterbelts. The third phase (1980–1989) saw the retreat of Communist power from the Maritime Realm. China, after ten years of radical Communism and chaos of the Cultural Revolution (1966–1976), broke away from the Continental Eurasian Realm to establish a new East Asian realm. Soviet influence declined in the Middle East, Sub-Saharan Africa, and Latin America. In 1989 the Berlin Wall fell, and Eastern Europe began to establish democratic governments. In 1991 the Soviet Union broke apart as its constituent republics became independent states. Compression zones are areas of conflict, but they are not contested by major powers.

-131-

Map **107**   International Conflicts in the Post–World War II World

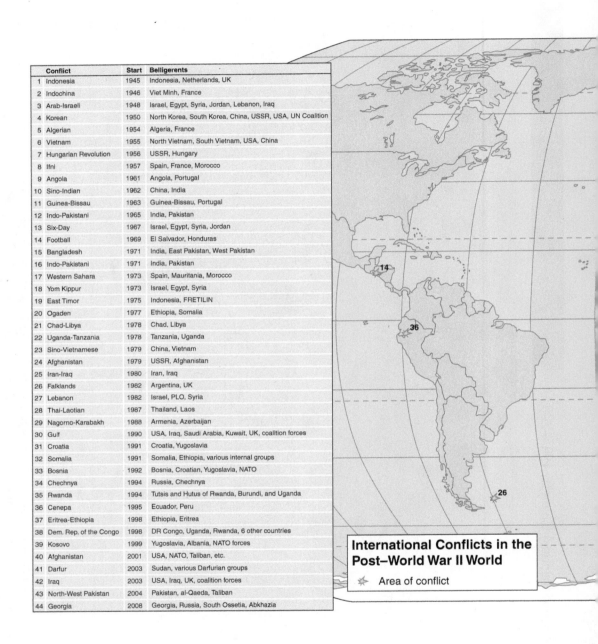

	Conflict	Start	Belligerents
1	Indonesia	1945	Indonesia, Netherlands, UK
2	Indochina	1946	Viet Minh, France
3	Arab-Israeli	1948	Israel, Egypt, Syria, Jordan, Lebanon, Iraq
4	Korean	1950	North Korea, South Korea, China, USSR, USA, UN Coalition
5	Algerian	1954	Algeria, France
6	Vietnam	1955	North Vietnam, South Vietnam, USA, China
7	Hungarian Revolution	1956	USSR, Hungary
8	Ifni	1957	Spain, France, Morocco
9	Angola	1961	Angola, Portugal
10	Sino-Indian	1962	China, India
11	Guinea-Bissau	1963	Guinea-Bissau, Portugal
12	Indo-Pakistani	1965	India, Pakistan
13	Six-Day	1967	Israel, Egypt, Syria, Jordan
14	Football	1969	El Salvador, Honduras
15	Bangladesh	1971	India, East Pakistan, West Pakistan
16	Indo-Pakistani	1971	India, Pakistan
17	Western Sahara	1973	Spain, Mauritania, Morocco
18	Yom Kippur	1973	Israel, Egypt, Syria
19	East Timor	1975	Indonesia, FRETILIN
20	Ogaden	1977	Ethiopia, Somalia
21	Chad-Libya	1978	Chad, Libya
22	Uganda-Tanzania	1978	Tanzania, Uganda
23	Sino-Vietnamese	1979	China, Vietnam
24	Afghanistan	1979	USSR, Afghanistan
25	Iran-Iraq	1980	Iran, Iraq
26	Falklands	1982	Argentina, UK
27	Lebanon	1982	Israel, PLO, Syria
28	Thai-Laotian	1987	Thailand, Laos
29	Nagorno-Karabakh	1988	Armenia, Azerbaijan
30	Gulf	1990	USA, Iraq, Saudi Arabia, Kuwait, UK, coalition forces
31	Croatia	1991	Croatia, Yugoslavia
32	Somalia	1991	Somalia, Ethiopia, various internal groups
33	Bosnia	1992	Bosnia, Croatian, Yugoslavia, NATO
34	Chechnya	1994	Russia, Chechnya
35	Rwanda	1994	Tutsis and Hutus of Rwanda, Burundi, and Uganda
36	Cenepa	1995	Ecuador, Peru
37	Eritrea-Ethiopia	1998	Ethiopia, Eritrea
38	Dem. Rep. of the Congo	1998	DR Congo, Uganda, Rwanda, 6 other countries
39	Kosovo	1999	Yugoslavia, Albania, NATO forces
40	Afghanistan	2001	USA, NATO, Taliban, etc.
41	Darfur	2003	Sudan, various Darfurian groups
42	Iraq	2003	USA, Iraq, UK, coalition forces
43	North-West Pakistan	2004	Pakistan, al-Qaeda, Taliban
44	Georgia	2008	Georgia, Russia, South Ossetia, Abkhazia

International Conflicts in the Post–World War II World

✳ Area of conflict

The Korean War and the Vietnam War dominated the post–World War II period in terms of international military conflict. But numerous smaller conflicts have taken place, with fewer numbers of belligerents and with fewer battle and related casualties. These smaller international conflicts have been mostly territorial conflicts, reflecting the continual readjustment of political boundaries and loyalties brought about by the end of colonial empires and the dissolution of the U.S.S.R. Many of these conflicts were not wars in the more traditional sense,

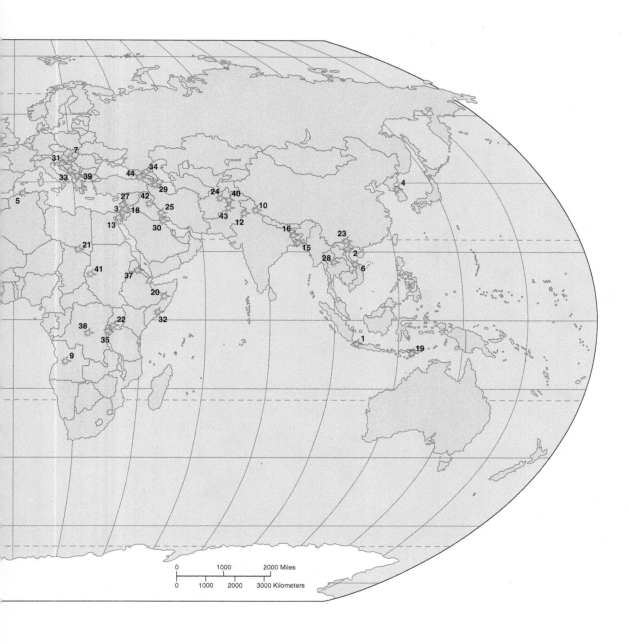

in which two or more countries formally declare war on one another, severing diplomatic ties and devoting their entire national energies to the war effort. Rather, many of these conflicts were and are undeclared wars, sometimes fought between rival groups within the same country with outside support from other countries. The aftermath of the September 11, 2001, terrorist attacks on the United States indicate the dawn of yet another type of international conflict, namely a "war" fought between traditional nation-states and non-state actors.

Map **108** The Geopolitical World in the Twenty-First Century

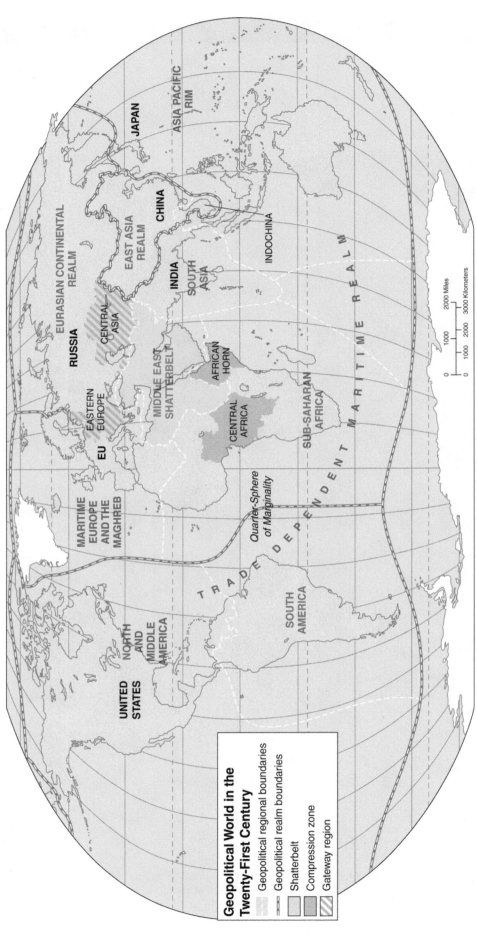

Geopolitical World in the Twenty-First Century

Geopolitical regional boundaries
Geopolitical realm boundaries
Shatterbelt
Compression zone
Gateway region

UNITED STATES

NORTH AND MIDDLE AMERICA

MARITIME EUROPE AND THE MAGHREB

EU

EASTERN EUROPE

RUSSIA

EURASIAN CONTINENTAL REALM

CENTRAL ASIA

JAPAN

ASIA PACIFIC RIM

EAST ASIA REALM

CHINA

MIDDLE EAST SHATTERBELT

INDIA

SOUTH ASIA

INDOCHINA

AFRICAN HORN

CENTRAL AFRICA

SUB-SAHARAN AFRICA

SOUTH AMERICA

Quarter-Sphere of Marginality

T R A D E   D E P E N D E N T   M A R I T I M E   R E A L M

0   1000   2000   3000 Kilometers
0   1000   2000 Miles

In the geostrategic structure of the world, the largest territorial units are realms. They are shaped by circulation patterns that link people, goods, and ideas. Realms are shaped by maritime and continental influences. Today's Atlantic and Pacific Trade-Dependent Maritime Realm has been shaped by international exchange over the oceans and their interior seas as mercantilism, capitalism, and industrialization gave rise to maritime-oriented states and to economic and political colonialism. The world's leading trading and economic powers are part of this realm. The Eurasian Continental Realm, centered around Russia, is inner-oriented, less influenced by outside economic or cultural forces, and politically closed, even after the fall of Communism. Expansion of NATO in Europe

has increased its feeling of being "hemmed in." East Asia has mixed Maritime and Continental influences. China has traditionally been continental, but reforms that began in the late 1970s increased the importance of its maritime-oriented southern coasts. Even so, its trade volume is still low, and it maintains a hold on inland areas such as Tibet and Xinjiang. Realms are subdivided into regions, some dependent on others, as South America is on North America. Regions located between powerful realms or regions may be shatterbelts (internally divided and caught up in competition between Great Powers) or gateways (facilitating the flow of ideas, goods, and people between regions).

-134-

Map 109 The United Nations

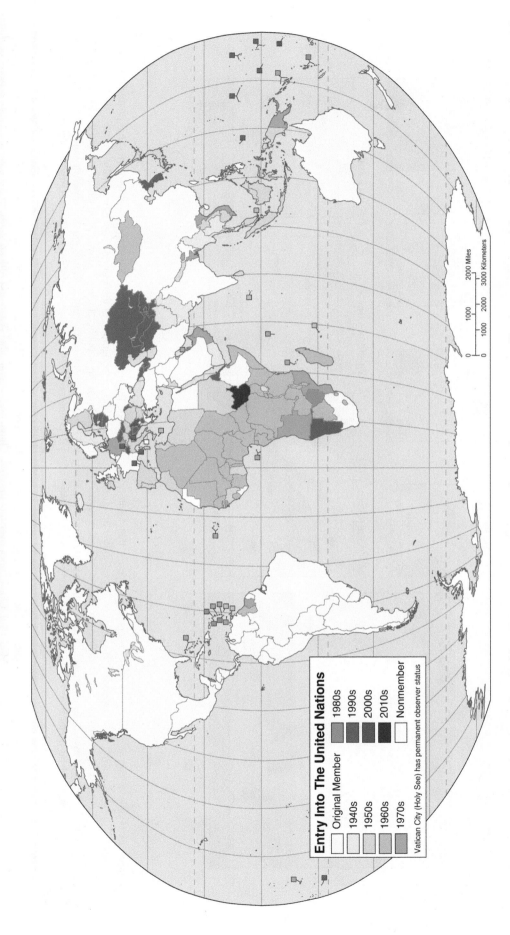

Entry Into The United Nations

- Original Member
- 1940s
- 1950s
- 1960s
- 1970s
- 1980s
- 1990s
- 2000s
- 2010s
- Nonmember

Vatican City (Holy See) has permanent observer status

0   1000   2000 Miles
0 1000 2000 3000 Kilometers

The United Nations was formed in 1945 after World War II to maintain international peace and security and to promote cooperation involving economic, social, cultural, and humanitarian problems. Originally consisting of 51 member states, the organization has grown to 192 member states in 2009. Most of the African continent and the smaller Caribbean states entered the UN during the 1960s and 1970s following the end of European colonial rule. The 1990s saw the entry of several countries following the dissolution of the Soviet Union and the breakup of Yugoslavia. Montenegro became the most recent country to join the UN in 2006 following its separation from Serbia.

China was represented by the government of the Republic of China at the creation of the United Nations. Following the Communist victory during the Chinese Civil War, the government fled to the island of Taiwan. UN representation was maintained by the Republic of China government until 1971 when the government of the People's Republic of China (mainland China) was recognized as the representatives of China to the organization. Western Sahara is not a member of the UN as its sovereignty status is in dispute. Much of the territory of Western Sahara is controlled by Morocco. Kosovo declared its independence in 2008, which was recognized by the United States and 91 other countries. As of 2009, its entry into the UN has been blocked by Russia, which has not recognized its independence from Serbia. The Holy See (Vatican City) and Palestine hold status as observers.

Map 110 Is It a Country?

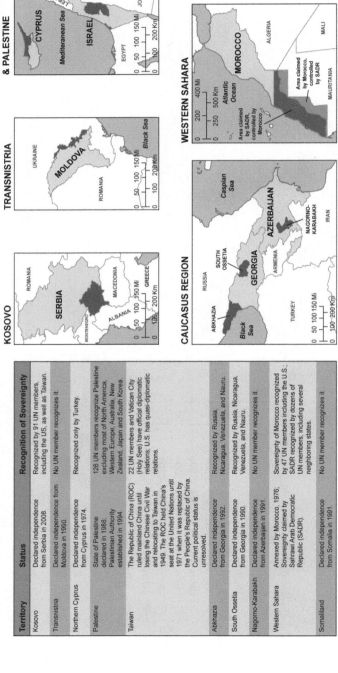

Territory	Status	Recognition of Sovereignty
Kosovo	Declared independence from Serbia in 2008.	Recognized by 91 UN members, including the US, as well as Taiwan.
Transnistria	Declared independence from Moldova in 1990.	No UN member recognizes it.
Northern Cyprus	Declared independence from Cyprus in 1974.	Recognized only by Turkey.
Palestine	State of Palestine declared in 1988; Palestinian Authority established in 1994.	126 UN members recognize Palestine excluding most of North America, Western Europe, Australia, New Zealand, Japan and South Korea.
Taiwan	The Republic of China (ROC) ruled mainland China until losing the Chinese Civil War and relocating to Taiwan in 1949. The ROC held China's seat at the United Nations until 1971 when it was replaced by the People's Republic of China. Current political status is unresolved.	22 UN members and Vatican City (Holy See) have official diplomatic relations; U.S. has quasi-diplomatic relations.
Abkhazia	Declared independence from Georgia in 1992.	Recognized by Russia, Nicaragua, Venezuela, and Nauru.
South Ossetia	Declared independence from Georgia in 1990.	Recognized by Russia, Nicaragua, Venezuela, and Nauru.
Nagorno-Karabakh	Declared independence from Azerbaijan in 1991	No UN member recognizes it.
Western Sahara	Annexed by Morocco, 1976. Sovereignty claimed by Sahrawi Arab Democratic Republic (SADR).	Sovereignty of Morocco recognized by 47 UN members including the U.S.; SADR recognized by dozens of UN members, including several neighboring states.
Somaliland	Declared independence from Somalia in 1991.	No UN member recognizes it.

When a group of people within a territory declares independence, does that territory become a country? The short answer is "it depends." This question begs a follow-up question: What makes a country a country? The criteria defining a country, or, more appropriately, a *state*, are fairly straightforward. To begin with, a state must have territory and a resident population. There must be political, economic, and social organization. A state possesses *sovereignty*. Generally, sovereignty can be thought of as having complete control over one's territory and possessing the right to defend oneself against external aggression. Finally, a state is *recognized* as a country by other states. Today, that recognition generally is manifested by entry in to the United Nations (Map 109). One cannot overstate the importance of international recognition, because it does not always happen automatically or quickly. For example, international recognition and subsequent entry of Macedonia into the UN was delayed because of objections by Greece to the name of the new country. Even though it is a member of the UN, Israel is not recognized as a state by several UN

members. The map above illustrates ten examples of territories that possess varying levels of international recognition of their sovereignty. None of the territories is a member of the UN, although the Republic of China (Taiwan) held China's seat in the UN until 1971, and the Palestine Liberation Organization was granted observer status to the UN in 1974 and 130 countries recognize the state of Palestine. Some of the territories have declared independence but have received no recognition (Somaliland, Transnistria) or limited recognition (Abkhazia, Nagorno-Karabakh, South Ossetia, Turkish Republic of North Cyprus). Kosovo has been recognized by 92 countries, including the United States, but not by Serbia, Russia, China, and several other Asian and African states. In the case of Taiwan, it never declared independence from mainland China. Rather, it views itself as the legitimate government of all of China, while the People's Republic of China views it as a breakaway province. A handful of countries have official diplomatic relations with the Republic of China. Much of the rest of the world, including the United States, has quasi-diplomatic relations with the island.

Map 111   United Nations Regions and Sub-Regions

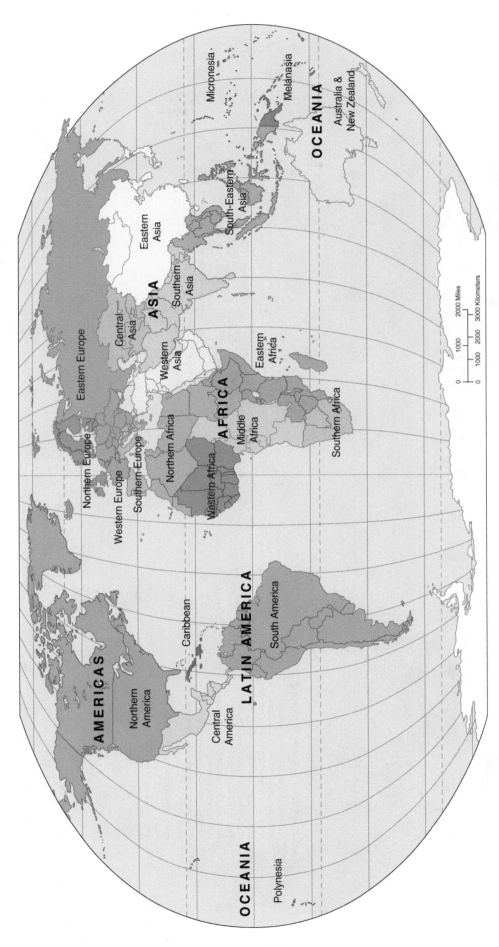

Where in the world is a country? At first, the response to that question seems fairly straightforward: China is in Asia, Angola is in Africa, and so on. The more one learns about the world, however, the more one finds that not everyone groups the world's countries the same way, nor do they use standard terminology. For example, is Russia in Europe? Is it in Asia? Is it in both? Some terms commonly used in Western culture tend not to be applicable in a global context. Using "Middle East" to describe the countries of the Arabian Peninsula and the eastern Mediterranean makes sense when viewing the world from the United Kingdom, but one would be hard-pressed to characterize these countries as "East" or in the "Middle" of countries to the east if one were in Japan. The map above presents the classification of the world's countries by the United Nations. How well do the names of the regions and sub-regions conform to the names with which you are familiar?

Map 112   Countries with Nuclear Weapons

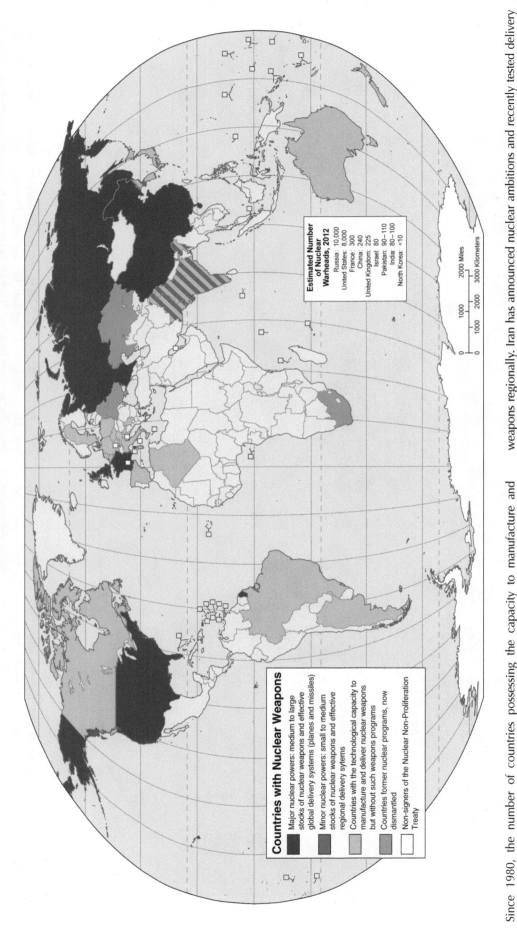

Countries with Nuclear Weapons

Major nuclear powers: medium to large stocks of nuclear weapons and effective global delivery systems (planes and missiles)

Minor nuclear powers: small to medium stocks of nuclear weapons and effective regional delivery systems

Countries with the technological capacity to manufacture and deliver nuclear weapons but without such weapons programs

Countries former nuclear programs, now dismantled

Non-signers of the Nuclear Non-Proliferation Treaty

Estimated Number of Nuclear Warheads, 2012

Russia:	10,000
United States:	8,000
France:	300
China:	240
United Kingdom:	225
Israel:	80
Pakistan:	90–110
India:	80–100
North Korea:	<10

Since 1980, the number of countries possessing the capacity to manufacture and deliver nuclear weapons has grown dramatically, increasing the chances of accidental or intentional nuclear exchanges. In addition to the traditional nuclear powers of the United States, Russia, China, the United Kingdom, and France, must now be added Israel, India, and Pakistan as countries that, without possessing the large stocks of weapons of the major powers, nor the extensive delivery systems of the United States and Russia, still have effective regional (and possibly global) delivery systems and medium stocks of warheads. Countries such as Kazakhstan, Ukraine, Georgia, and Belarus that were created out of what had been the Soviet Union did have some nuclear capacity in the 1991–1995 period but have since had all nuclear weapons removed from their territories. North Korea has recently announced the re-suspension of its nuclear weapons programs, although it may possess a small stock of nuclear warheads along with the capacity to deliver those weapons regionally. Iran has announced nuclear ambitions and recently tested delivery systems. Until the overthrow of the Baathist regime of Saddam Hussein by a U.S.-led military coalition in 2003, Iraq also had nuclear ambitions. The proliferation of nuclear states threatens global security, and the objective of the Nuclear Non-Proliferation Treaty was to reduce the chances for expanding nuclear arsenals worldwide. This treaty has been partially successful in that a number of countries in the developed world certainly have the capacity to manufacture and deliver nuclear weapons but have chosen not to do so. These countries include Canada, European countries other than the United Kingdom and France, South Korea, Japan, Australia, New Zealand, and Brazil and Argentina in South America. On the other side of the coin, the still-possible intent of North Korea to emerge as a nuclear power may force countries such as South Korea and Japan to re-think their positions as non-nuclear countries.

Map 113   Military Expenditures

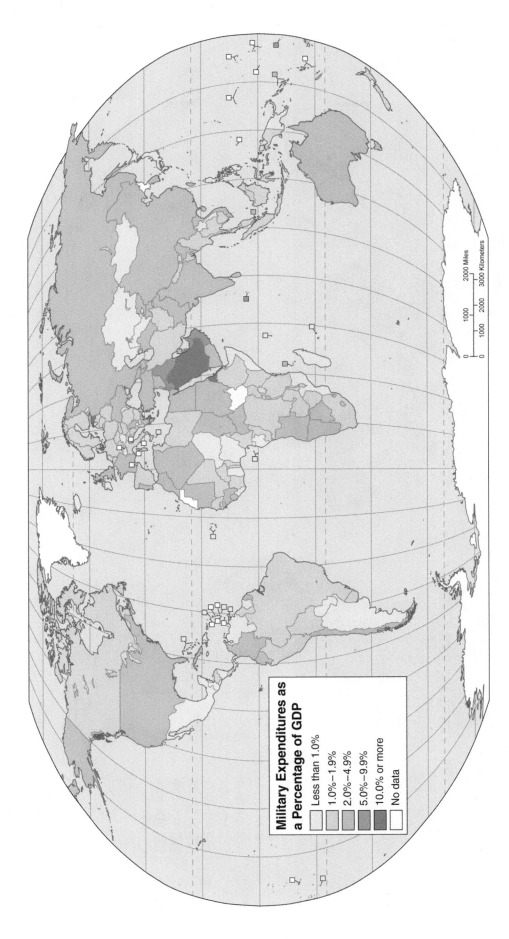

Military Expenditures as a Percentage of GDP

- Less than 1.0%
- 1.0%–1.9%
- 2.0%–4.9%
- 5.0%–9.9%
- 10.0% or more
- No data

Many countries devote a significant proportion of their total central governmental expenditures to defense: weapons, personnel, and research and development of military hardware. A glance at the map reveals that there are a number of regions in which defense expenditures are particularly high, reflecting the degree of past and present political tension between countries. The clearest example is the Middle East. The steady increase in military expenditures by developing countries is one of the most alarming (and least well-known) worldwide defense issues. Where the end of the Cold War has meant a

substantial reduction of military expenditures for the countries in North America and Europe and for Russia, in many of the world's developing countries military expenditures have risen between 15 percent and 20 percent per year for the past few years, averaging out to 7.5 percent per year for the past quarter century. Even though many developing countries still spend less than 5 percent of their gross national product on defense, these funds could be put to different uses in such human development areas as housing, land reform, health care, and education.

Map 114　Distribution of Minority Populations

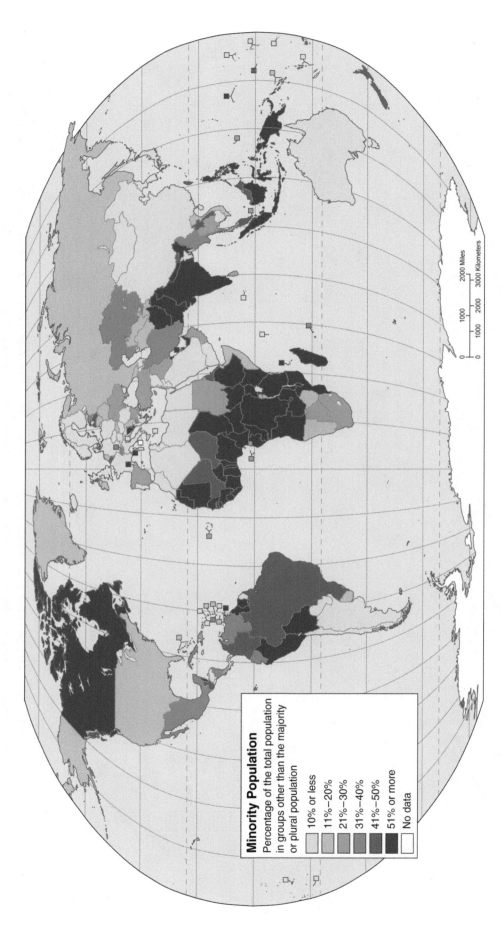

Minority Population
Percentage of the total population in groups other than the majority or plural population

- 10% or less
- 11%–20%
- 21%–30%
- 31%–40%
- 41%–50%
- 51% or more
- No data

The presence of minority ethnic, national, or racial groups within a country's population can add a vibrant and dynamic mix to the whole. Plural societies with a high degree of cultural and ethnic diversity should, according to some social theorists, be among the world's most healthy. Unfortunately, the reality of the situation is quite different from theory or expectation. The presence of significant minority populations played an important role in the disintegration of the Soviet Union; the continuing existence of minority populations within the new states formed from former Soviet republics threatens the viability and stability of those young political units. In Africa, national boundaries were drawn by colonial powers without regard for the geographical distribution of ethnic groups, and the continuing tribal conflicts that have resulted hamper both economic and political development. Even in the most highly developed regions of the world, the presence of minority ethnic populations poses significant problems: witness the separatist movement in Canada, driven by the desire of some French-Canadians to be independent of the English majority, and the continuing ethnic conflict between Flemish-speaking and Walloon-speaking Belgians. This map, by arraying states on a scale of homogeneity to heterogeneity, indicates areas of existing and potential social and political strife.

-140-

Map 115   Marginalized Minorities: Declining Indigenous Populations

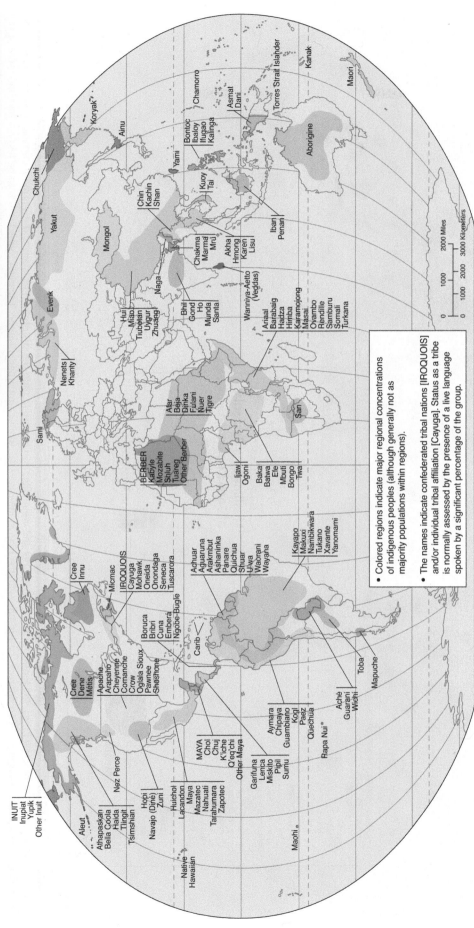

- Colored regions indicate major regional concentrations of indigenous peoples (although generally not as majority populations within regions).

- The names indicate confederated tribal nations [IROQUOIS] and/or individual tribal affiliation [Cayuga]. Status as a tribe is normally assessed by the presence of a live language is normally assessed by the presence of a live language spoken by a significant percentage of the group.

The world's indigenous (native) populations, those with the longest historical association with a geographic region, have nearly always been marginalized by colonialism, migration, economic development, or other external forces. Often confined to specific territories such as reservations or preserves, indigenous peoples have strived to preserve their languages, their cultures, their very identities—and often have sought to regain control of their ancestral lands as they have defined them. It has been estimated that, of the approximately 5,000 indigenous cultures remaining, fewer than half will survive through the end of the twenty-first century. There are many reasons for this: departure from the traditional life and homeland for urban areas and jobs; populations and population growth rates that have dropped below critical mass for maintenance; the scourge of substance abuse, particularly alcohol; disease rates that are and have been higher among locationally marginal populations; because they are and have been marginal, they may not have had the same levels of exposure to certain diseases as more densely-crowded populations. The best historical example of this is the massive decline in American Indian populations (in both North and South America) after European colonization. American Indians simply did not have the built-in immunities against diseases such as smallpox, measles, chicken pox, or even the common cold as did Europeans and Africans and mortality rates were enormous among many populations—often exceeding 90%. A similar process occurred in the latter half of the twentieth century as the Amazon Basin was rapidly occupied by representatives of mixed Old World populations (European and African) and Old World diseases took a heavy toll among indigenous peoples of the Amazon. The loss of indigenous culture is not just an academic issue but one that bears upon the survival of humanity. For it is the indigenous culture, living close to their natural environment, who may have the knowledge of local biology that could aid in the identification of substances that could assist in the eradication of major diseases among non-indigenous populations.

Map 116   The Political Geography of a Global Religion: The Islamic World

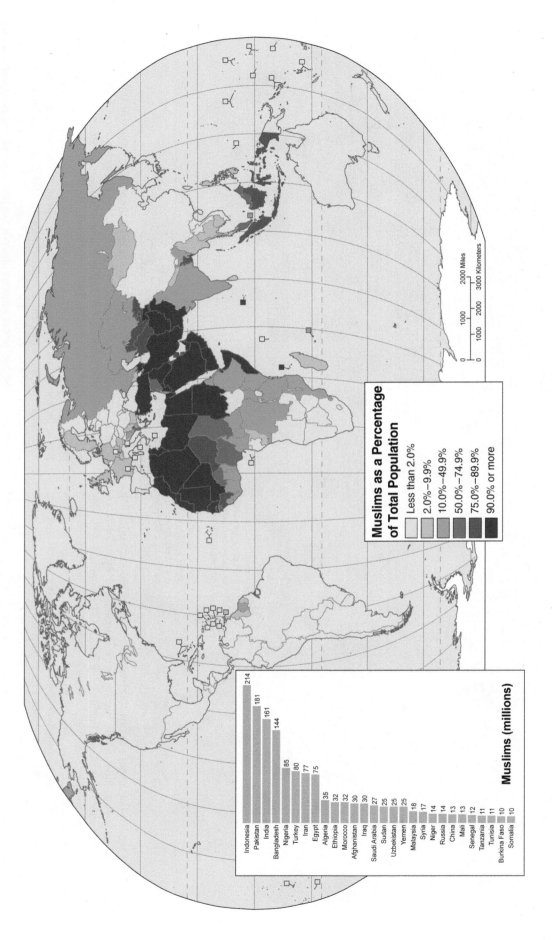

Muslims as a Percentage of Total Population

- Less than 2.0%
- 2.0%–9.9%
- 10.0%–49.9%
- 50.0%–74.9%
- 75.0%–89.9%
- 90.0% or more

Bar chart — **Muslims (millions)**:

Country	Muslims (millions)
Indonesia	214
Pakistan	181
India	161
Bangladesh	144
Nigeria	85
Turkey	80
Iran	77
Egypt	75
Algeria	35
Ethiopia	32
Morocco	32
Afghanistan	30
Iraq	30
Saudi Arabia	27
Sudan	25
Uzbekistan	25
Yemen	25
Malaysia	18
Syria	17
Niger	14
Russia	14
China	13
Mali	13
Senegal	12
Tanzania	11
Tunisia	11
Burkina Faso	10
Somalia	10

Scale: 0 1000 2000 Miles; 0 1000 2000 3000 Kilometers

Islam, as a religion, does not promote conflict. The term *jihad*, often mistranslated to mean "holy war," in fact refers to the struggle to find God and to promote the faith. In spite of the beneficent nature of Islamic teachings, the tensions between Muslims and adherents of other faiths often flare into warfare. A comparison of this map with the map of international conflict will show a disproportionate number of wars in that portion of the world where Muslims are either majority or significant minority populations. The reasons for this are based more in the nature of government, cultures, and social structure, than in the tenets of the faith of Islam. Nevertheless, the spatial correlations cannot be ignored. Similarly, terrorist incidents falling considerably short of open armed warfare are spatially consistent with the distribution of Islam and even more consistent with the presence of Islamic fundamentalism or "Islamism," which tends to be less tolerant and more aggressive than the mainstream of the religion. Terrorism is also consistent with those areas where the legacy of colonialism or the persistent presence of non-Islamic cultures intrude into the Islamic world.

-142-

Map 117   World Refugees: Country of Origin

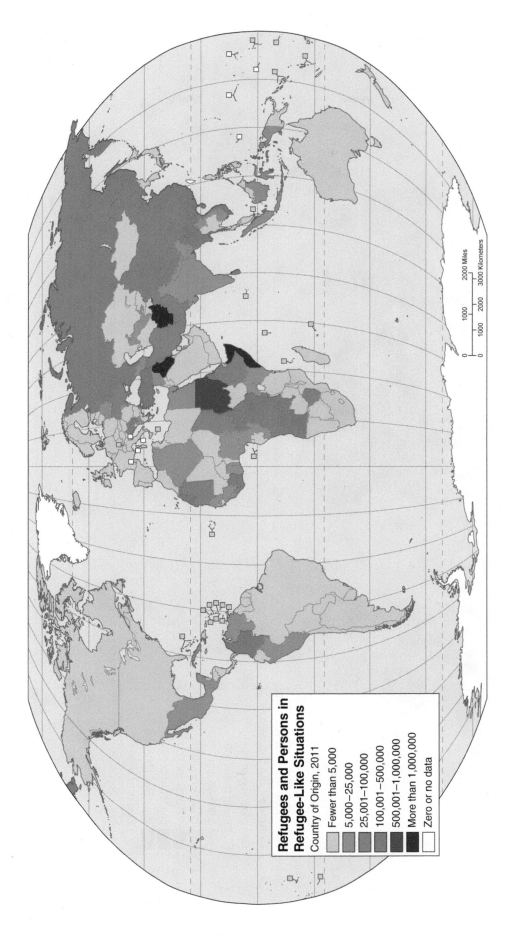

Refugees and Persons in Refugee-Like Situations

Country of Origin, 2011

Fewer than 5,000
5,000–25,000
25,001–100,000
100,001–500,000
500,001–1,000,000
More than 1,000,000
Zero or no data

0   1000   2000 Miles
0   1000   2000   3000 Kilometers

Refugees are persons who have been driven from their homes and seek refuge in another country. Although there are many reasons why people flee their home country, the vast majority are fleeing armed conflict. In such cases, there may be a mass exodus from the country involving tens or perhaps hundreds of thousands of persons. Most refugees flee into neighboring countries. Because armed conflict is oftentimes short-lived, the number of refugees in the world changes from year to year, sometimes substantially. The refugee population is recognized by international agencies and is monitored by the United Nations High Commissioner for Refugees (UNHCR).

-143-

Map 118 World Refugees: Country of Asylum

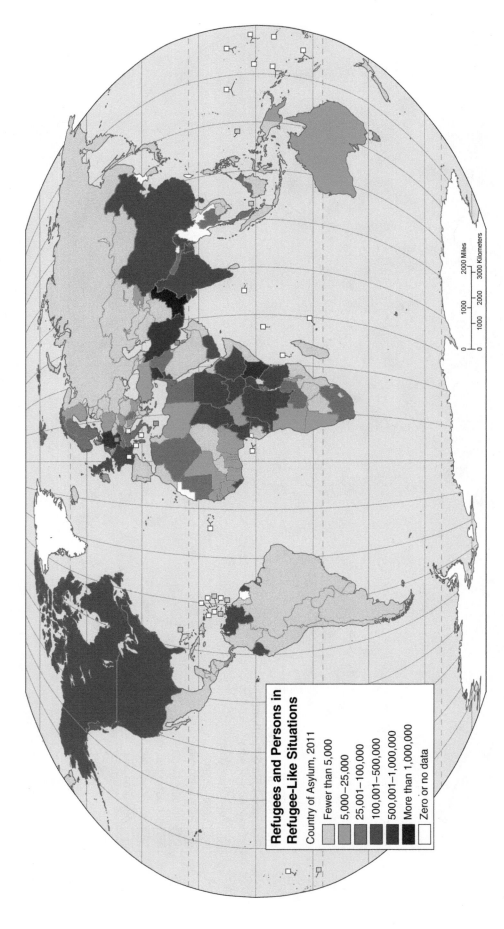

Refugees and Persons in Refugee-Like Situations

Country of Asylum, 2011

- Fewer than 5,000
- 5,000–25,000
- 25,001–100,000
- 100,001–500,000
- 500,001–1,000,000
- More than 1,000,000
- Zero or no data

0   1000   2000   2000 Miles
0 1000 2000   3000 Kilometers

When refugees flee their country, they most commonly flee to a neighboring country, and not every country is equally equipped to handle such an influx of persons. During such times, international agencies often financially reward the countries of refuge for their willingness to take in externally displaced persons. For most of the host countries, the challenge of hosting a large refugee population is a short-term problem. As we have seen in recent decades, the burden of hosting massive numbers of refugees can be a destabilizing force for the country of refuge.

Map 119   Internally Displaced Persons

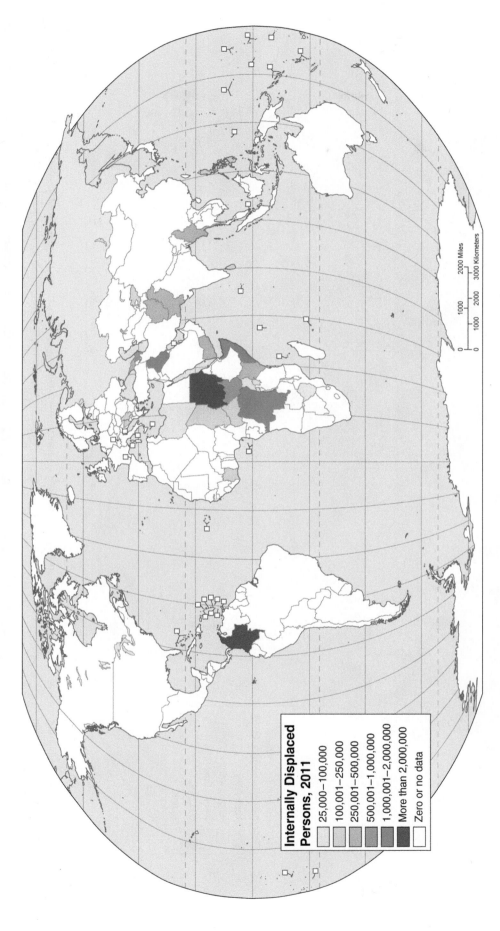

Internally Displaced Persons, 2011

- 25,000–100,000
- 100,001–250,000
- 250,001–500,000
- 500,001–1,000,000
- 1,000,001–2,000,000
- More than 2,000,000
- Zero or no data

0   1000   2000 Miles

0   1000   2000   3000 Kilometers

Internally displaced persons (IDPs) are those who flee their homes because of conflict, persecution, or disaster and seek refuge in another location within their country. With the exception of not leaving their country, they are essentially the same as refugees. International organizations offer the same assistance to internally displaced persons as is provided to refugees. The actual number of IDPs is more difficult to assess than refugee data. Not only do IDP populations fluctuate, there likely are a large number of displaced persons who flee to the larger cities in the countries rather than to the camps established by international relief organizations. In early 2011, the countries with the greatest number of IPDs were Colombia, Iraq, Democratic Republic of the Congo, Somalia, and Sudan.

Map 120 Abuse of Public Trust

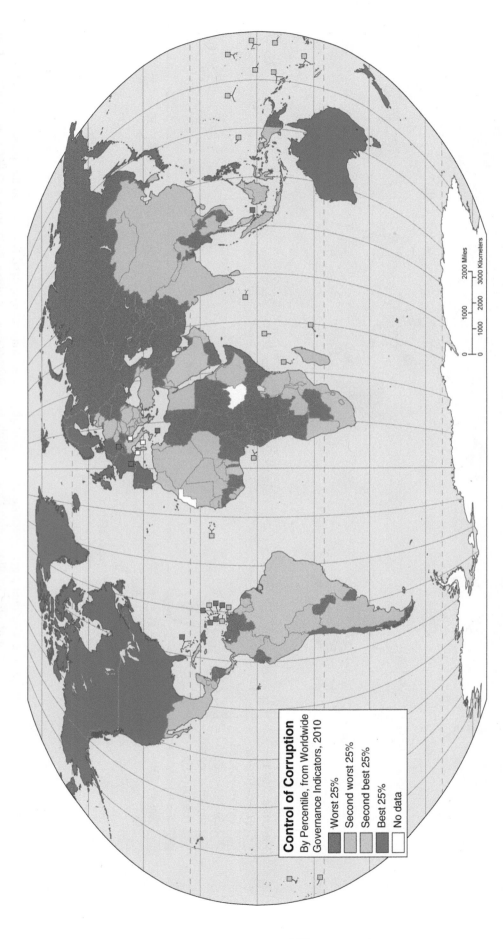

Control of Corruption
By Percentile, from Worldwide Governance Indicators, 2010

- Worst 25%
- Second worst 25%
- Second best 25%
- Best 25%
- No data

Abusing the public trust is simply another way of saying "corruption in government." In many parts of the world, corruption in the government is not an aberration but a way of life. Normally, although not always, governmental corruption is an indication of a weak and ineffective government, one that negatively affects such public welfare issues as public health, sanitation, education, and the provision of social services. It also tends to impact the cost of doing business and, thereby, drives away the foreign capital so badly needed in many African and Asian countries for economic development. Corruption is not automatic in poor countries, nor are rich countries free from it. But there is a general correlation between abuse of the public trust and lower levels of per capita income—excepting such countries as the Baltic states and Chile that have reached high standards of governance without joining the ranks of the wealthy countries. Studies by the World Bank have shown that countries that address issues of corruption and clean up the operations of their governments increase national incomes as much as four or five times. In those countries striving to attain governments that function according to a rule of law—rather than a rule of abusing the public trust—such important demographic measures as child mortality drop by as much as 75 percent. Clearly, good government and good business and higher incomes and better living conditions for the general public all go hand-in-hand.

Map 121   Political and Civil Liberties

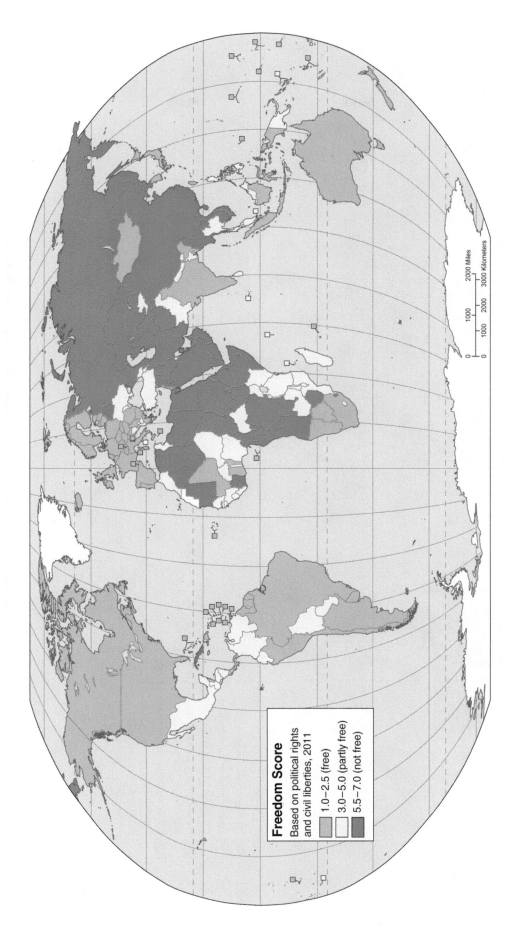

Freedom Score

Based on political rights and civil liberties, 2011

- 1.0–2.5 (free)
- 3.0–5.0 (partly free)
- 5.5–7.0 (not free)

Although measures of political and civil liberty are somewhat difficult to obtain and assess, there are some generally accepted standards that can be evaluated: open elections and competitive political parties, the rule of law, freedoms of speech and press, judicial systems separate from other branches of government, and limits on the power of elected or appointed governmental officials. Interstingly, there appear to be correlations between "degrees of freedom" and such other characteristics of a state as per capita wealth, environmental quality, and healthy economic growth—characteristics that may be mutually contradictory. There is no empirical evidence of a causal link between democratic institutions and consumption; on the other hand, there is clear evidence of a positive relationship between wealth and consumption. Therefore, the three variables are closely correlated and should be used in assessing the nature of the state in any part of the world.

Map 122 Human Rights Abuse

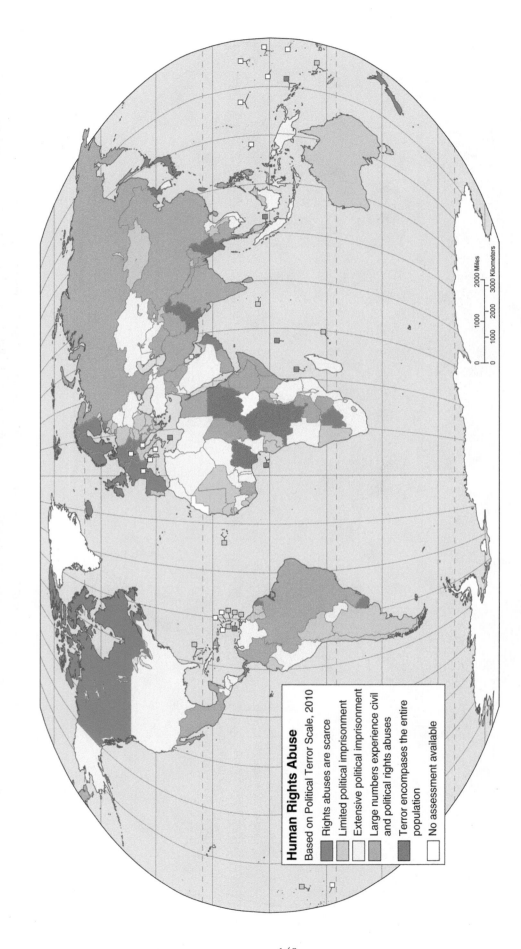

Human Rights Abuse
Based on Political Terror Scale, 2010

- Rights abuses are scarce
- Limited political imprisonment
- Extensive political imprisonment
- Large numbers experience civil and political rights abuses
- Terror encompases the entire population
- No assessment available

0   1000   2000 Miles
0   1000   2000   3000 Kilometers

The Political Terror Scale measures levels of political violence and terror that a country experiences in a particular year. The score is calculated using data from the U.S. State Department Country Reports on Human Rights and yearly country reports from Amnesty International. At the lowest end of the scale, torture or political murder are scarce and the rule of law dominates. Higher scores indicate increasing pervasiveness of human rights abuses. At the highest level, a country's entire population is affected by political terror, genocide, or other crimes against humanity. Countries with high scores typically are dictatorships or totalitarian states whose leaders oftentimes carry out political terror through secret police forces or death squads. In many politically unstable countries, the Political Terror Scale may fluctuate from year to year. In the early 2000s countries such as Liberia, Colombia, and Rwanda ranked much higher than in 2010. Conversely, Central African Republic, and Thailand have seen increases in the score since the early 2000s. North Korea, Democratic Republic of the Congo, Somalia, and Sudan have had very high scores since the turn of the century.

Map 123   Women's Rights

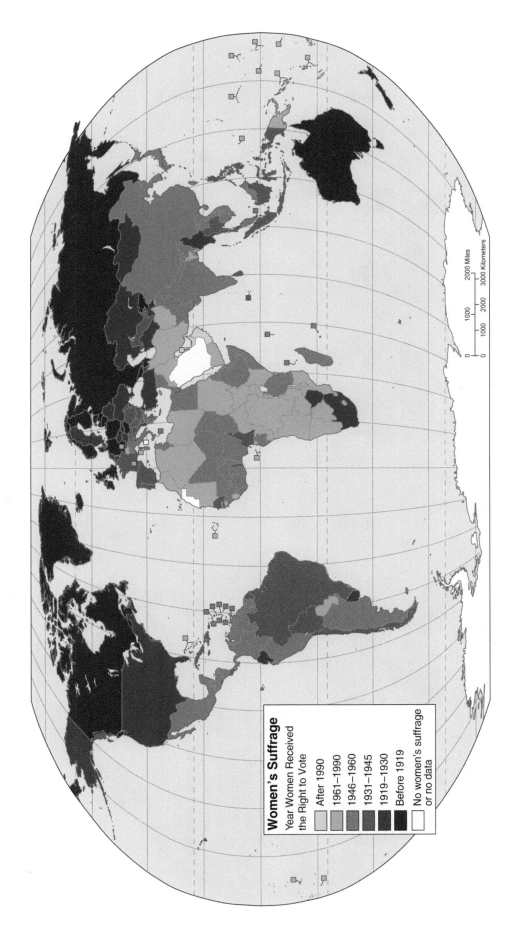

Women's Suffrage

Year Women Received
the Right to Vote

- After 1990
- 1961–1990
- 1946–1960
- 1931–1945
- 1919–1930
- Before 1919
- No women's suffrage
  or no data

The "rights" referred to in this map refer primarily to the right to vote. But where women have the right to vote in free elections, the other fundamental rights tend to become available as well: the right to own property, the right to an education, the right to leave a domestic alliance without fear of retribution, or the right to be treated as a human being rather than property. But the time lag between women receiving the right to vote and their attainment of other fundamental human rights does not occur immediately or,

in many cases, even relatively quickly. On the map, the most recent countries to grant suffrage to women are in Africa and Southwest Asia. In these regions, women still do not have access to many of the basic rights of what we would consider to be a civilized life. Bahrain, Oman, Kuwait, and the United Arab Emirates have all granted women the right to vote since 2000. Women will be allowed to vote in municipal elections in Saudi Arabia starting in 2015.

-149-

Map 124 Capital Punishment

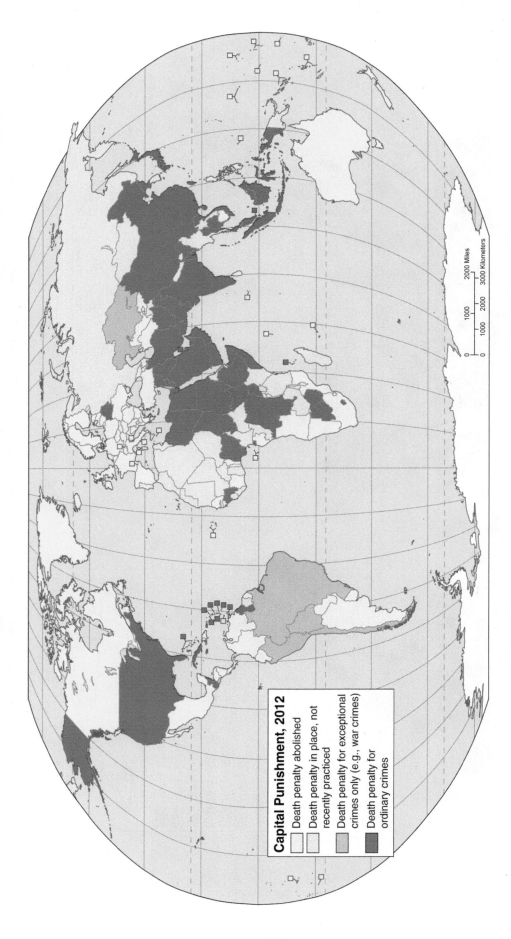

The most basic human right is life itself. Many have argued that one cannot consider the condition of human rights in countries without including analysis of the extent to which capital punishment—where a person is put to death by the state as punishment for a crime—is practiced. Nearly 100 countries have abolished the practice outright and nearly 40 have ceased using the death penalty as punishment. In eight countries—Bolivia, Brazil, Chile, El Salvador, Fiji, Israel, Kazakhstan, and Peru—capital punishment is reserved for exceptional circumstances such as war crimes. Since 2000, twenty-five countries have abolished the death penalty. In the United States, eighteen states and the District of Columbia have no death penalty statute.

Map 125   Human Trafficking

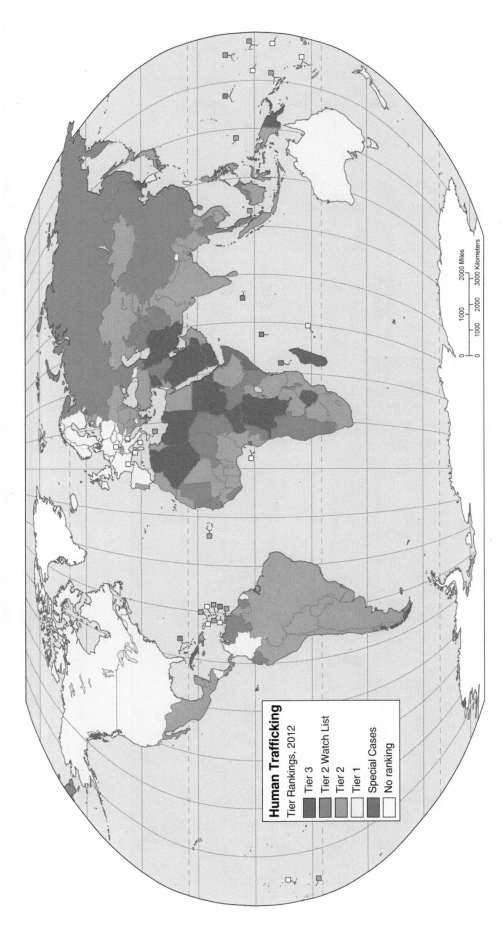

Human trafficking is the trade of human beings for the purposes of compelling them to service against their will. The term is broad and can include many activities, such as forced labor, sex trafficking, bonded labor (when a person uses services to pay off a loan), involuntary domestic servitude, forced child labor, child military conscription, and child sex trafficking. In 2000, the United Nations adopted the Protocol to Prevent, Suppress and Punish Trafficking in Persons, Especially Women and Children in an attempt to prevent and combat trafficking in persons. To date, 117 countries have adopted the protocol. That same year the United States passed the Victims of Trafficking and Violence Protection Act, which was reauthorized in 2003 as the Trafficking Victims Protection Act (TVPA). The U.S. State Department ranks the countries of the world based on the degree to which they meet

standards of TVPA and places the countries into three tiers. Tier 1 countries are those that comply with the TVPA minimum standards. Tier 2 countries do not fully comply with the TVPA's minimum standards, but are making significant efforts to bring themselves into compliance with those standards. Tier 3 countries do not fully comply with the minimum standards and are not making significant efforts to do so. The State Department places some of the Tier 2 countries on a Watch List. In these countries, the absolute number of victims of severe forms of trafficking is very significant or is significantly increasing, or there is a failure to provide evidence of increasing efforts to combat severe forms of trafficking in persons from the previous year. Somalia has been identified as a "Special Case"—a country without the government infrastructure to effectively prevent human trafficking.

Map 126   Flashpoints, 2013

Map 126a

Drug-Related Homicides
2006 - 2011

Fewer than 100
101–500
501–1,000
1,001–2,500
2,501–5,000
More than 5,000

Mexico: Although drug-related crime has been present in Mexico for decades, the last few years has seen a dramatic increase in narco-related violence. The surge in violence came as a result of Mexican drug cartels supplanting Colombian drug cartels for control of illegal drug trafficking into the United States. These cartels have become heavily armed and not only are capable of battling government troops, but also have targeted police officers, politicians, journalists, and entertainers for kidnapping and murder. Although there have been many drug cartels operating in Mexico, power recently has begun to consolidate among those operating in the cities of Ciudad Juaréz and Tijuana as well as the states of Michoacán, Sinaloa, and Tamaulipas. Mexico consistently ranks in the top ten most

dangerous countries for journalists to work in. Nearly 40,000 persons died in drug-related violence between 2006 and 2011 with more than half of those deaths occurring in 2010 and 2011 alone. All of this presents problems both for Mexico and for the United States. At nearly 2,000 miles in length, the border between Mexico and the United States is the most frequently crossed international border in the world. Given the likelihood of continued demand for foreign narcotics in the United States in the near future, narco-related violence, and the ever-increasing attention to U.S.-Mexican border issues, Mexico will likely continue to be a flashpoint for the foreseeable future.

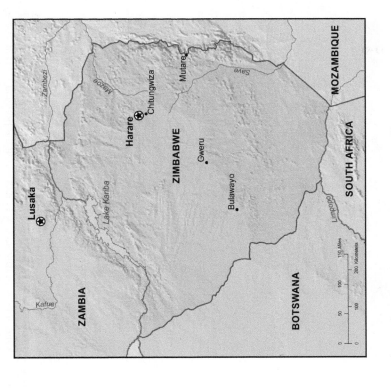

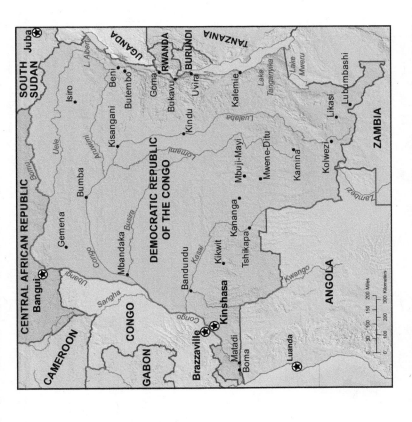

Dem. Rep. Congo: The war in the Democratic Republic of the Congo (formerly Zaire) has preoccupied United Nations and African diplomats since 1999. Troops from Zimbabwe, Angola, Sudan, Chad, and Namibia joined with the president of the Congo, Laurent Kabila, against his former allies Rwanda, Burundi, and Uganda, which each backed several separate Congolese rebel groups. The origins of the conflict lie in the overthrow of longtime dictator Mobutu Sese Seko by Kabila's army in 1997 after a year of civil war. Kabila's failure to call elections or stabilize the country's economy led to further rounds of rebellion in the huge but fractious nation—rebellion supported by the economic and military assistance of neighboring Rwanda, Burundi, and Uganda. After Kabila's assassination in 2001 and the succession of his son, Joseph, to the presidency, accord seemed to have been reached, and the various conflicting parties agreed to withdraw troops in 2002. But in early 2003, new fighting flared along the country's eastern border, threatening a new and broadened war and the addition of more deaths to the more than 5 million since 1998. Diplomats called the conflict "Africa's first world war." With fighting continuing in the east, fears are that the Congo conflict could destabilize the entire southern half of the continent, leading to massive refugee flows and abject poverty.

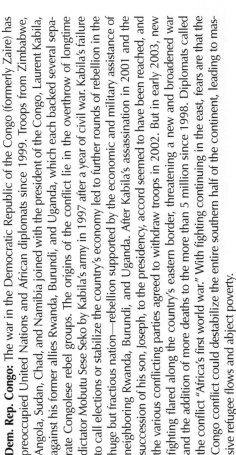

Zimbabwe: When Zimbabwe achieved independence in 1980, Robert Mugabe assumed leadership of the country—a position he has held ever since. In 2000, Mugabe instituted a highly controversial land reform program, appropriating white-owned farms and giving them to tribal leaders. A country that had Africa's highest literacy rate and was one of the continent's leaders in agricultural production, Zimbabwe quickly deteriorated into conditions of abject poverty and famine. Hyperinflation was rampant in the early 2000s, peaking at a rate of more than 11 million percent and bringing the country to the brink of economic collapse. In the wake of a disputed presidential election in 2008, Mugabe agreed to share power with opposition candidate Morgan Tsvangirai. Although Zimbabwe does not have the tribal conflicts that beset so many African nations (most Zimbabweans are Shona), the feelings between the supporters of the two rival political factions run deeply enough that, if the power-sharing arrangement does not work and the economy continues to deteriorate, the country could be plunged into a civil war.

Map 126e

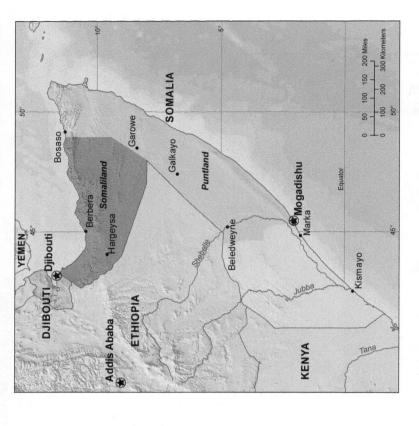

Map 126d

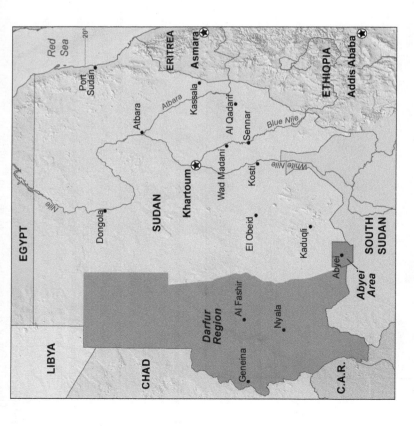

Sudan: Since Sudan achieved independence in 1956, military regimes favoring Islamic-oriented governments have dominated national politics. These regimes have embroiled the country in a civil war for nearly all of the past half-century. These wars have been rooted in the attempts of northern economic, political, and social interests dominated by Muslims to control territories occupied by non-Muslim, non-Arab southern Sudanese, such as the Dinka tribal groups. Since 1983, war and famine have resulted in more than 2 million deaths and more than 4 million people displaced. The current regime is a mixture of military elite and an Islamist party that came to power in a 1989 coup. After more than five decades of conflict with the ruling north, South Sudan gained its independence in 2011. However, clashes have persisted along the newly formed border with South Sudan, particularly in the contested area of Abyei, which is claimed by both Sudan and South Sudan. Additionally, the western region of Darfur has seen a protracted conflict between separatist forces and the Sudanese government, displacing more than two million persons.

Somalia: With the ouster of the government led by Mohamed Said Barre in January 1991, turmoil, factional fighting, and anarchy have followed in Somalia, with several separate governments arising in different parts of the country. The northern clans declared an independent Republic of Somaliland. Although not recognized by any government, it has maintained a stable existence. Puntland, the central portion of Somalia, from the Horn of Africa to the coast of the Indian Ocean and the border with Ethiopia, has been a self-governing autonomous state since 1998. In 2004 a new UN-backed government, the Transitional Federal Government (TFG), was created for the entire country, but the government has not been able to gain effective control. In 2006 and 2007 a push by central government forces, backed up by units of the Ethiopia regular army, succeeded in driving Islamic extremist forces out of the Mogadishu region. Ethiopia withdrew in 2009 and a series of political reforms followed. Compounding the political turmoil in the country, a severe drought occurred in the southern part of the country resulting in severe famine. Tens of thousands of Somalis died and nearly one million fled to refugee camps in Kenya and Ethiopia. In 2011, an agreement was reached to transition the country to a unified central government. In August 2012, a new constitution was overwhelmingly approved by the National Constituent Assembly.

Map 126f

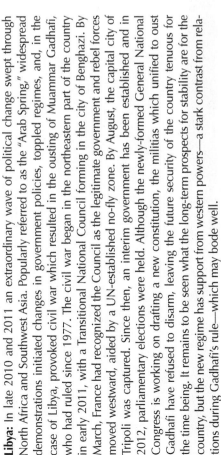

Map 126g

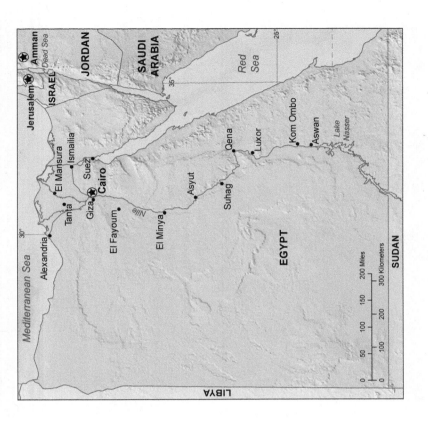

Libya: In late 2010 and 2011 an extraordinary wave of political change swept through North Africa and Southwest Asia. Popularly referred to as the "Arab Spring," widespread demonstrations initiated changes in government policies, toppled regimes, and, in the case of Libya, provoked civil war which resulted in the ousting of Muammar Gadhafi, who had ruled since 1977. The civil war began in the northeastern part of the country in early 2011, with a Transitional National Council forming in the city of Benghazi. By March, France had recognized the Council as the legitimate government and rebel forces moved westward, aided by a UN-established no-fly zone. By August, the capital city of Tripoli was captured. Since then, an interim government has been established and in 2012, parliamentary elections were held. Although the newly-formed General National Congress is working on drafting a new constitution, the militias which unified to oust Gadhafi have refused to disarm, leaving the future security of the country tenuous for the time being. It remains to be seen what the long-term prospects for stability are for the country, but the new regime has support from western powers—a stark contrast from relations during Gadhafi's rule—which may bode well.

Egypt: At the same time that Libya devolved into civil war, Egypt saw a regime change that was just as dramatic as what occurred in its neighbor to the west. Unlike Libya, however, change occurred largely through widespread, non-violent civil resistance. As the revolutionary wave of the "Arab Spring" spread, millions of Egyptians took to the streets to protest long-held grievances against President Hosni Mubarak, who had ruled since 1981. By the end of 2011, Mubarak had resigned and control of the country was assumed by the Supreme Council of Armed Forces who dissolved parliament, suspended the constitution, and announced elections in early 2012. In June of that year, Mohamed Morsi, a candidate of the pro-Islamist Muslim Brotherhood, became the first civilian president of the country since 1952. As Egypt moves forward, observers point to two potential flash-points: First, will Islamism define government policy and if so, what will be the reaction of the population? Second, what will be the role of the Egyptian military if President Morsi attempts to strengthen the role of the president at the expense of power previously held by the military?

Map 126h

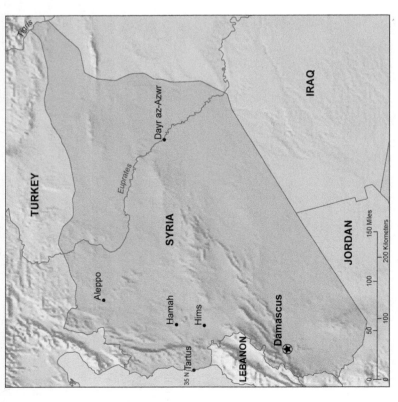

Syria: Like many of its neighboring countries, Syria saw widespread demonstrations during 2011. Unlike many of the other countries of "Arab Spring," Syria's internal conflict continued well beyond when the others completed either major policy change or regime change. In Egypt, President Mubarak stepped down. In Libya, civil war toppled the regime of Gadhafi. In Syria, however, the Baathist regime, with the support of the military, actively engaged the uprising with far more determination than was seen in Libya. Mass protests inspired by those in Tunisia, Egypt, and Libya have been met with tanks and widespread arrests. There are accusations that the Syrian military has opened fire on civilians. In 2012, large protests against the Assad regime erupted in Aleppo, Syria's second-largest city, where fighting between rebels and government forces forced as many as 100,000 persons to flee.

Map 126i

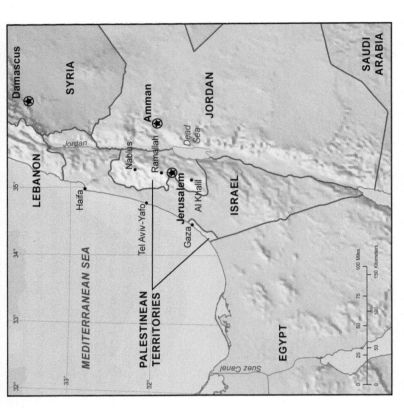

Israel and Its Neighbors: The modern state of Israel was created out of the former British Protectorate of Palestine, inhabited primarily by Muslim Arabs, after World War II. Conflict between Arabs and Israeli Jews has been a constant ever since. Much of the present tension revolves around the West Bank area, not part of the original Israeli state but taken from Jordan, an Arab country, in the Six-Day War of 1967. Many Palestinians had settled this part of Jordan after the creation of Israel and remain as a majority population in the West Bank region today. Israel has established many agricultural settlements within the region since 1967, angering Palestinian Arabs. For Israel, the West Bank is the region of ancient Judea and this region, won in battle, will not be ceded back to Palestinian Arabs without protracted or severe military action. The West Bank, inhabited by nearly 400,000 Israeli settlers and 4 million Palestinians, is also the location of most of the suicide bombings carried out by Islamic militant groups from 2001 to 2009. By early 2008, the Gaza Strip had emerged as the most critical flashpoint in the area. The Israeli government and the Palestinian Authority had agreed to resume peace talks with the goal being a peace agreement by the end of the year. But in late 2008 and early 2009, Israeli troops responded to rocket attacks by Hamas, a leading Palestinian political party, by attacking Gaza in force. In late 2010, talks between Israel and the Palestinians had commenced, but were called off by the end of the year.

Map 126j

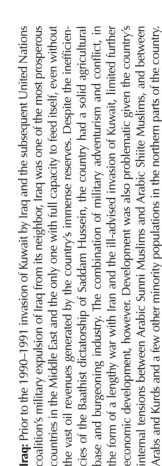

Map 126k

Iraq: Prior to the 1990–1991 invasion of Kuwait by Iraq and the subsequent United Nations coalition's military expulsion of Iraq from its neighbor, Iraq was one of the most prosperous countries in the Middle East and the only one with full capacity to feed itself, even without the vast oil revenues generated by the country's immense reserves. Despite the inefficiencies of the Baathist dictatorship of Saddam Hussein, the country had a solid agricultural base and burgeoning industry. The combination of military adventurism and conflict, in the form of a lengthy war with Iran and the ill-advised invasion of Kuwait, limited further economic development, however. Development was also problematic given the country's internal tensions between Arabic Sunni Muslims and Arabic Shiite Muslims, and between Arabs and Kurds and a few other minority populations in the northern parts of the country. Elections were held in 2009 and a new government was installed in 2010. In 2010, the United States ended combat operations with full withdrawal of troops at the end of 2011.

Kurdistan: Where Turkey, Iran, and Iraq meet in the high mountain region of the Tauros and Zagros mountains, a nation of 25 million people exists. This nation is "Kurdistan," but the Kurds, the occupants of this area for more than 3,000 years, have no state, and receive much less attention than other stateless nations such as the Palestinians. Following the 1991 Gulf War between Iraq and a U.S.-led coalition of European and Arabic states, the United Nations demarcated a Kurdish "Security Zone" in northern Iraq. From 1991 to 2003, the Security Zone was anything but secure as Iraqi militants from the south and Turks from the north infringed on Kurdish territory, and internal militant extremist groups, such as the Kurdish Workers' Party, staged periodic attacks on rival villages. During the 2003 U.S.-led invasion of Iraq that eliminated the Baathist regime of Saddam Hussein, the Kurds played an important role in securing the northern portions of Iraq for the American-British coalition and fought alongside elements of the Iraqi army from cities such as Mosul and Kirkuk. Rich in oil and history, Kurdistan will probably remain as a nation without a state, shared by Iraq, Turkey, and Iran—none of which is likely to give up substantial portions of territory for the establishment of a Kurdish state. In 2012, the portion of northern Iraq under Kurdish control was the most stable of that war-torn country.

Map 126m

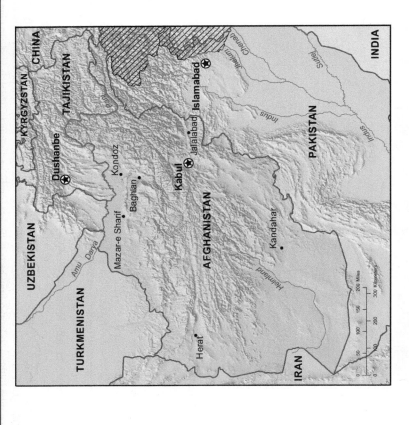

Afghanistan: In the aftermath of the tragic September 11, 2001, terrorist attacks, the United States (backed to varying degrees by its allies) declared a massive and global "war on terrorism," targeting terrorist groups and states that provide "safe harbor" to them. Front and center in this war was the Taliban regime of Islamic extremists that controlled about 95 percent of Afghanistan and harbored the al-Qaeda terrorist network. U.S. and British forces, aided by members of the Northern Alliance of Afghan rebels, expelled the Taliban government in 2002, and in 2003, Hamid Karzai became the first democratically elected president of the country. Despite the imposition of democracy, Afghanistan still is plagued by warlords in remote areas of the country who refuse to recognize the legally constituted government. In addition, significant pockets of resistance from remnants of the former Taliban regime and from al-Qaeda forces are engaged in ongoing military conflict with American and Pakistani troops along the Afghanistan-Pakistan border. Taliban resurgence, particularly based in Pakistan, grew throughout 2008 and into 2009, partly as the result of an ineffective and corrupt central government that, for all practical purposes, controls only the region of the capital city, Kabul.

Map 126l

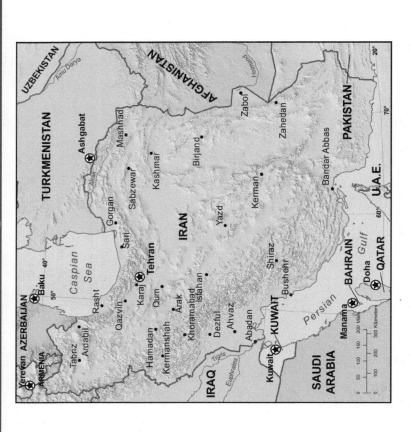

Iran: Iran has been at odds with the United States and Israel since the overthrow of the pro-Western government of the Shah and the installation of a fundamentalist Islamic republic in 1979. The Iranians took a large number of American hostages following the political revolution and held them for over a year. During an eight-year war between Iran and its nearest neighbor, Iraq, the United States provided military aid to Iraq, further straining the relations between Iran and the West. Iran's promise to continue development of nuclear facilities that could lead to the development of nuclear weapons has heightened distrust of Iran in the West. More likely, however, is that once Iran has developed facilities capable of producing weapons-grade plutonium, Israel will carry out the same type of preemptive strikes it has previously used on Iraq and Syria. Iran is a large and important country, poorly understood by the United States. In 2009, hard-line President Mahmoud Ahmadinejad won re-election in a highly contested and controversial vote. Millions of Iranians took to the street to protest, but Supreme Leader Ayatollah Khamenei endorsed Ahmadinejad as the winner and declared the protests illegal. Ahmadinejad has pushed for construction of an atomic power station, declaring production of nuclear fuel to be an "inalienable right." In response, the UN imposed sanctions on Iran in 2010. With an ancient imperial tradition, Iran is the historical core of Shiah Islam (nearly 90 percent of Iranians are Shiite Muslims), and it possesses enormous reserves of oil and natural gas.

Map 126n

Map 126o

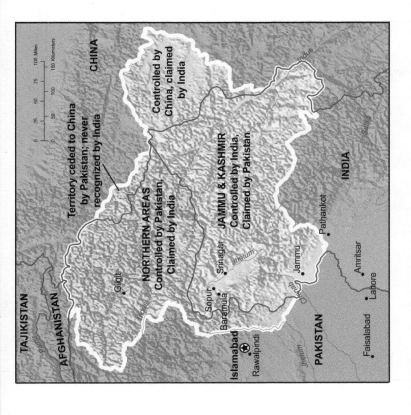

Pakistan: Pakistan's importance to the West and its potential as a flashpoint represent one of the most critical threats to world peace in the opening years of the twenty-first century. Pakistan is one of the ten most populous countries in the world; it possesses both nuclear weapons and a delivery system (as does its nearest neighbor and chief antagonist, India); and it teeters on the brink of being either a Western-style representative democracy or an Islamic fundamentalist state. Strategically, Pakistan lies at the western end of the core of the Muslim world (although large Muslim populations exist to the south and east in India, Malaysia, and Indonesia) and is immediately adjacent to U.S. military operations against the Taliban in Afghanistan. Physically, Pakistan is an incredibly rugged country, mixing a large river floodplain (the Indus) with high mountain country, with peaks in excess of 25,000 feet in elevation in the northwest. Culturally, the country is a mixture of different linguistic and ethnic groups. The government has tried to encourage the use of Urdu as the national language, but less than 10 percent of Pakistanis speak Urdu as their primary language. Pakistan's influence as a precarious U.S. ally in the war on terror has been tested recently. Terror attacks in Mumbai, India, in 2008 and 2011 have been linked to Islamic terror groups operating primarily in Pakistan and in 2011, al-Qaeda leader Osama bin Laden was shot and killed at a compound in the northern Pakistani city of Abbottabad.

Kashmir Region: When the British withdrew from South Asia in 1947, the former states of British India were asked to decide whether they wanted to become part of a new Hindu India or a Muslim Pakistan. In the state of Jammu and Kashmir, the rulers were Hindu and the majority population was Muslim. The maharajah (prince) of Kashmir opted to join India, but an uprising of the Muslim majority precipitated a war between India and Pakistan over control of this high mountain region. In 1949 a cease-fire line was established by the UN, leaving most of the territory of Jammu and Kashmir in Indian hands. Since then, Pakistan and India have waged intermittent skirmishes over the disputed territory that holds the headwaters of the Indus River, a life-giving stream to desert Pakistan. While Jammu and Kashmir refers specifically to the state in northern India, "Kashmir" is used to describe the larger area of contention that includes Pakistan's northern areas and Azad Kashmir, as well as territory controlled by China. In 1999, extremist Muslim groups demanding independence escalated the periodic battles into a full-fledged, if small, war between two of Asia's major powers—both possessing nuclear weapons. The specter of nuclear exchange caused both Pakistan and India to back down, and although the area remains disputed, military activity has quieted somewhat. Given that there has been no resolution to the situation, Kashmir will continue to be at the leading edge of the simmering feud between Hindu and Muslim populations that has been part of South Asian politics since independence from Great Britain and the partition into separate states in 1947.

Map 126p

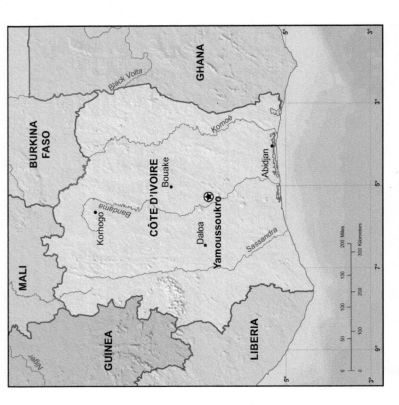

Map 126q

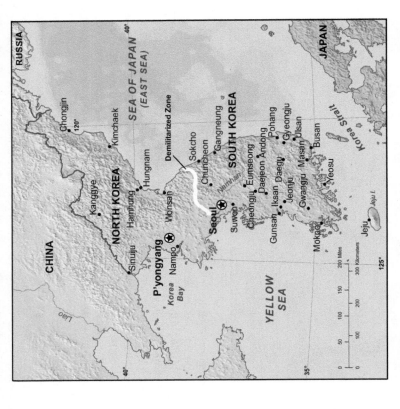

Côte d'Ivoire: Until relatively recently, its close ties to France following independence in 1960 and its development of cocoa production for export (along with significant foreign investment) made Côte d'Ivoire (Ivory Coast) one of the most prosperous of the tropical African states. Since 1999, however, political turmoil has disrupted economic development. In December 1999, a military coup in Côte d'Ivoire overthrew the elected government. In 2000, an election resulted in Laurent Gbagbo assuming the presidency. Another coup, this one a failure, was launched in 2002 with rebel forces claiming the northern half of the country. A unity government was established in 2003 with elections to be held in 2005. In Gbagbo held to the presidency, however, and the elections were postponed until 2010. In this election, the country's electoral commission announced that Gbagbo had lost to former prime minister Alassane Ouattara, a result recognized internationally. The Ivorian constitutional council, however, ruled that Gbagbo was the winner. In the months that followed, the country has plunged into civil war. In 2011, Ouattara, with the support of UN and French forces, was able to finally oust Gbagbo from his stronghold in the country's largest city, Abidjan. UN forces will remain in the country until this most recent crisis is finally resolved.

Korean Peninsula: Although active military conflict has not existed since the 1950s, the Korean peninsula remains an important flashpoint. Since the end of the "Korean War," South Korea has flourished economically. North Korea, on the other hand, adopted a policy of diplomatic and economic self-reliance, becoming one of the world's most authoritarian and isolated states. Yet, North Korea so mismanaged and misallocated its resources that, by the mid-1990s, the country was unable to feed itself. An estimated 2 million North Koreans have died in the past decade as a result of severe food shortages. It continues to expend resources to maintain one of the world's largest armies. In 2006, North Korea announced the development of nuclear weapons and delivery systems designed to "protect" against American aggression. In 2009, North Korea conducted a nuclear test, which was followed by the test firing of several short-range missiles and a long-range rocket. Peninsular tensions reached another high in 2010 when the South Korean warship *Cheonan* exploded and sank. South Korea blamed North Korea, then ceased all cross-border trade. The world's attention refocused on the peninsula in December 2011 following the death of North Korean dictator Kim Jong-Il. His son, Kim Jong-Un, was named his successor, and he continued the rhetoric of his father and grandfather warning in 2012 of a "sacred war" with South Korea and the United States.

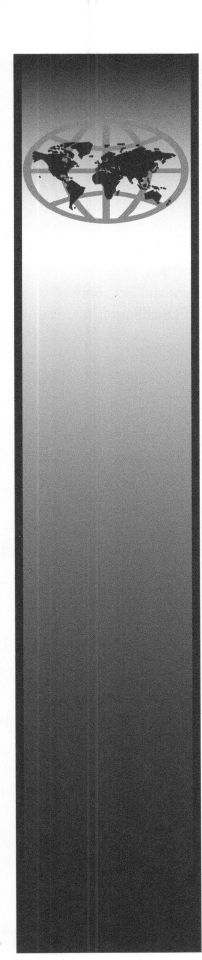

Unit VII

World Regions

Map 127   North America: Political Divisions

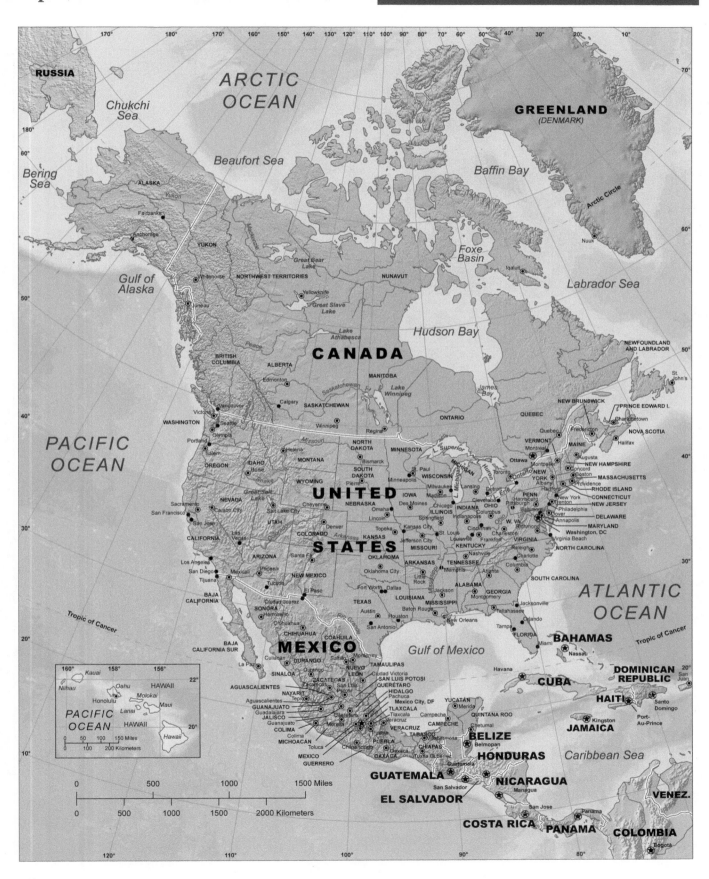

Map **128**   North America: Physical Features

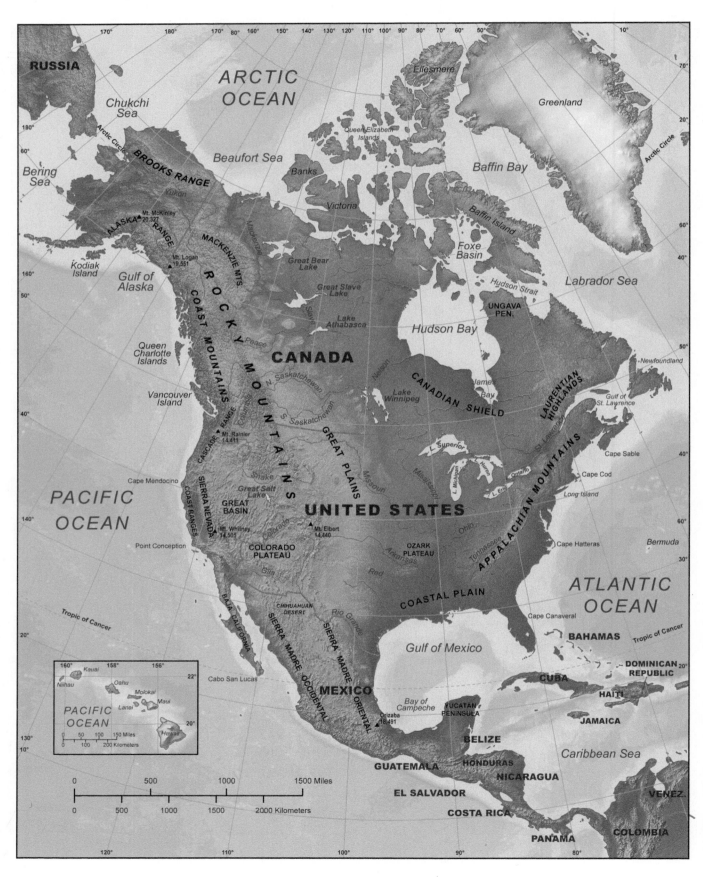

Map **129**   North America: Environment and Economy

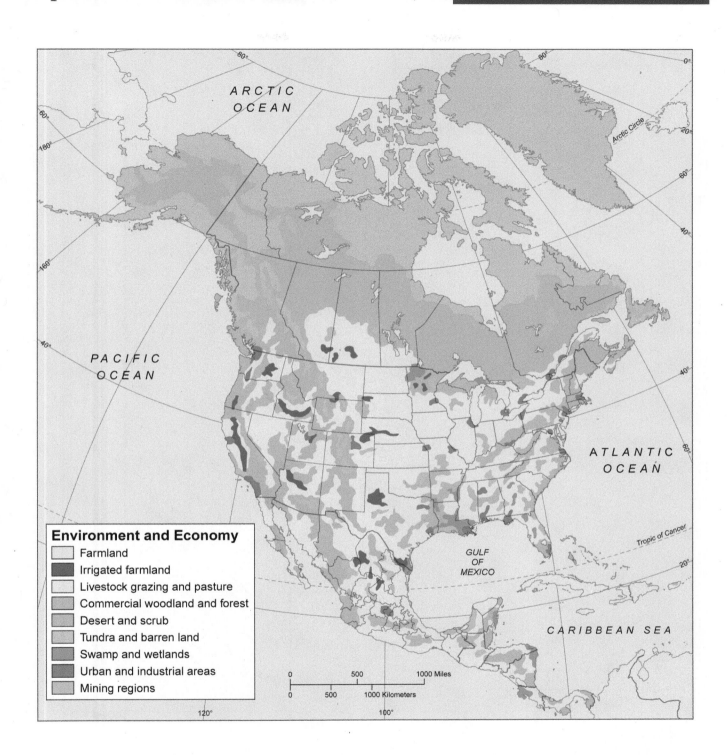

Environment and Economy

- Farmland
- Irrigated farmland
- Livestock grazing and pasture
- Commercial woodland and forest
- Desert and scrub
- Tundra and barren land
- Swamp and wetlands
- Urban and industrial areas
- Mining regions

The use of land in North America represents a balance among agriculture, resource extraction, and manufacturing that is unmatched. The United States, as the world's leading industrial power, is also the world's leader in commercial agricultural production. Canada, despite its small population, is a ranking producer of both agricultural and industrial products, and Mexico has begun to emerge from its developing nation status to become an important industrial and agricultural nation as well. The countries of Middle America and the Caribbean are just beginning the transition from agriculture to modern industrial economies. Part of the basis for the high levels of economic productivity in North America is environmental: a superb blend of soil, climate, and raw materials. But just as important is the cultural and social mix of the plural societies of North America, a mix that historically aided the growth of the economic diversity necessary for developed economies.

Map 130　North America: Population Distribution

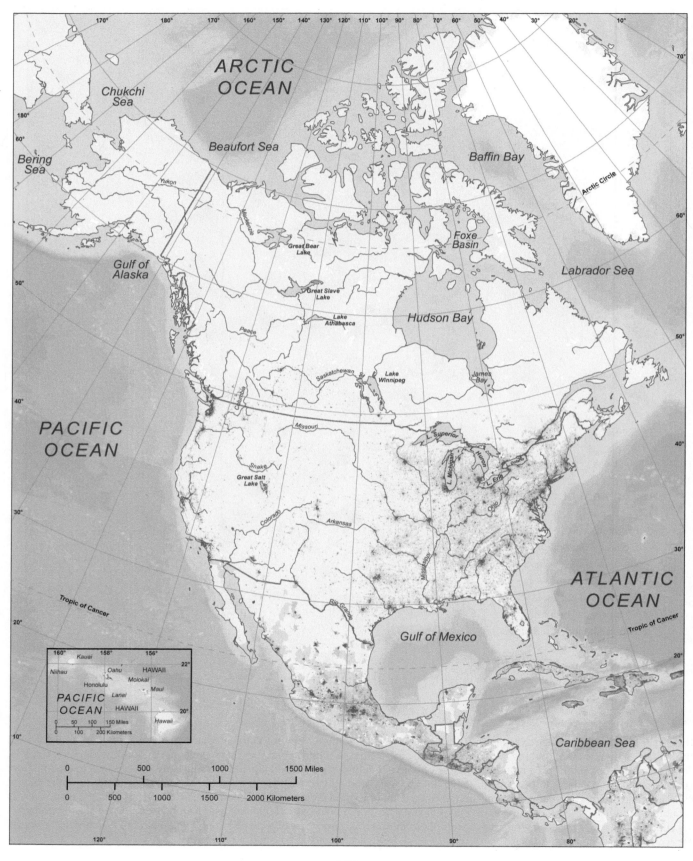

Map 131 Canada

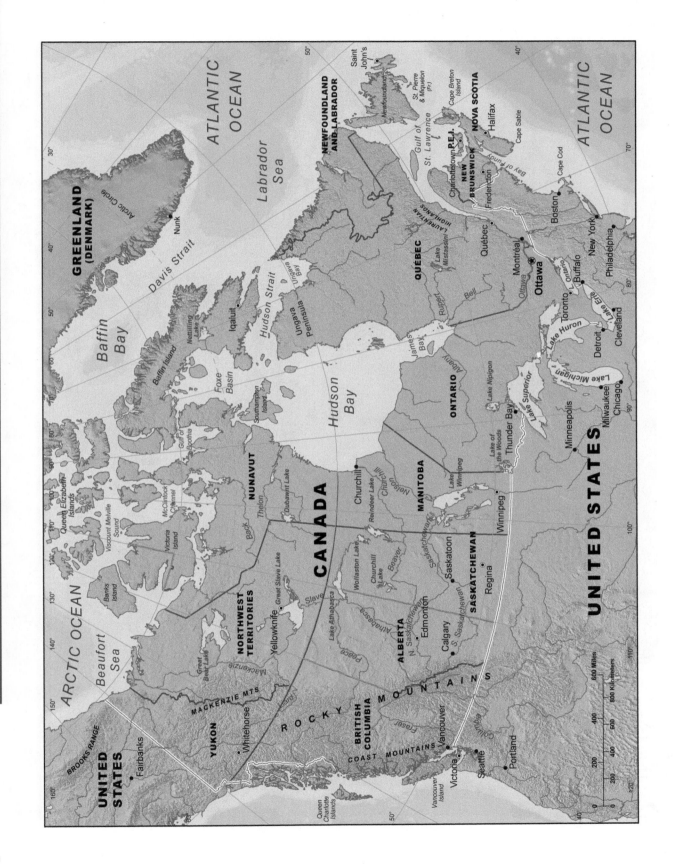

Map 132 United States

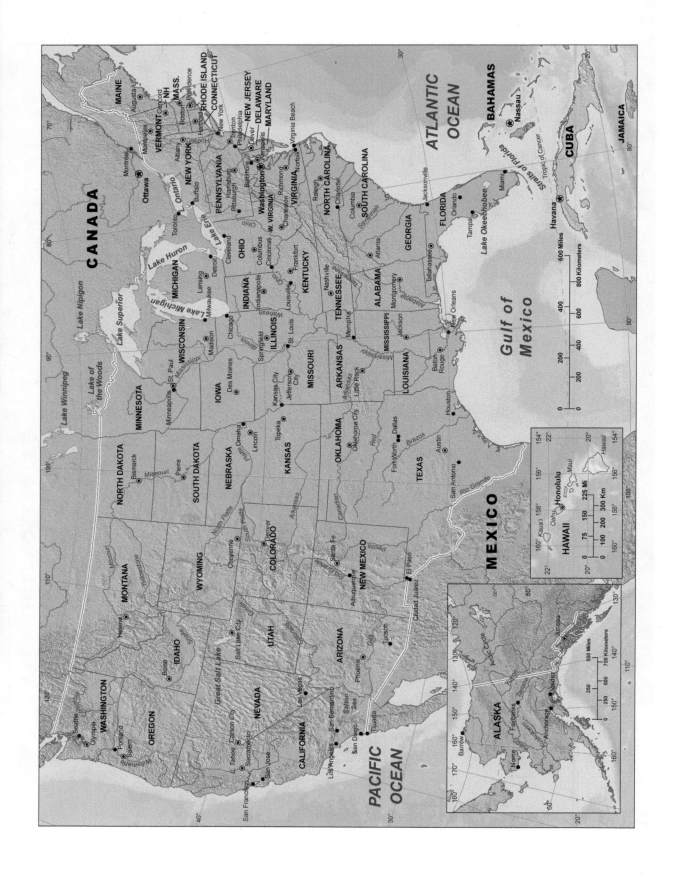

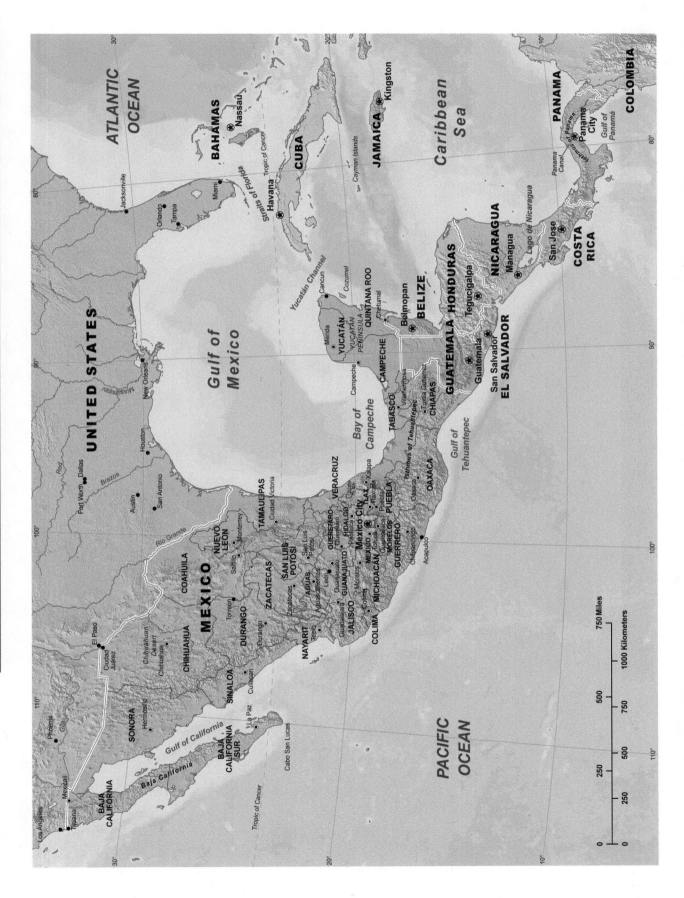

Map 133   Middle America

Map 134   The Caribbean

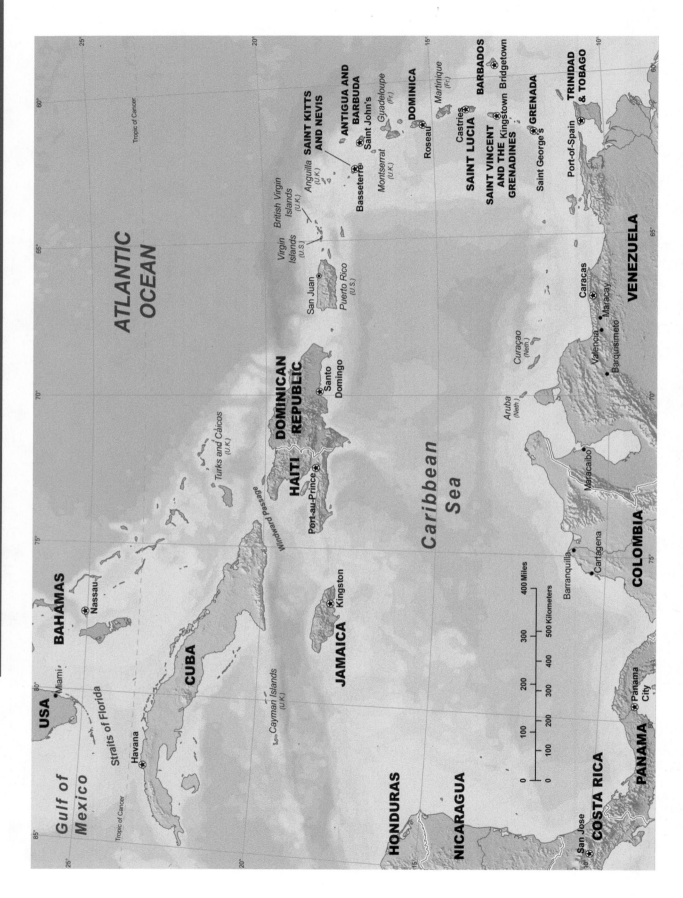

Map 135   South America: Political Divisions

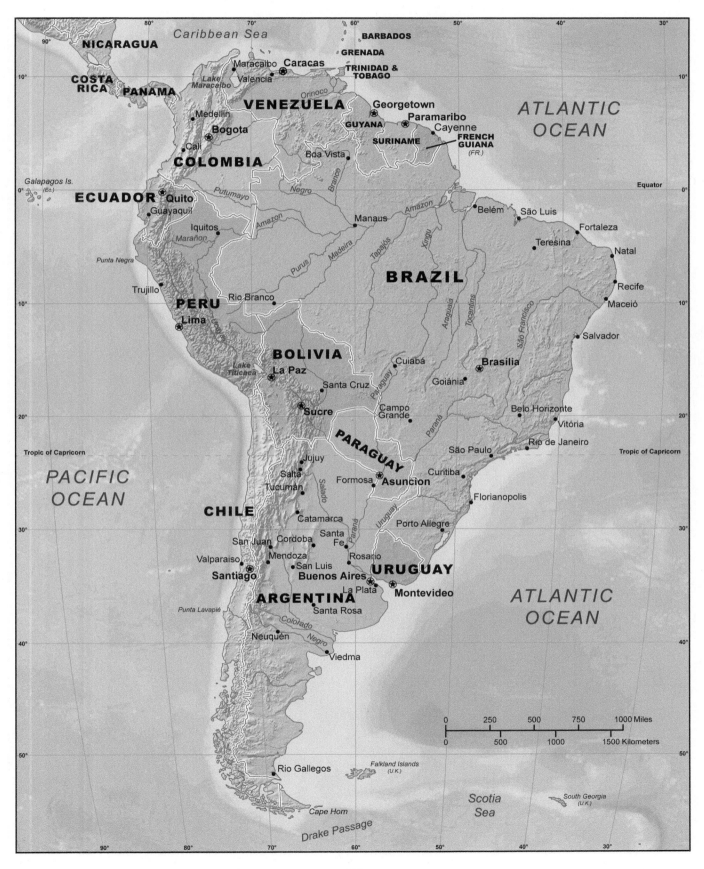

Map 136   South America: Physical Features

Map 137 South America: Environment and Economy

Environment and Economy
- Farmland
- Irrigated farmland
- Livestock grazing and pasture
- Commercial woodland and forest
- Desert and scrub
- Tundra and barren land
- Swamp and wetlands
- Urban and industrial areas
- Mining regions

South America is a region just beginning to emerge from a colonial-dependency economy in which raw materials flowed from the continent to more highly developed economic regions. With the exception of Brazil, Argentina, Chile, and Uruguay, most of the continent's countries still operate under the traditional mode of exporting raw materials in exchange for capital that tends to accumulate in the pockets of a small percentage of the population. The land use patterns of the continent are, therefore, still dominated by resource extraction and agriculture. A problem posed by these patterns is that little of the continent's land area is actually suitable for either commercial forestry or commercial crop agriculture without extremely high environmental costs. Much of the agriculture, then, is based on high value tropical crops that can be grown in small areas profitably, or on extensive livestock grazing. Even within the forested areas of the Amazon Basin where forest clearance is taking place at unprecedented rates, much of the land use that replaces forest is grazing.

Map 138   South America: Population Distribution

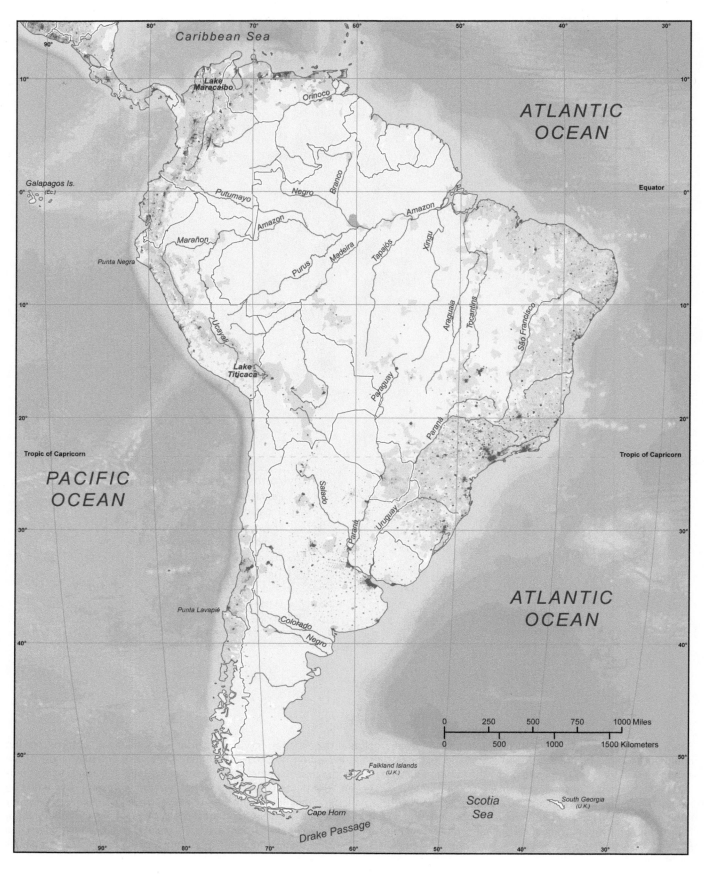

Map 139   Europe: Political Divisions

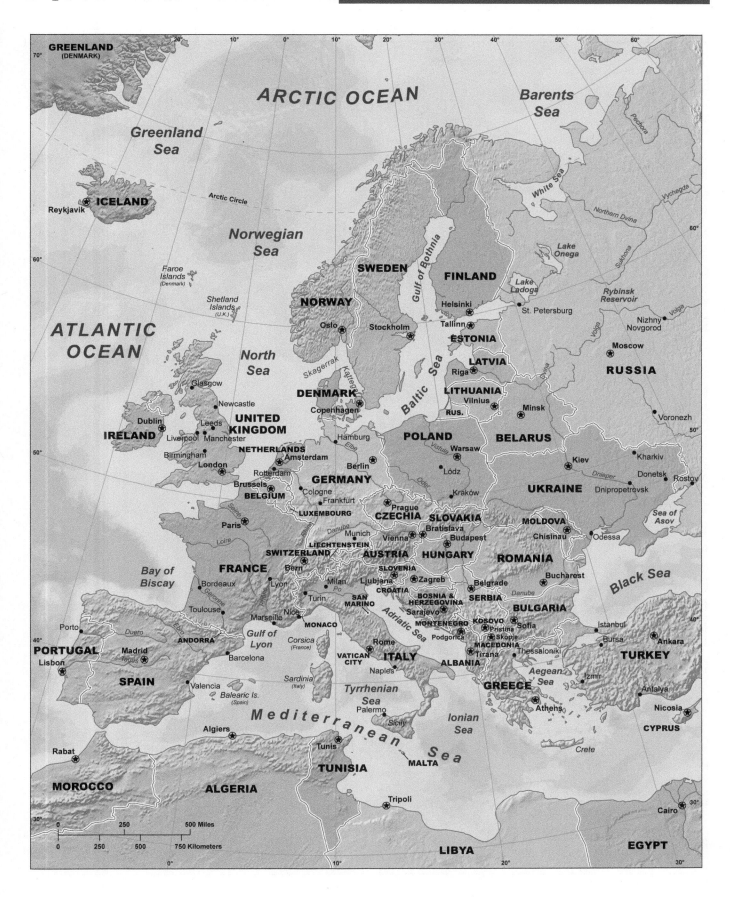

Map **140**   Europe: Physical Features

Map 141   Europe: Environment and Economy

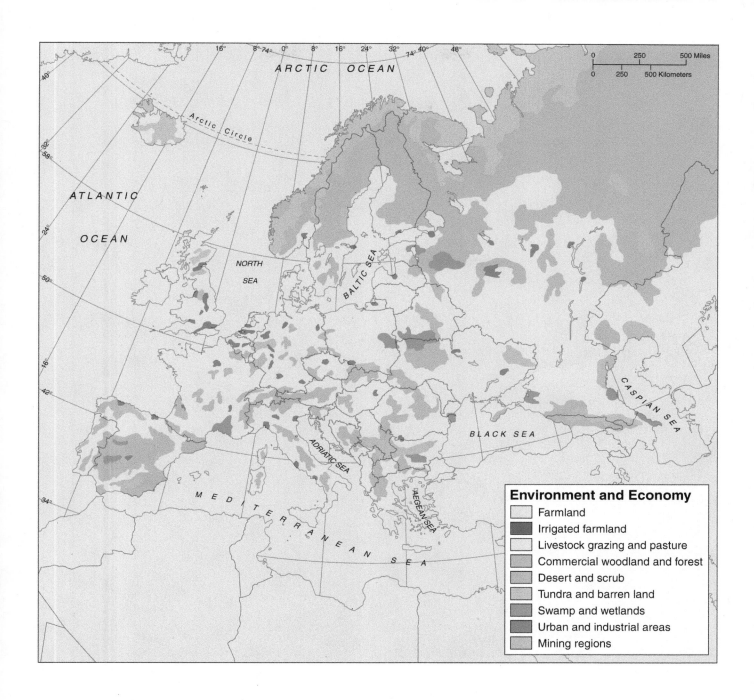

Environment and Economy
- Farmland
- Irrigated farmland
- Livestock grazing and pasture
- Commercial woodland and forest
- Desert and scrub
- Tundra and barren land
- Swamp and wetlands
- Urban and industrial areas
- Mining regions

More than any other continent, Europe bears the imprint of human activity—mining, forestry, agriculture, industry, and urbanization. Virtually all of western and central Europe's natural forest vegetation is gone, lost to clearing for agriculture beginning in prehistory, to lumbering that began in earnest during the Middle Ages, or more recently, to disease and destruction brought about by acid precipitation. Only in the far north and the east do some natural stands remain. The region is the world's most heavily industrialized and the industrial areas on the map represent only the largest and most significant. Not shown are the industries that are found in virtually every small town and village and smaller city throughout the industrial countries for Europe. Europe also possesses abundant raw materials and a very productive agricultural base. The mineral resources have long been in a state of active exploitation and the mining regions shown on the map are, for the most part, old regions in upland areas that are somewhat less significant now than they may have been in the past. Agriculturally, the northern European plain is one of the world's great agricultural regions but most of Europe contains decent land for agriculture.

-177-

Map 142   Europe: Population Distribution

Map 143   European Political Boundaries, 1914–1948

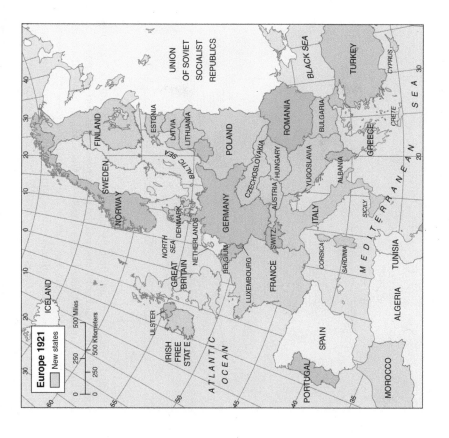

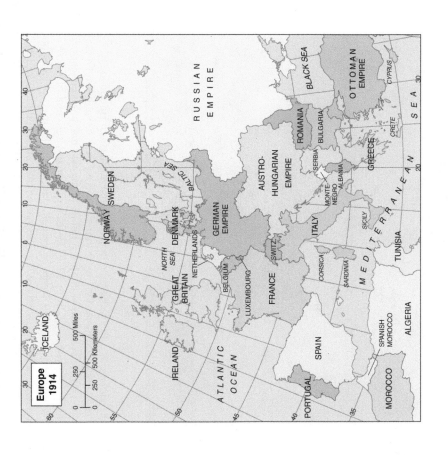

In 1914, on the eve of the First World War, Europe was dominated by the United Kingdom and France in the west, the German Empire and the Austro-Hungarian Empire in central Europe, and the Russian Empire in the east. Battle lines for the conflict that began in 1914 were drawn when the United Kingdom, France, and the Russian Empire joined together as the Triple Entente. In the view of the Germans, this coalition was designed to encircle Germany and its Austrian ally, which, along with Italy, made up the Triple Alliance. The German and Austrian fears were heightened from 1912–1914 when a Russian-sponsored "Balkan League" pushed the Ottoman Turkish Empire from Europe, leaving behind the weak and mutually antagonistic Balkan states Serbia and Montenegro. In August 1914, Germany and Austria-Hungary attacked in several directions and World War I began. Four years later, after massive loss of life and destruction, the central European empires were defeated. The victorious French, English, and Americans (who had entered the war in 1917) restructured the map of Europe in 1919, carving nine new states out of the remains of the German and Austro-Hungarian empires and the westernmost portions of the Russian Empire which, by the end of the war, was deep in the Revolution that deposed the czar and brought the Communists to power in a new Union of Soviet Socialist Republics.

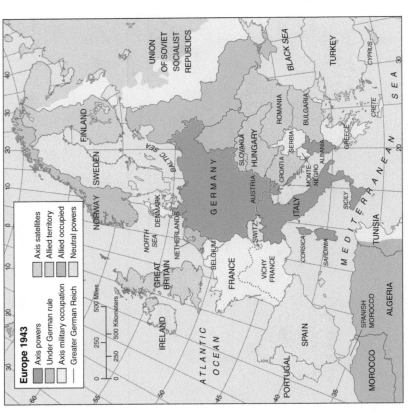

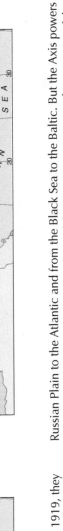

Europe 1948
NATO states
Warsaw Pact states

Europe 1943
Axis powers
Under German rule
Axis military occupation
Greater German Reich
Axis satellites
Allied territory
Allied occupied
Neutral powers

When the victorious Allies redrew the map of central and eastern Europe in 1919, they caused as many problems as they were trying to solve. The interval between the First and Second World Wars was really just a lull in a long war that halted temporarily in 1918 and erupted once again in 1939. Defeated Germany, resentful of the terms of the 1918 armistice and 1919 Treaty of Versailles and beset by massive inflation and unemployment at home, overthrew the Weimar republican government in 1933 and installed the National Socialist (Nazi) party led by Adolf Hitler in Berlin. Hitler quickly began making good on his promises to create a "thousand year realm" of German influence by annexing Austria and the Czech region of Czechoslovakia and allying Germany with a fellow fascist state in Mussolini's Italy. In September 1939 Germany launched the lightning-quick combined infantry, artillery, and armor attack known as **der Blitzkrieg** and took Poland to the east and, in quick succession, the Netherlands, Belgium, and France to the west. By 1943 the greater German Reich extended from the

Russian Plain to the Atlantic and from the Black Sea to the Baltic. But the Axis powers of Germany and Italy could not withstand the greater resources and manpower of the combined United Kingdom–United States–USSR–led Allies and, in 1945, Allied armies occupied Germany. Once again, the lines of the central and eastern European map were redrawn. This time, a strengthened Soviet Union took back most of the territory the Russian Empire had lost at the end of the First World War. Germany was partitioned into four occupied sectors (English, French, American, and Russian) and later into two independent countries, the Federal Republic of Germany (West Germany) and the German Democratic Republic (East Germany). Although the Soviet Union's territory stopped at the Polish, Hungarian, Czechoslovakian, and Romanian borders, the eastern European countries (Poland, East Germany, Czechoslovakia, Hungary, Romania, Yugoslavia, Albania, and Bulgaria) became Communist between 1945 and 1948 and were separated from the West by the Iron Curtain.

-180-

Map 144   Europe: Political Changes, 1989–2009

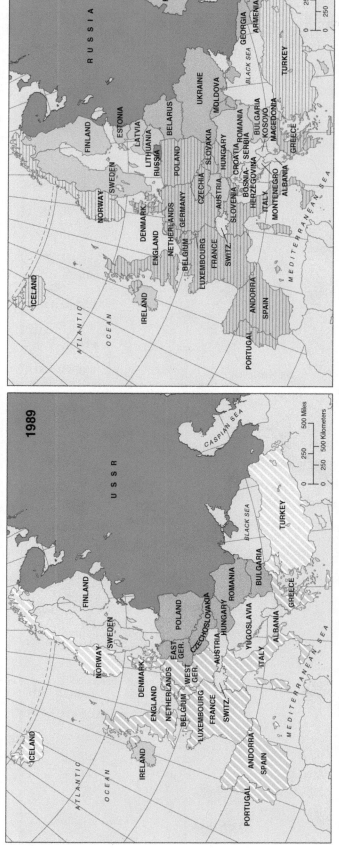

Europe: Political Changes 1989–2009

1989

Union of Soviet Socialist Republics

Warsaw Pact Countries (excluding the USSR)

North Atlantic Treaty Organization Countries

European Community (formerly the EEC)

2009

Former Republics of the USSR, now independent countries

Russian Federation

NATO Countries, 2005

Associated with NATO / petitioned for entry

European Union (formerly the EC)

During the last decade of the twentieth century, one of the most remarkable series of political geographic changes of the last 500 years took place. The bipolar East-West structure that had characterized Europe's political geography since the end of the Second World War altered in the space of a very few years. In the mid-1980s, as Soviet influence over eastern and central Europe weakened, those countries began to turn to the capitalist West. Between 1989, when the country of Hungary was the first Soviet satellite to open its borders to travel, and 1991, when the Soviet Union dissolved into 15 independent countries, abrupt change in political systems occurred. The result is a new map of Europe that includes a number of countries not present on the map of 1989. These countries have emerged as the result of reunification, separation, or independence from the former Soviet Union and Yugoslavia. The new political structure has been accompanied by growing economic cooperation.

Map 145   Western Europe

Map **146**   Eastern Europe

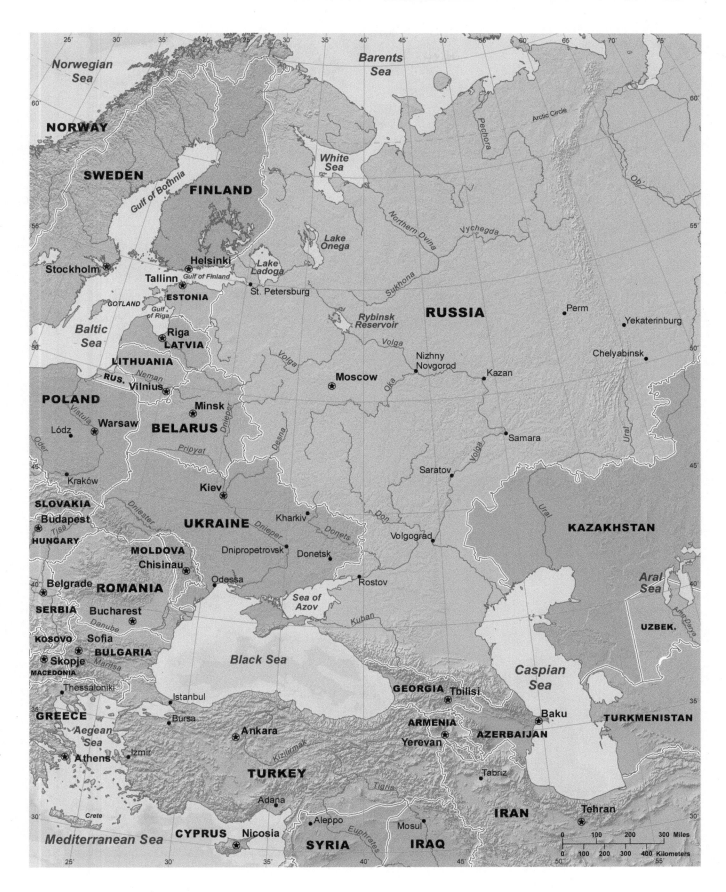

Map 147   Africa: Political Divisions

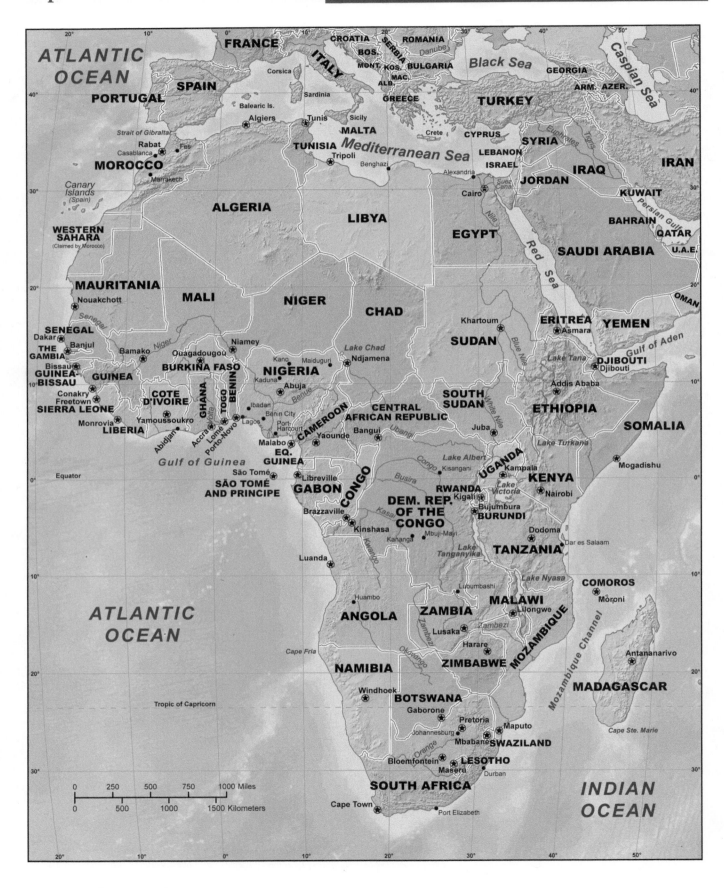

Map 148   Africa: Physical Features

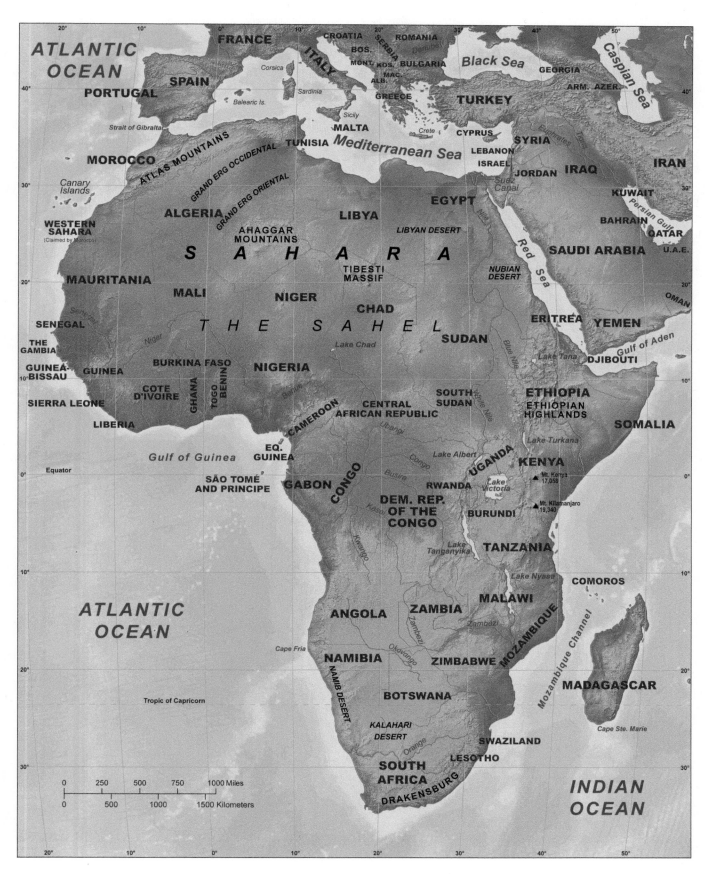

Map 149 Africa: Population Distribution

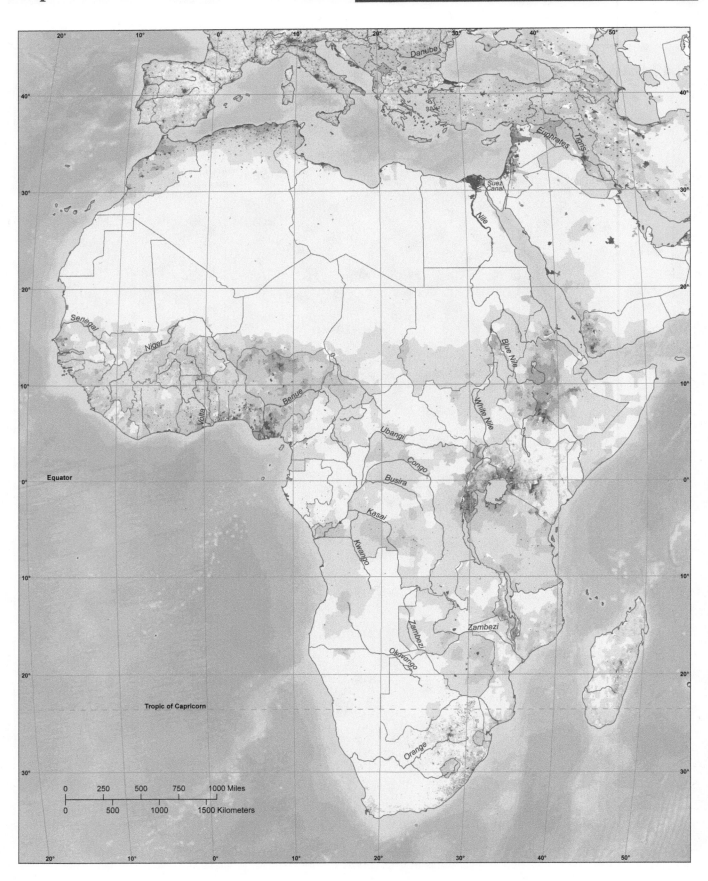

Map 150   Africa: Colonialism to Independence

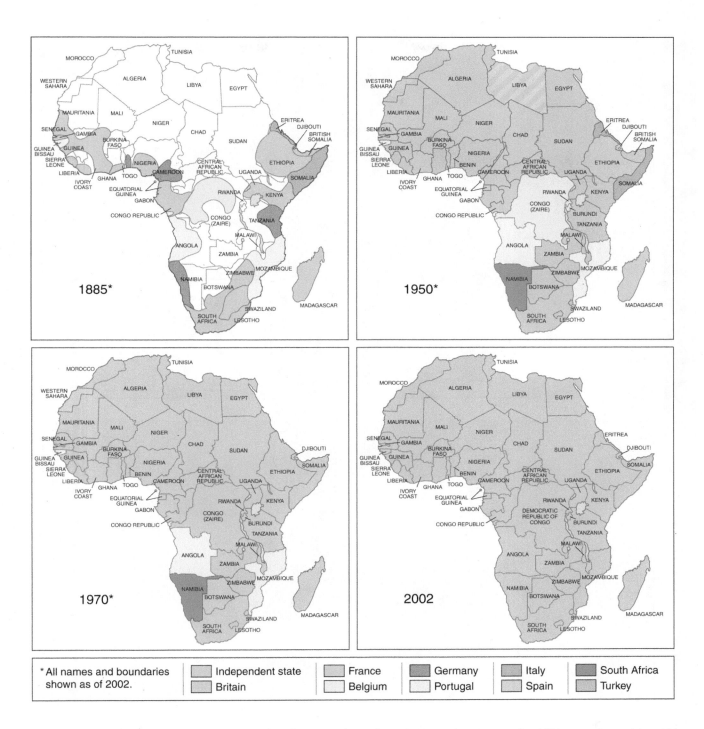

1885*

1950*

1970*

2002

*All names and boundaries shown as of 2002.

Independent state | Britain | France | Belgium | Germany | Portugal | Italy | Spain | South Africa | Turkey

In few parts of the world has the transition from colonialism to independence been as abrupt as on the African continent. Most African states did not become colonies until the nineteenth century and did not become independent until the twentieth, nearly all of them after World War II. Much of the colonial power in Africa is social and economic. The African colony provided the mother country with raw materials in exchange for marginal economic returns, and many African countries still exist in this colonial dependency relationship. An even more important component of the colonial legacy of Europe in Africa is geopolitical. When the world's colonial powers joined at the Conference of Berlin in 1884, they divided up Africa to fit their own needs, drawing boundary lines on maps without regard for terrain or drainage features, or for tribal/ethnic linguistic, cultural, economic, or political borders. Traditional Africa was enormously disrupted by this process. After independence, African countries retained boundaries that are legacies of the colonial past; and African countries today are beset by internal problems related to tribal and ethnic conflicts, the disruption of traditional migration patterns, and inefficient spatial structures of market and supply.

Map 151   African Cropland and Dryland Degradation

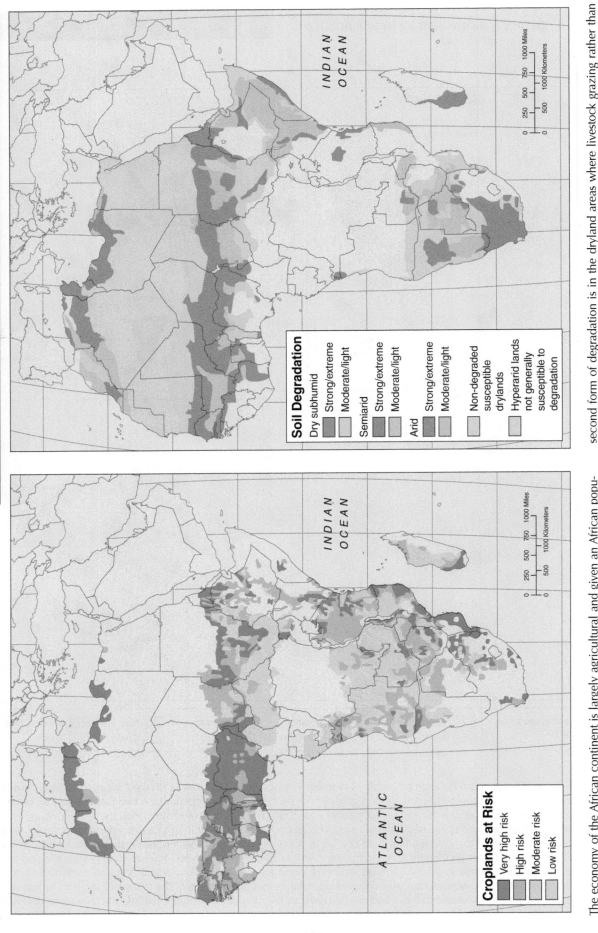

Soil Degradation

Dry subhumid
- Strong/extreme
- Moderate/light

Semiarid
- Strong/extreme
- Moderate/light

Arid
- Strong/extreme
- Moderate/light

- Non-degraded susceptible drylands
- Hyperarid lands not generally susceptible to degradation

INDIAN OCEAN

0   250   500   750   1000 Miles
0   500   1000 Kilometers

Croplands at Risk
- Very high risk
- High risk
- Moderate risk
- Low risk

ATLANTIC OCEAN

INDIAN OCEAN

0   250   500   750   1000 Miles
0   500   1000 Kilometers

The economy of the African continent is largely agricultural and given an African population of more than 1 billion people, much of the agricultural environment is degraded. Two forms of degradation exist. The first of these is in cropland areas where susceptible tropical soils and high population densities have produced major cropland degradation, largely in the form of loss of fertility. Irrigation and the advent of artificial fertilizer have not helped soils to restore their natural chemical balances after generations of misuse. The second form of degradation is in the dryland areas where livestock grazing rather than cropping is the dominant agricultural form. Populations of domesticated stock that are too large for the carrying capacity of the environment, exacerbated by the view of livestock as wealth and worsened by increasing human populations, have all contributed to the conversion of semi-arid grasslands into desert.

-188-

Map 152 Northern Africa

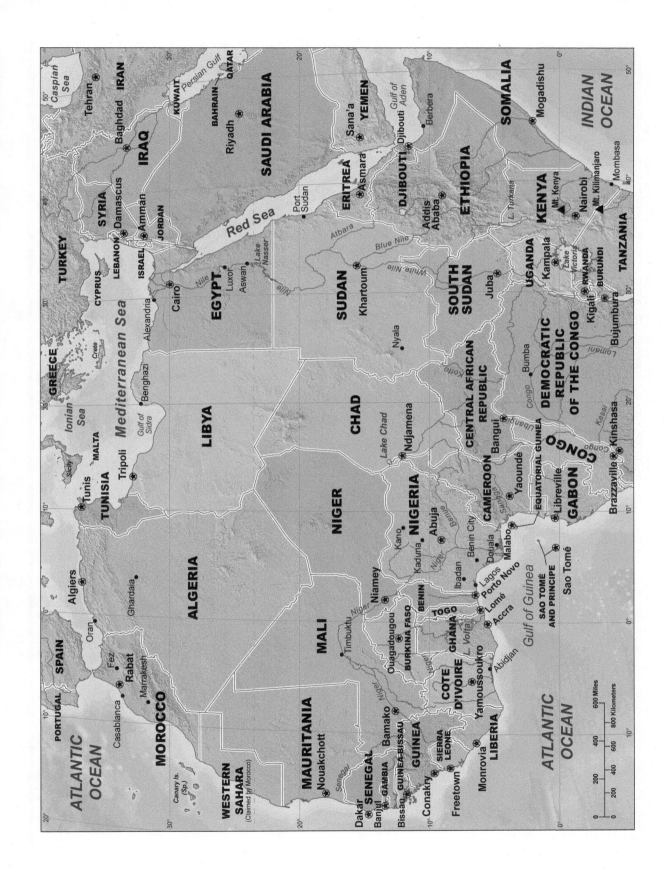

Map 153 Southern Africa

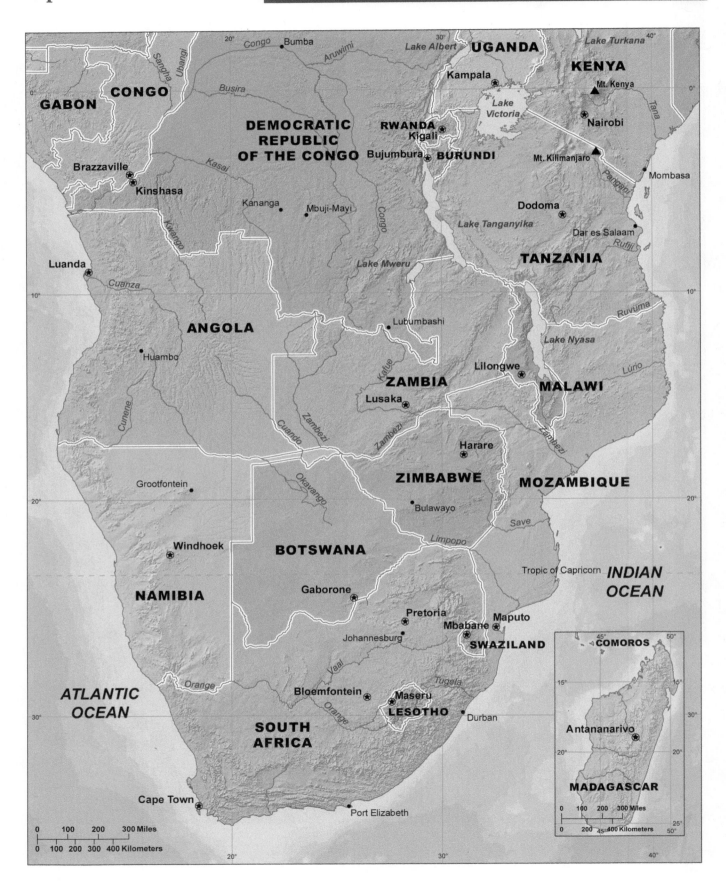

Map 154   Asia: Political Divisions

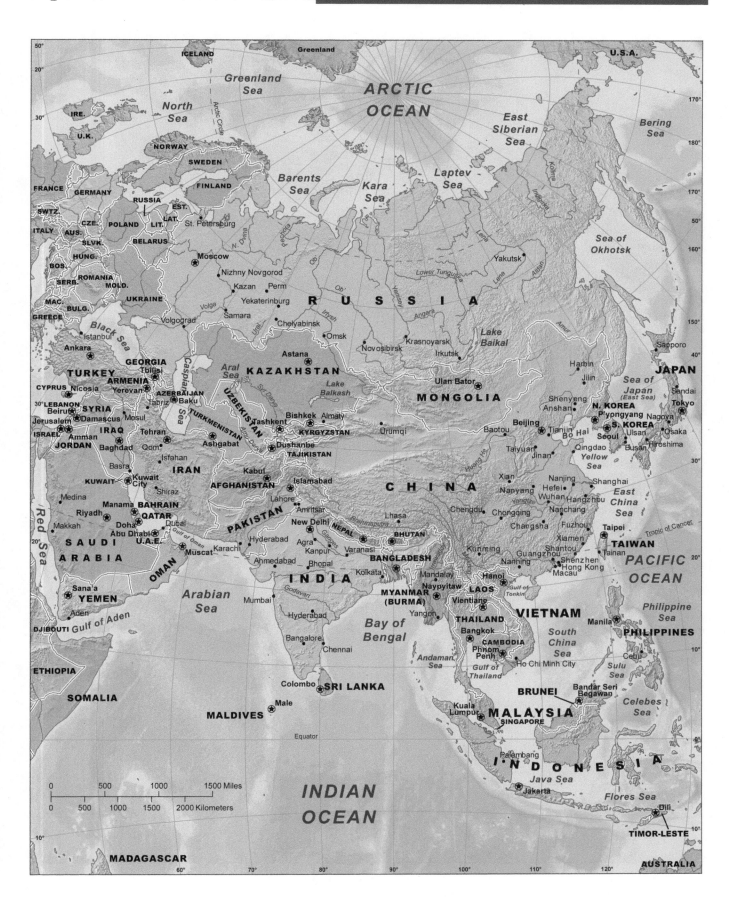

Map 155 Asia: Physical Features

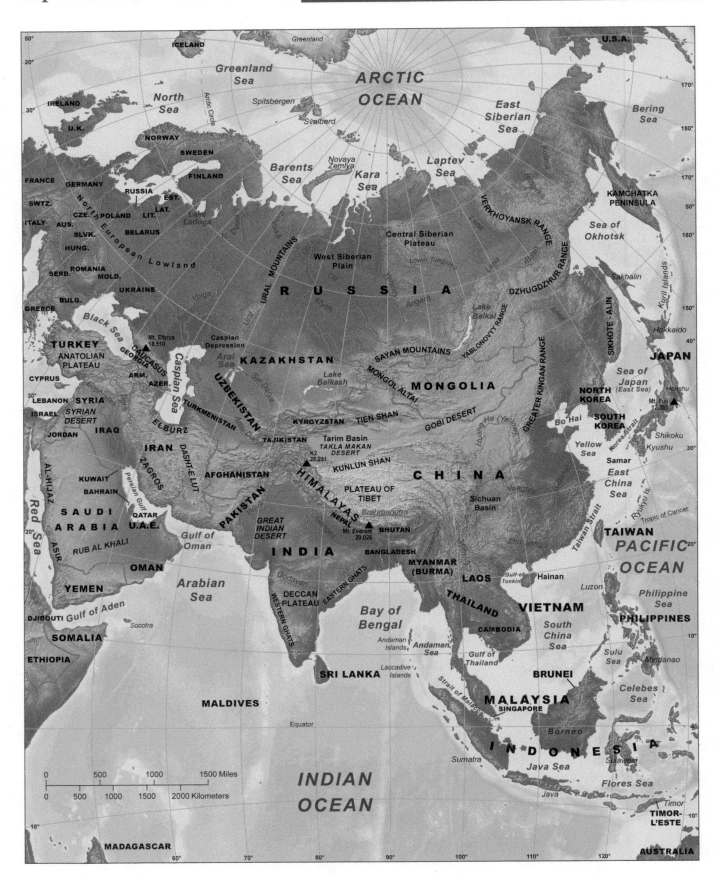

Map 156   Asia: Environment and Economy

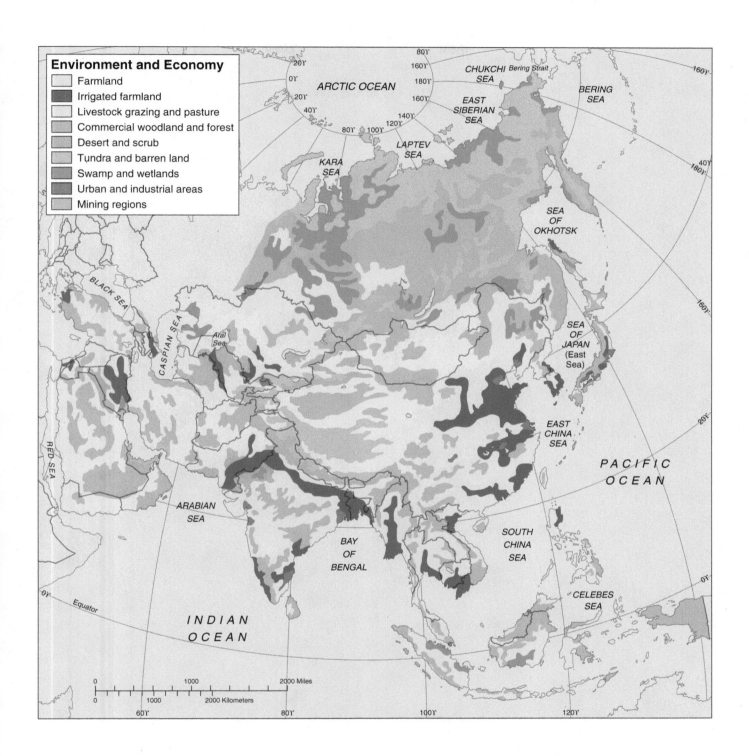

Environment and Economy
- Farmland
- Irrigated farmland
- Livestock grazing and pasture
- Commercial woodland and forest
- Desert and scrub
- Tundra and barren land
- Swamp and wetlands
- Urban and industrial areas
- Mining regions

Asia is a land of extremes of land use with some of the world's most heavily industrialized regions, barren and empty areas, and productive and densely populated farm regions. Asia is a region of rapid industrial growth. Yet Asia remains an agricultural region with three out of every four workers engaged in agriculture. Asian commercial agriculture and intensive subsistence agriculture is characterized by irrigation. Some of Asia's irrigated lands are desert requiring additional water. But most of the Asian irrigated regions have sufficient precipitation for crop agriculture and irrigation is a way of coping with seasonal drought—the wet-and-dry cycle of the monsoon—often gaining more than one crop per year on irrigated farms. Agricultural yields per unit area in many areas of Asia are among the world's highest. Because the Asian population is so large and the demands for agricultural land so great, Asia is undergoing rapid deforestation and some areas of the continent have only small remnants of a once-abundant forest reserve.

Map 157 Asia: Population Distribution

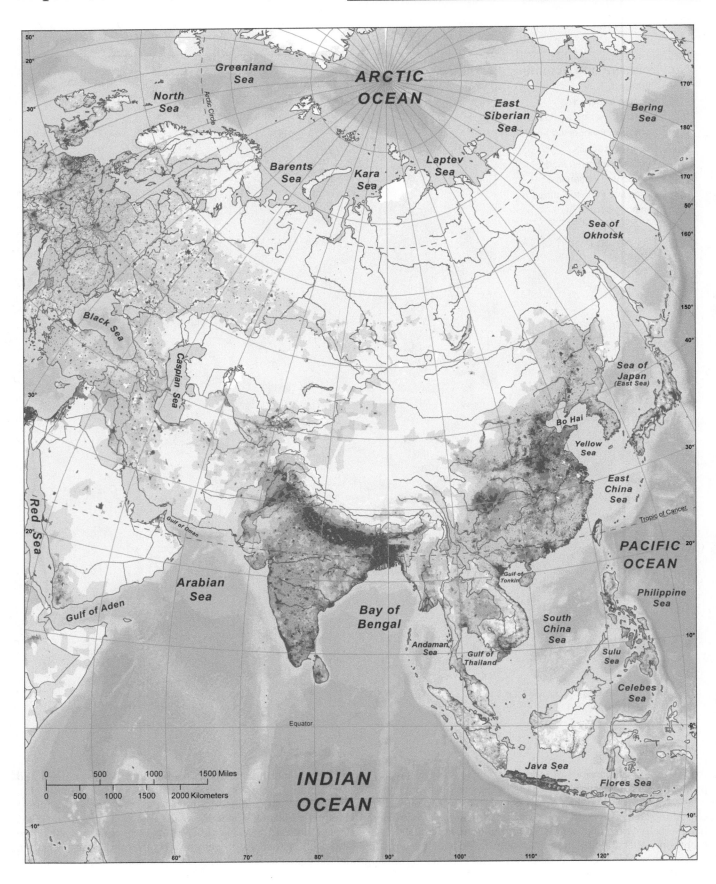

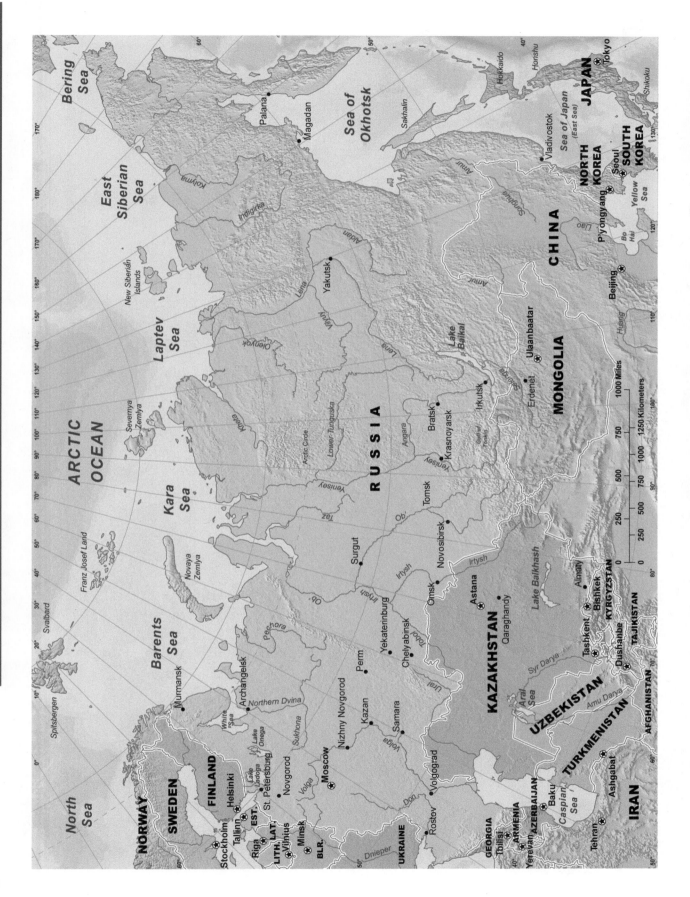

Map **158** Northern Asia

ARCTIC OCEAN

Bering Sea

East Siberian Sea

Laptev Sea

Kara Sea

Barents Sea

White Sea

North Sea

Sea of Okhotsk

Sea of Japan (East Sea)

Yellow Sea

Bo Hai

RUSSIA

CHINA

MONGOLIA

KAZAKHSTAN

UZBEKISTAN

TURKMENISTAN

AFGHANISTAN

IRAN

KYRGYZSTAN

TAJIKISTAN

JAPAN

NORTH KOREA

SOUTH KOREA

NORWAY

SWEDEN

FINLAND

EST.

LAT.

LITH.

BLR.

UKRAINE

GEORGIA

ARMENIA

AZERBAIJAN

Tokyo

Hokkaido

Honshu

Shikoku

Vladivostok

Seoul

P'yongyang

Beijing

Ulaanbaatar

Erdenet

Almaty

Bishkek

Tashkent

Dushanbe

Ashgabat

Tehran

Baku

Yerevan

Tbilisi

Rostov

Volgograd

Astana

Qaraghandy

Omsk

Novosibirsk

Tomsk

Surgut

Yekaterinburg

Chelyabinsk

Perm

Kazan

Samara

Nizhny Novgorod

Moscow

Novgorod

St. Petersburg

Helsinki

Stockholm

Tallinn

Riga

Vilnius

Minsk

Murmansk

Archangelsk

Krasnoyarsk

Bratsk

Irkutsk

Yakutsk

Magadan

Palana

Svalbard

Spitsbergen

Franz Josef Land

Novaya Zemlya

Severnya Zemlya

New Siberian Islands

Sakhalin

Lake Baikal

Lake Ladoga

Lake Onega

Lake Balkhash

Aral Sea

Caspian Sea

Arctic Circle

Kolyma

Indigirka

Lena

Aldan

Vilyuy

Olenyok

Lower Tunguska

Yenisey

Yenisey

Angara

Ob'

Ob'

Taz

Irtysh

Irtysh

Tobol

Ural

Volga

Volga

Don

Dnieper

Pechora

Northern Dvina

Sukhona

Syr Darya

Amu Darya

Amur

Amur

Songhua

Liao

Huang

Gulf of Tomin

1000 Miles

1250 Kilometers

1000

750

500

250

0

1000

750

500

250

0

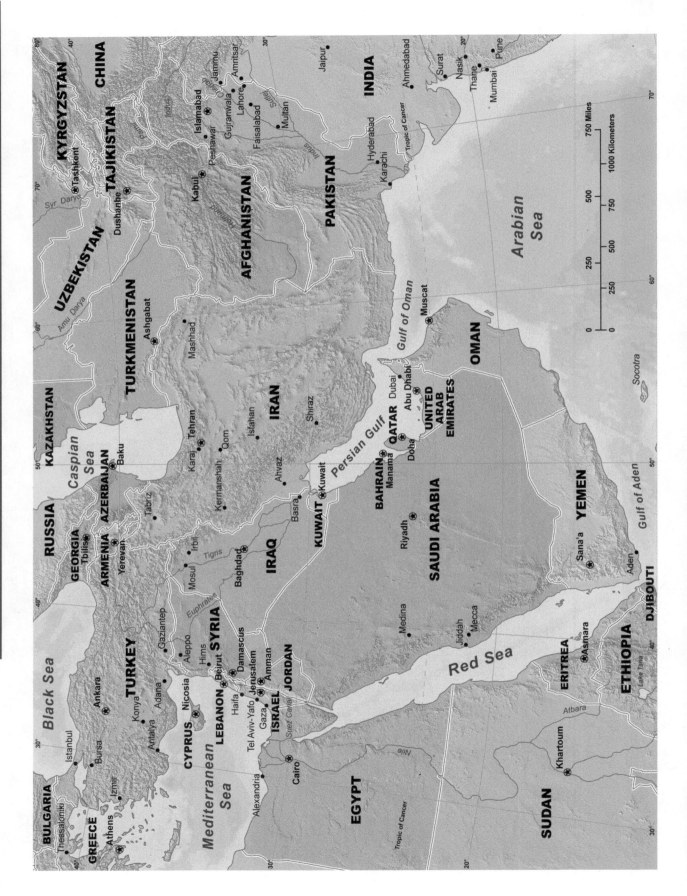

Map **159** Southwestern Asia

Map **160**   South Asia

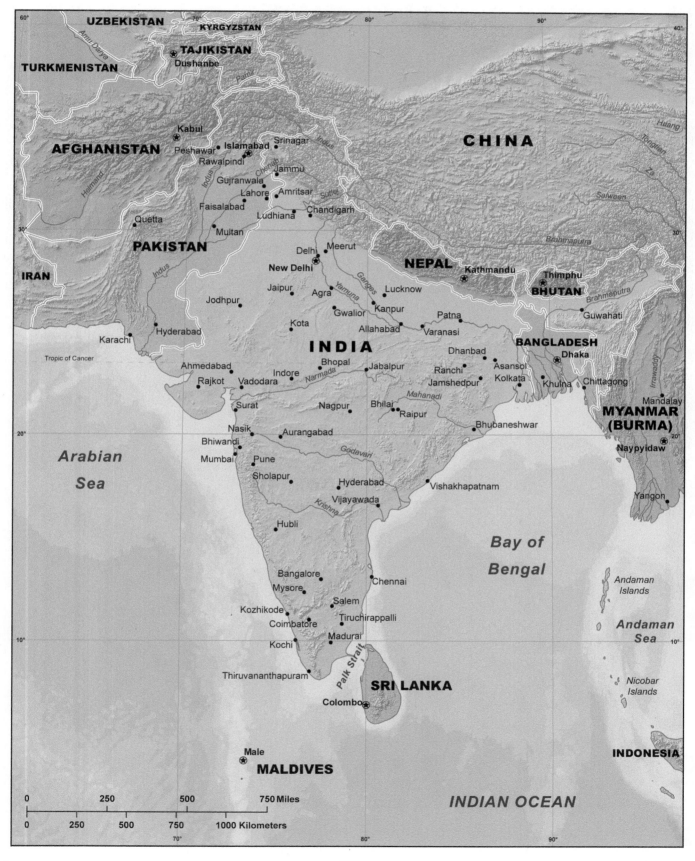

Map 161 East Asia

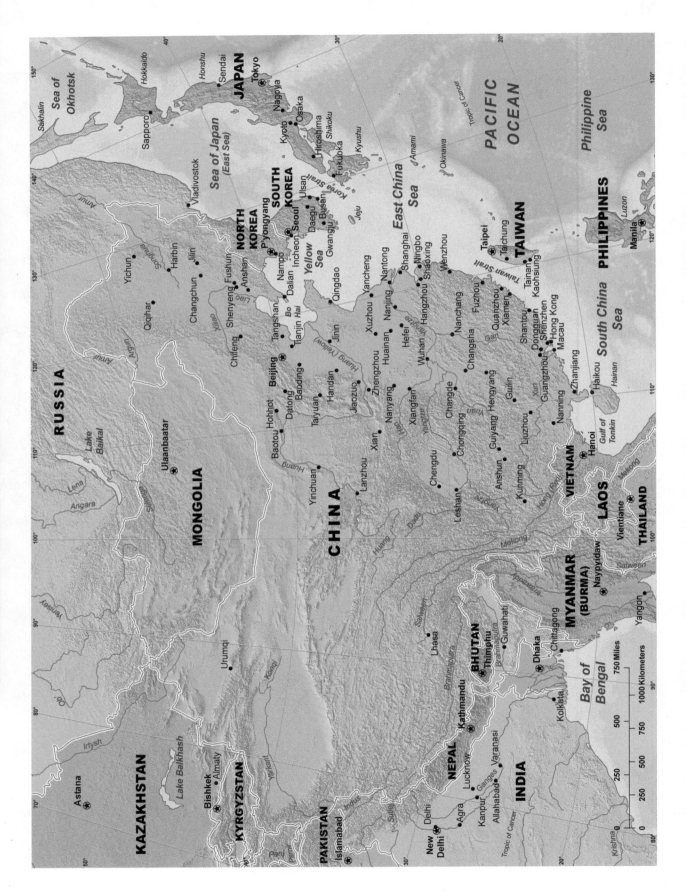

Map **162** Southeast Asia

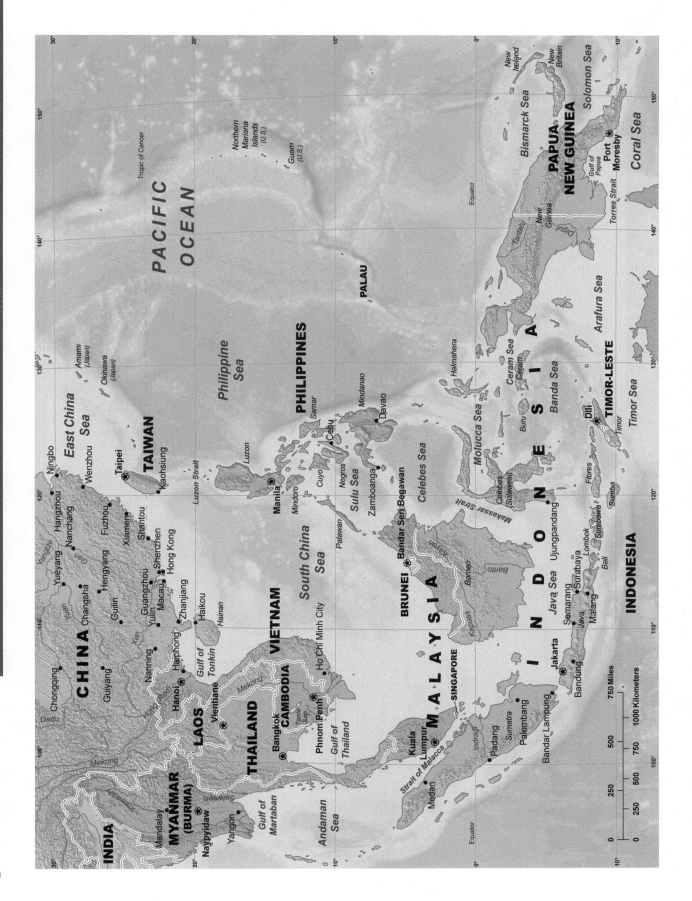

Map 163 Australasia and Oceania: Political Divisions

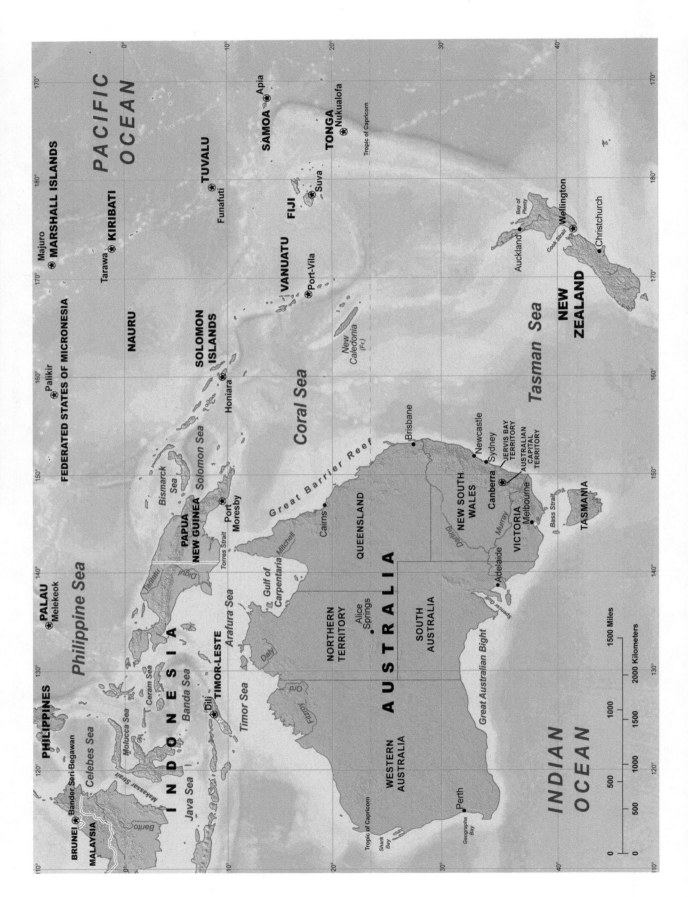

Map **164** Australasia and Oceania: Physical Features

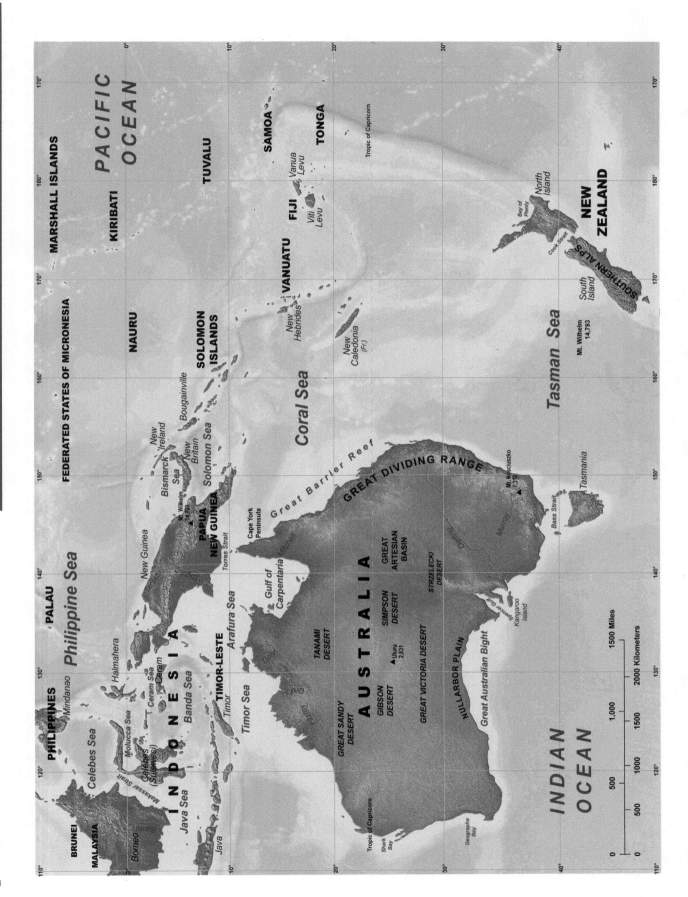

Map 165   Australasia and Oceania: Environment and Economy

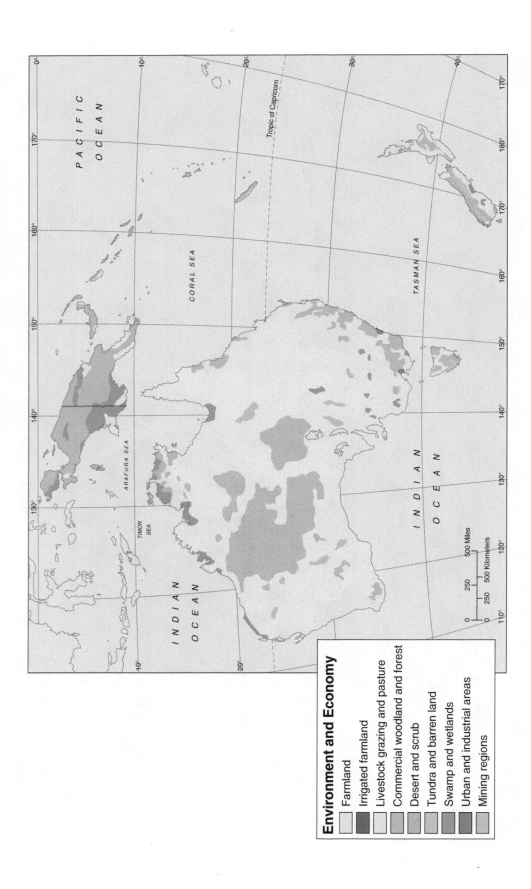

Environment and Economy

- Farmland
- Irrigated farmland
- Livestock grazing and pasture
- Commercial woodland and forest
- Desert and scrub
- Tundra and barren land
- Swamp and wetlands
- Urban and industrial areas
- Mining regions

Australasia is dominated by the world's smallest and most uniform continent. Flat, dry and mostly hot, Australia has the simplest of land use patterns: where rainfall exists so does agricultural activity. Two agricultural patterns dominate the map: livestock grazing, primarily sheep, and wheat farming, although some sugar cane production exists in the north and some cotton is grown elsewhere. Only about 6 percent of the continent consists of arable land so the areas of wheat farming, dominant as they may be in the context of Australian agriculture, are small. Australia also supports a healthy mineral resource economy, with iron and copper and precious metals making up the bulk of the extraction. Elsewhere in the region, tropical forests dominate Papua New Guinea, with some subsistence agriculture and livestock. New Zealand's temperate climate with abundant precipitation supports a productive livestock industry and little else besides tourism—which is an important economic element throughout the remainder of the region as well.

Map 166 Australasia and Oceania: Population Distribution

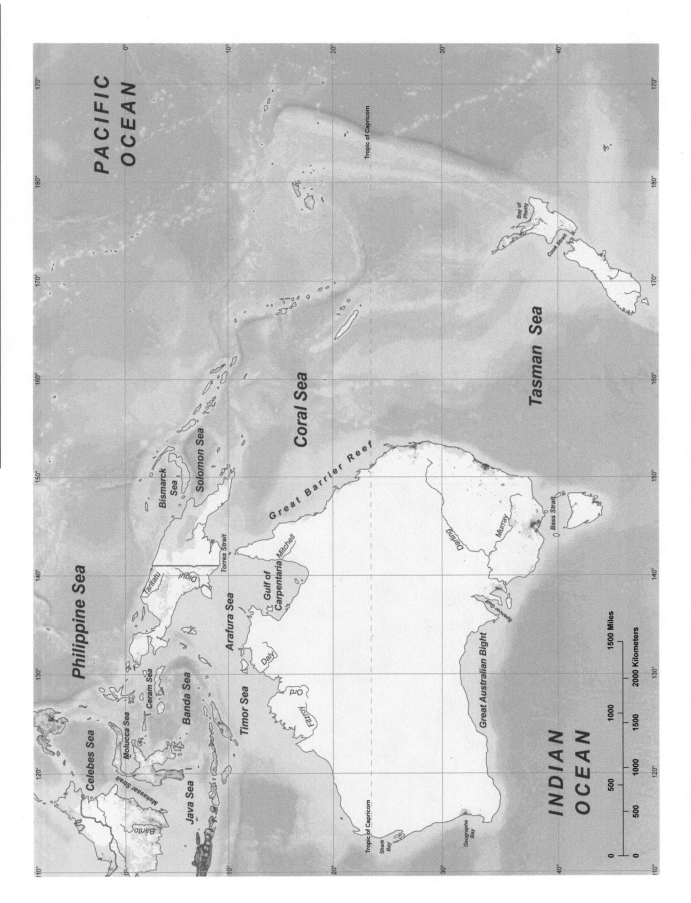

Map **167** The Arctic

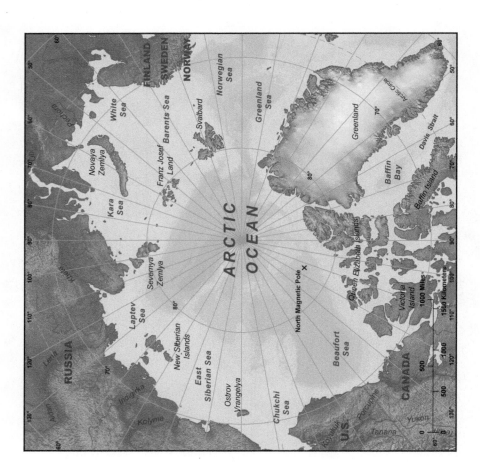

Map **168** Antarctica

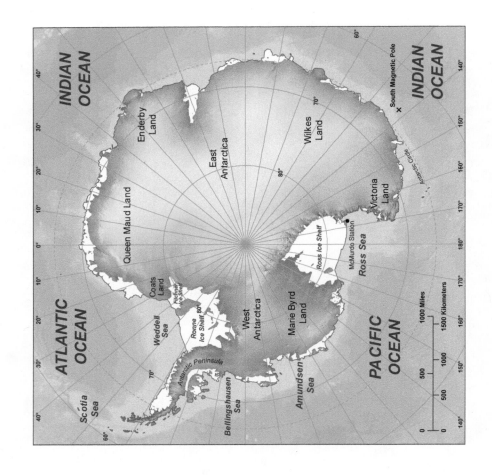

Unit VIII

Country and
Dependency Profiles

Country and Dependency Profiles

Afghanistan

Afghanestan

Official Name: Islamic Republic of Afghanistan
Capital: Kabul
Area: 250,001 sq mi (647,500 sq km)
Population: 30,419,928
Major Language(s): Dari, Pashto
Major Religion(s): Sunni Islam
Currency: afghani

Andorra

Official Name: Principality of Andorra
Capital: Andorra la Vella
Area: 181 sq mi (468 sq km)
Population: 85,082
Major Language(s): Catalan, Spanish, French
Major Religion(s): Roman Catholic Christianity
Currency: euro

Albania

Shqiperia

Official Name: Republic of Albania
Capital: Tirana
Area: 11,100 sq mi (28,748 sq km)
Population: 3,002,859
Major Language(s): Albanian
Major Religion(s): Sunni Islam
Currency: lek

Angola

Official Name: Republic of Angola
Capital: Luanda
Area: 481,354 sq mi (1,246,700 sq km)
Population: 18,056,072
Major Language(s): Portuguese, Bantu languages
Major Religion(s): Indigenous beliefs, Christianity
Currency: kwanza

Algeria

Al Jaza'ir

Official Name: People's Democratic Republic of Algeria
Capital: Algiers
Area: 919,595 sq mi (2,381,740 sq km)
Population: 35,406,303
Major Language(s): Arabic, French, Berber dialects
Major Religion(s): Sunni Islam
Currency: Algerian dinar

Anguilla

(overseas territory of the United Kingdom)

Official Name: Anguilla
Capital: Valley, The
Area: 39 sq mi (102 sq km)
Population: 15,423
Major Language(s): English
Major Religion(s): Protestant Christianity
Currency: East Caribbean dollar

American Samoa

(territory of the United States)

Official Name: Territory of American Samoa
Capital: Pago Pago
Area: 77 sq mi (199 sq km)
Population: 68,061
Major Language(s): Samoan
Major Religion(s): Protestant Christianity
Currency: U.S. dollar

Antigua and Barbuda

Official Name: Antigua and Barbuda
Capital: Saint John's
Area: 171 sq mi (443 sq km)
Population: 89,018
Major Language(s): English
Major Religion(s): Protestant Christianity
Currency: East Caribbean dollar

Argentina

Official Name: Argentine Republic
Capital: Buenos Aires
Area: 1,068,302 sq mi (2,766,890 sq km)
Population: 42,192,494
Major Language(s): Spanish
Major Religion(s): Roman Catholic Christianity
Currency: Argentine peso

Austria

Oesterreich

Official Name: Republic of Austria
Capital: Vienna
Area: 32,382 sq mi (83,870 sq km)
Population: 8,219,743
Major Language(s): German
Major Religion(s): Roman Catholic Christianity
Currency: euro

Armenia

Hayastan

Official Name: Republic of Armenia
Capital: Yerevan
Area: 11,484 sq mi (29,743 sq km)
Population: 2,970,495
Major Language(s): Armenian
Major Religion(s): Armenian Apostolic Christianity
Currency: dram

Azerbaijan

Azarbaycan

Official Name: Republic of Azerbaijan
Capital: Baku
Area: 33,436 sq mi (86,600 sq km)
Population: 9,493,600
Major Language(s): Azerbaijani
Major Religion(s): Shia Islam
Currency: manats

Aruba

(part of the Kingdom of the Netherlands)

Official Name: Aruba
Capital: Oranjestad
Area: 75 sq mi (193 sq km)
Population: 107,635
Major Language(s): Papiamento, Dutch
Major Religion(s): Roman Catholic Christianity
Currency: Arubian guilder/florin

Bahamas, The

Official Name: Commonwealth of the Bahamas
Capital: Nassau
Area: 5,382 sq mi (13,940 sq km)
Population: 316,182
Major Language(s): English
Major Religion(s): Protestant Christianity
Currency: Bahamian dollar

Australia

Official Name: Commonwealth of Australia
Capital: Canberra
Area: 2,967,909 sq mi (7,686,850 sq km)
Population: 22,015,576
Major Language(s): English, Aboriginal languages
Major Religion(s): Christianity
Currency: Australian dollar

Bahrain

Al Bahrayn

Official Name: Kingdom of Bahrain
Capital: Manama
Area: 257 sq mi (665 sq km)
Population: 1,248,348
Major Language(s): Arabic, English
Major Religion(s): Sunni Islam
Currency: Bahraini dollar

Country and Dependency Profiles

Bangladesh

Banladesh

Official Name: People's Republic of Bangladesh
Capital: Dhaka
Area: 55,599 sq mi (144,000 sq km)
Population: 161,083,804
Major Language(s): Bangla, English
Major Religion(s): Sunni Islam
Currency: taka

Barbados

Official Name: Barbados
Capital: Bridgetown
Area: 166 sq mi (431 sq km)
Population: 287,733
Major Language(s): English
Major Religion(s): Protestant Christianity
Currency: Barbadian dollar

Belarus

Byelarus'

Official Name: Republic of Belarus
Capital: Minsk
Area: 80,155 sq mi (207,600 sq km)
Population: 9,542,883
Major Language(s): Belarusian, Russian
Major Religion(s): Eastern Orthodox Christianity
Currency: Belarusian ruble

Belgium

Belgique/Belgie

Official Name: Kingdom of Belgium
Capital: Brussels
Area: 11,787 sq mi (30,528 sq km)
Population: 10,438,353
Major Language(s): Dutch, French, German
Major Religion(s): Roman Catholic Christianity
Currency: euro

Belize

Official Name: Belize
Capital: Belmopan
Area: 8,867 sq mi (22,966 sq km)
Population: 327,719
Major Language(s): Spanish, Creole, English
Major Religion(s): Roman Catholic Christianity
Currency: Belizian dollar

Benin

Official Name: Republic of Benin
Capital: Porto-Novo
Area: 43,483 sq mi (112,620 sq km)
Population: 9,598,787
Major Language(s): French, Fon, Yoruba, other African languages
Major Religion(s): Indigenous beliefs, Christianity
Currency: CFA franc

Bermuda

(overseas territory of the United Kingdom)

Official Name: Bermuda
Capital: Hamilton
Area: 20 sq mi (53 sq km)
Population: 69,080
Major Language(s): English
Major Religion(s): Protestant Christianity
Currency: Bermudian dollar

Bhutan

Druk Yul

Official Name: Kingdom of Bhutan
Capital: Thimphu
Area: 18,147 sq mi (47,000 sq km)
Population: 716,896
Major Language(s): Dzongkha, Tibetan dialects, Nepalese dialects
Major Religion(s): Buddhism
Currency: ngultrum

Bolivia

Official Name: Plurinational State of Bolivia
Capital: La Paz (administrative), Sucre (constitutional)
Area: 424,164 sq mi (1,098,580 sq km)
Population: 10,290,003
Major Language(s): Spanish, Quechua, Aymara
Major Religion(s): Roman Catholic Christianity
Currency: boliviano

British Virgin Islands

Overseas territory of the United Kingdom

Official Name: British Virgin Islands
Capital: Road Town
Area: 58 sq mi (151 sq km)
Population: 31,148
Major Language(s): English
Major Religion(s): Protestant Christianity
Currency: U.S. dollar

Bosnia and Herzegovina

Bosna i Hercegovina

Official Name: Bosnia and Herzegovina
Capital: Sarajevo
Area: 19,772 sq mi (51,209 sq km)
Population: 4,622,292
Major Language(s): Bosnian, Croatian, Serbian
Major Religion(s): Christianity, Islam
Currency: konvertibilna mark

Brunei

Official Name: Brunei Darussalam
Capital: Bandar Seri Begawan
Area: 2,228 sq mi (5,770 sq km)
Population: 408,786
Major Language(s): Malay, English
Major Religion(s): Sunni Islam
Currency: Bruneian dollar

Botswana

Official Name: Republic of Botswana
Capital: Gaborone
Area: 231,804 sq mi (600,370 sq km)
Population: 2,098,018
Major Language(s): Setswana, Kalanga, English
Major Religion(s): Christianity
Currency: pula

Bulgaria

Balgariya

Official Name: Republic of Bulgaria
Capital: Sofia
Area: 42,823 sq mi (110,910 sq km)
Population: 7,037,935
Major Language(s): Bulgarian
Major Religion(s): Eastern Orthodox Christianity
Currency: leva

Brazil

Brasil

Official Name: Federative Republic of Brazil
Capital: Brasilia
Area: 3,286,488 sq mi (8,511,965 sq km)
Population: 205,716,890
Major Language(s): Portuguese, many Amerindian languages
Major Religion(s): Roman Catholic Christianity
Currency: real

Burkina Faso

Official Name: Burkina Faso
Capital: Ouagadougou
Area: 105,869 sq mi (274,200 sq km)
Population: 17,275,115
Major Language(s): French, many African languages
Major Religion(s): Sunni Islam
Currency: CFA franc

Country and Dependency Profiles

Burma (Myanmar)

Myanma Naingngandaw

Official Name: Union of Burma
Capital: Rangoon
Area: 261,970 sq mi (678,500 sq km)
Population: 54,584,650
Major Language(s): Burmese
Major Religion(s): Buddhism
Currency: kyat

Canada

Official Name: Canada
Capital: Ottawa
Area: 3,855,103 sq mi (9,984,670 sq km)
Population: 34,300,083
Major Language(s): English, French, many Amerindian languages
Major Religion(s): Roman Catholic Christianity
Currency: Canadian dollar

Burundi

Official Name: Republic of Burundi
Capital: Bujumbura
Area: 10,745 sq mi (27,830 sq km)
Population: 10,557,259
Major Language(s): Kirundi, French, Swahili
Major Religion(s): Roman Catholic Christianity
Currency: Burundi franc

Cape Verde

Cabo Verde

Official Name: Republic of Cape Verde
Capital: Praia
Area: 1,557 sq mi (4,033 sq km)
Population: 523,568
Major Language(s): Portuguese
Major Religion(s): Roman Catholic Christianity
Currency: escudo

Cambodia

Kampuchea

Official Name: Kingdom of Cambodia
Capital: Phnom Penh
Area: 69,900 sq mi (181,040 sq km)
Population: 14,952,665
Major Language(s): Khmer
Major Religion(s): Buddhism
Currency: riel

Cayman Islands

(overseas territory of the United Kingdom)

Official Name: Cayman Islands
Capital: George Town
Area: 101 sq mi (262 sq km)
Population: 52,560
Major Language(s): English
Major Religion(s): Christianity
Currency: Caymanian dollar

Cameroon

Cameroun

Official Name: Republic of Cameroon
Capital: Yaounde
Area: 183,568 sq mi (475,440 sq km)
Population: 20,129,878
Major Language(s): English, French, many African languages
Major Religion(s): Indigenous beliefs, Christianity
Currency: CFA franc

Central African Republic

Official Name: Central African Republic
Capital: Bangui
Area: 240,535 sq mi (622,984 sq km)
Population: 5,057,208
Major Language(s): French, Sangho
Major Religion(s): Indigenous beliefs, Christianity
Currency: CFA franc

Chad

Tchad/Tshad

Official Name: Republic of Chad
Capital: N'Djamena
Area: 495,755 sq mi (1,284,000 sq km)
Population: 10,975,648
Major Language(s): French, Arabic, many African languages
Major Religion(s): Sunni Islam
Currency: CFA franc

Chile

Official Name: Republic of Chile
Capital: Santiago
Area: 292,260 sq mi (756,950 sq km)
Population: 17,067,369
Major Language(s): Spanish
Major Religion(s): Roman Catholic Christianity
Currency: Chilean peso

China

Zhongguo

Official Name: People's Republic of China
Capital: Beijing
Area: 3,705,407 sq mi (9,596,960 sq km)
Population: 1,343,239,923
Major Language(s): Chinese (Mandarin), Yue (Cantonese), many other dialects
Major Religion(s): Officially atheist
Currency: Renminbi yuan

Colombia

Official Name: Republic of Colombia
Capital: Bogotá
Area: 439,736 sq mi (1,138,910 sq km)
Population: 45,239,079
Major Language(s): Spanish
Major Religion(s): Roman Catholic Christianity
Currency: Colombian peso

Comoros

Komori/Comores/Juzur al Qamar

Official Name: Union of the Comoros
Capital: Moroni
Area: 838 sq mi (2,170 sq km)
Population: 737,284
Major Language(s): Arabic, French
Major Religion(s): Sunni Islam
Currency: Comoran franc

Congo, Democratic Republic of

Congo (Kinshasa)

Official Name: Democratic Republic of the Congo
Capital: Kinshasa
Area: 905,345 sq mi (2,344,858 sq km)
Population: 73,599,190
Major Language(s): French, Lingala, Kingwana
Major Religion(s): Christianity, Islam, Indigenous beliefs
Currency: Congolese franc

Congo, Republic of

Congo (Brazzaville)

Official Name: Republic of the Congo
Capital: Brazzaville
Area: 132,047 sq mi (342,000 sq km)
Population: 4,366,266
Major Language(s): French, Lingala, Monokutuba
Major Religion(s): Christianity, Animist
Currency: Congolese franc

Costa Rica

Official Name: Republic of Costa Rica
Capital: San Jose
Area: 19,730 sq mi (51,100 sq km)
Population: 4,636,348
Major Language(s): Spanish
Major Religion(s): Roman Catholic Christianity
Currency: colón

Country and Dependency Profiles

Côte d'Ivoire

Official Name: Republic of Côte d'Ivoire
Capital: Yamoussoukro
Area: 124,503 sq mi (322,460 sq km)
Population: 21,952,093
Major Language(s): French, many African languages
Major Religion(s): Sunni Islam, Christianity
Currency: CFA franc

Cyprus

Kypros/Kibris

Official Name: Republic of Cyprus
Capital: Nicosia
Area: 3,571 sq mi (9,250 sq km)
Population: 1,138,071
Major Language(s): Greek, Turkish
Major Religion(s): Eastern Orthodox Christianity
Currency: euro

Croatia

Hrvatska

Official Name: Republic of Croatia
Capital: Zagreb
Area: 21,831 sq mi (56,542 sq km)
Population: 4,489,409
Major Language(s): Croatian
Major Religion(s): Roman Catholic Christianity
Currency: kuna

Czechia (Czech Republic)

Cesko

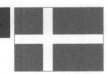

Official Name: Czech Republic
Capital: Prague
Area: 30,450 sq mi (78,866 sq km)
Population: 10,177,300
Major Language(s): Czech
Major Religion(s): Roman Catholic Christianity
Currency: koruny

Cuba

Official Name: Republic of Cuba
Capital: Havana
Area: 42,803 sq mi (110,860 sq km)
Population: 11,075,244
Major Language(s): Spanish
Major Religion(s): Roman Catholic Christianity
Currency: Cuban peso

Denmark

Danmark

Official Name: Kingdom of Denmark
Capital: Copenhagen
Area: 16,639 sq mi (43,094 sq km)
Population: 5,543,453
Major Language(s): Danish
Major Religion(s): Protestant Christianity
Currency: Danish kroner

Curaçao

Constituent country within the Kingdom of the Netherlands

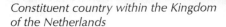

Official Name: Curaçao
Capital: Willemstad
Area: 17 sq mi (44 sq km)
Population: 145,834
Major Language(s): Papiamentu, Dutch
Major Religion(s): Roman Catholic Christianity
Currency: Netherlands Antilles guilder

Djibouti

Jibuti

Official Name: Republic of Djibouti
Capital: Djibouti
Area: 8,880 sq mi (23,000 sq km)
Population: 774,389
Major Language(s): French, Arabic
Major Religion(s): Sunni Islam
Currency: Djiboutian franc

Dominica

Official Name: Commonwealth of Dominica
Capital: Roseau
Area: 291 sq mi (754 sq km)
Population: 73,126
Major Language(s): English
Major Religion(s): Roman Catholic Christianity
Currency: East Caribbean dollar

Dominican Republic

La Dominicana

Official Name: Dominican Republic
Capital: Santo Domingo
Area: 18,815 sq mi (48,730 sq km)
Population: 10,088,598
Major Language(s): Spanish
Major Religion(s): Roman Catholic Christianity
Currency: Dominican peso

Ecuador

Official Name: Republic of Ecuador
Capital: Quito
Area: 109,483 sq mi (283,560 sq km)
Population: 15,223,680
Major Language(s): Spanish, many Amerindian languages
Major Religion(s): Roman Catholic Christianity
Currency: U.S. dollar

Egypt

Misr

Official Name: Arab Republic of Egypt
Capital: Cairo
Area: 386,662 sq mi (1,001,450 sq km)
Population: 83,688,164
Major Language(s): Arabic
Major Religion(s): Sunni Islam
Currency: Egyptian pound

El Salvador

Official Name: Republic of El Salvador
Capital: San Salvador
Area: 8,124 sq mi (21,040 sq km)
Population: 6,090,646
Major Language(s): Spanish
Major Religion(s): Roman Catholic Christianity
Currency: U.S. dollar

Equatorial Guinea

Guinea Ecuatorial/Guinee equatoriale

Official Name: Republic of Equatorial Guinea
Capital: Malabo
Area: 10,831 sq mi (28,051 sq km)
Population: 685,991
Major Language(s): Spanish, French, Fang, Bubi
Major Religion(s): Roman Catholic Christianity
Currency: CFA franc

Eritrea

Ertra

Official Name: State of Eritrea
Capital: Asmara
Area: 46,842 sq mi (121,320 sq km)
Population: 6,086,495
Major Language(s): Tigrinya, Arabic
Major Religion(s): Sunni Islam, Christianity
Currency: nakfa

Estonia

Eesti

Official Name: Republic of Estonia
Capital: Tallinn
Area: 17,462 sq mi (45,226 sq km)
Population: 1,274,709
Major Language(s): Estonian, Russian
Major Religion(s): Protestant Christianity
Currency: euro

Country and Dependency Profiles

Ethiopia

Ityop'iya

Official Name: Federal Democratic Republic of Ethiopia
Capital: Addis Ababa
Area: 435,186 sq mi (1,127,127 sq km)
Population: 93,815,992
Major Language(s): Amarigna, Oromigna, English
Major Religion(s): Ethiopian Orthodox Christianity, Islam
Currency: birr

Falkland Islands (Islas Malvinas)

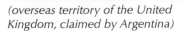

(overseas territory of the United Kingdom, claimed by Argentina)

Official Name: Falkland Islands (Islas Malvinas)
Capital: Stanley
Area: 4,700 sq mi (12,173 sq km)
Population: 3,140
Major Language(s): English
Major Religion(s): Christianity
Currency: British pound

Faroe Islands

Foroyar
(part of the Kingdom of Denmark)

Official Name: Faroe Islands
Capital: Torshavn
Area: 540 sq mi (1,399 sq km)
Population: 49,483
Major Language(s): Faroese
Major Religion(s): Protestant Christianity
Currency: Danish kroner

Fiji

Fiji/Viti

Official Name: Republic of Fiji
Capital: Suva
Area: 7,054 sq mi (18,270 sq km)
Population: 890,057
Major Language(s): English, Fijian
Major Religion(s): Christianity
Currency: Fijian dollar

Finland

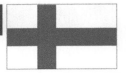

Suomi/Finland

Official Name: Republic of Finland
Capital: Helsinki
Area: 130,559 sq mi (338,145 sq km)
Population: 5,252,930
Major Language(s): Finnish
Major Religion(s): Protestant Christianity
Currency: euro

France

Official Name: French Republic
Capital: Paris
Area: 248,429 sq mi (643,427 sq km)
Population: 65,630,692
Major Language(s): French
Major Religion(s): Roman Catholic Christianity
Currency: euro

French Polynesia

(territorial overseas collectivity of France)

Official Name: French Polynesia
Capital: Papeete
Area: 1,609 sq mi (4,167 sq km)
Population: 274,512
Major Language(s): French, Polynesian
Major Religion(s): Protestant Christianity, Roman Catholic
Currency: XPF franc

Gabon

Official Name: Gabonese Republic
Capital: Libreville
Area: 103,347 sq mi (267,667 sq km)
Population: 1,608,321
Major Language(s): French, Fang
Major Religion(s): Christianity
Currency: CFA franc

Gambia, The

Official Name: Republic of The Gambia
Capital: Banjul
Area: 4,363 sq mi (11,300 sq km)
Population: 1,840,454
Major Language(s): English, Mandinka, Wolof, Fula
Major Religion(s): Sunni Islam
Currency: dalasis

Gilbraltar

(overseas territory of the United Kingdom)

Official Name: Gibraltar
Capital: Gibraltar
Area: 2.5 sq mi (6.5 sq km)
Population: 29,034
Major Language(s): English, Spanish
Major Religion(s): Roman Catholic Christianity
Currency: Gibraltar pound

Georgia

Sak'art'velo

Official Name: Georgia
Capital: T'bilisi
Area: 26,911 sq mi (69,700 sq km)
Population: 4,570,934
Major Language(s): Georgian, Russian
Major Religion(s): Eastern Orthodox Christianity
Currency: laris

Greece

Ellas or Ellada

Official Name: Hellenic Republic
Capital: Athens
Area: 50,942 sq mi (131,940 sq km)
Population: 10,767,827
Major Language(s): Greek
Major Religion(s): Eastern Orthodox Christianity
Currency: euro

Germany

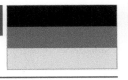

Deutschland

Official Name: Federal Republic of Germany
Capital: Berlin
Area: 137,847 sq mi (357,021 sq km)
Population: 81,305,856
Major Language(s): German
Major Religion(s): Christianity
Currency: euro

Greenland

Kalaallit Nunaat
(part of the Kingdom of Denmark)

Official Name: Greenland
Capital: Nuuk
Area: 836,330 sq mi (2,166,086 sq km)
Population: 57,695
Major Language(s): Greenlandic, Danish, English
Major Religion(s): Protestant Christianity
Currency: Danish kroner

Ghana

Official Name: Republic of Ghana
Capital: Accra
Area: 92,456 sq mi (239,460 sq km)
Population: 25,241,998
Major Language(s): English, many African languages
Major Religion(s): Christianity
Currency: cedi

Grenada

Official Name: Grenada
Capital: Saint George's
Area: 133 sq mi (344 sq km)
Population: 109,011
Major Language(s): English, French patois
Major Religion(s): Roman Catholic Christianity
Currency: East Caribbean dollar

Country and Dependency Profiles

Guam

(unicorportated territory of the United States)

Official Name: Territory of Guam
Capital: Hagatna
Area: 209 sq mi (541 sq km)
Population: 159,914
Major Language(s): English, Chamorro
Major Religion(s): Roman Catholic Christianity
Currency: U.S. dollar

Guinea-Bissau

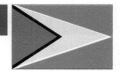

Guine-Bissau

Official Name: Republic of Guinea-Bissau
Capital: Bissau
Area: 13,946 sq mi (36,120 sq km)
Population: 1,628,603
Major Language(s): Portuguese, Crioulo
Major Religion(s): Sunni Islam, Indigenous beliefs
Currency: CFA franc

Guatemala

Official Name: Republic of Guatemala
Capital: Guatemala
Area: 42,043 sq mi (108,890 sq km)
Population: 14,099,032
Major Language(s): Spanish, many Amerindian languages
Major Religion(s): Roman Catholic Christianity
Currency: quetzal

Guyana

Official Name: Cooperative Republic of Guyana
Capital: Georgetown
Area: 83,000 sq mi (214,970 sq km)
Population: 741,908
Major Language(s): English, many Amerindian languages
Major Religion(s): Hinduism, Christianity
Currency: Guyanese dollar

Guernsey

(British crown dependency)

Official Name: Bailiwick of Guernsey
Capital: Saint Peter Port
Area: 30 sq mi (78 sq km)
Population: 65,345
Major Language(s): English
Major Religion(s): Protestant Christianity
Currency: British pound

Haiti

Haiti/Ayiti

Official Name: Republic of Haiti
Capital: Port-au-Prince
Area: 10,714 sq mi (27,750 sq km)
Population: 9,801,665
Major Language(s): French, Creole
Major Religion(s): Roman Catholic Christianity
Currency: gourde

Guinea

Guinee

Official Name: Republic of Guinea
Capital: Conakry
Area: 94,926 sq mi (245,857 sq km)
Population: 10,884,958
Major Language(s): French, many African languages
Major Religion(s): Sunni Islam
Currency: Guinean franc

Holy See (Vatican City)

Santa Sede (Citta del Vaticano)

Official Name: The Holy See (State of the Vatican City)
Capital: Vatican City
Area: 0.4 sq mi (1.1 sq km)
Population: 836
Major Language(s): Italian, Latin
Major Religion(s): Roman Catholic Christianity
Currency: euro

Honduras

Official Name: Republic of Honduras
Capital: Tegucigalpa
Area: 43,278 sq mi (112,090 sq km)
Population: 8,296,693
Major Language(s): Spanish, many Amerindian languages
Major Religion(s): Roman Catholic Christianity
Currency: lempira

Hungary

Magyarorszag

Official Name: Republic of Hungary
Capital: Budapest
Area: 35,919 sq mi (93,030 sq km)
Population: 9,958,453
Major Language(s): Hungarian
Major Religion(s): Roman Catholic Christianity
Currency: forint

Iceland

Island

Official Name: Republic of Iceland
Capital: Reykjavik
Area: 39,769 sq mi (103,000 sq km)
Population: 313,183
Major Language(s): Icelandic
Major Religion(s): Protestant Christianity
Currency: kronur

India

India/Bharat

Official Name: Republic of India
Capital: New Delhi
Area: 1,269,346 sq mi (3,287,590 sq km)
Population: 1,205,073,612
Major Language(s): Hindi, English, 17 other languages
Major Religion(s): Hinduism, Sunni Islam
Currency: Indian rupee

Indonesia

Official Name: Republic of Indonesia
Capital: Jakarta
Area: 741,100 sq mi (1,919,440 sq km)
Population: 248,216,193
Major Language(s): Bahasa Indonesia, many local dialects
Major Religion(s): Sunni Islam
Currency: rupiah

Iran

Official Name: Islamic Republic of Iran
Capital: Tehran
Area: 636,296 sq mi (1,648,000 sq km)
Population: 78,868,711
Major Language(s): Persian, Turkic dialects, Kurdish
Major Religion(s): Shia Islam
Currency: Iranian rial

Iraq

Al Iraq

Official Name: Republic of Iraq
Capital: Baghdad
Area: 168,754 sq mi (437,072 sq km)
Population: 31,129,225
Major Language(s): Arabic, Kurdish
Major Religion(s): Shia and Sunni Islam
Currency: New Iraqi dinar

Ireland

Eire

Official Name: Ireland
Capital: Dublin
Area: 27,135 sq mi (70,280 sq km)
Population: 4,722,028
Major Language(s): English, Irish
Major Religion(s): Roman Catholic Christianity
Currency: euro

Country and Dependency Profiles

Isle of Man

(British Crown Dependency)

Official Name: Isle of Man
Capital: Douglas
Area: 221 sq mi (572 sq km)
Population: 85,421
Major Language(s): English, Manx Gaelic
Major Religion(s): Protestant Christianity
Currency: Manx pound

Japan

Nihon/Nippon

Official Name: Japan
Capital: Tokyo
Area: 145,883 sq mi (377,835 sq km)
Population: 127,368,088
Major Language(s): Japanese
Major Religion(s): Shintoism
Currency: yen

Israel

Yisra'el

Official Name: State of Israel
Capital: Jerusalem (proclaimed), Tel Aviv (de facto)
Area: 8,019 sq mi (20,770 sq km)
Population: 7,590,758
Major Language(s): Hebrew, Arabic
Major Religion(s): Judaism
Currency: Israeli new shekel

Jersey

(British crown dependency)

Official Name: Bailiwick of Jersey
Capital: Saint Helier
Area: 45 sq mi (116 sq km)
Population: 94,949
Major Language(s): English
Major Religion(s): Protestant Christianity
Currency: British pound

Italy

Italia

Official Name: Italian Republic
Capital: Rome
Area: 116,306 sq mi (301,230 sq km)
Population: 61,261,254
Major Language(s): Italian, German, French
Major Religion(s): Roman Catholic Christianity
Currency: euro

Jordan

Al Urdun

Official Name: Hashemite Kingdom of Jordan
Capital: Amman
Area: 35,637 sq mi (92,300 sq km)
Population: 6,508,887
Major Language(s): Arabic
Major Religion(s): Sunni Islam
Currency: Jordanian dinar

Jamaica

Official Name: Jamaica
Capital: Kingston
Area: 4,244 sq mi (10,991 sq km)
Population: 2,889,187
Major Language(s): English, English patois
Major Religion(s): Protestant Christianity
Currency: Jamaica dollar

Kazakhstan

Astana

Official Name: Republic of Kazakhstan
Capital: Almaty
Area: 1,049,155 sq mi (2,717,300 sq km)
Population: 17,522,010
Major Language(s): Kazakh, Russian
Major Religion(s): Sunni Islam, Eastern Orthodox Christianity
Currency: tenge

Kenya

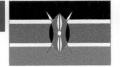

Official Name: Republic of Kenya
Capital: Nairobi
Area: 224,962 sq mi (582,650 sq km)
Population: 43,013,341
Major Language(s): English, Kiswahili
Major Religion(s): Christianity
Currency: Kenyan shilling

Kiribati

Official Name: Republic of Kiribati
Capital: Taraw
Area: 313 sq mi (811 sq km)
Population: 101,998
Major Language(s): I-Kiribati, English
Major Religion(s): Roman Catholic Christianity
Currency: Australian dollar

Korea, North
Choson

Official Name: Democratic People's Republic of Korea
Capital: P'yongyang
Area: 46,541 sq mi (120,540 sq km)
Population: 24,589,122
Major Language(s): Korean
Major Religion(s): Buddhism
Currency: North Korean won

Korea, South
Han'guk

Official Name: Republic of Korea
Capital: Seoul
Area: 38,023 sq mi (98,480 sq km)
Population: 48,860,500
Major Language(s): Korean
Major Religion(s): Buddhism
Currency: South Korean won

Kosovo
Kosova

Official Name: Republic of Kosovo
Capital: Pristina
Area: 4,203 sq mi (10,887 sq km)
Population: 1,836,529
Major Language(s): Albanian, Serbian
Major Religion(s): Sunni Islam
Currency: euro

Kuwait
Al Kuwayt

Official Name: State of Kuwait
Capital: Kuwait
Area: 6,880 sq mi (17,820 sq km)
Population: 2,646,314
Major Language(s): Arabic
Major Religion(s): Sunni Islam
Currency: Kuwaiti dinar

Kyrgyzstan

Official Name: Kyrgyz Republic
Capital: Bishkek
Area: 76,641 sq mi (198,500 sq km)
Population: 5,496,737
Major Language(s): Kyrgyz, Uzbek, Russian
Major Religion(s): Sunni Islam, Eastern Orthodox Christianity
Currency: som

Laos
Pathet Lao

Official Name: Lao People's Democratic Republic
Capital: Vientiane
Area: 91,429 sq mi (236,800 sq km)
Population: 6,586,266
Major Language(s): Lao, French, English
Major Religion(s): Buddhism
Currency: kip

Country and Dependency Profiles

Latvia

Latvija

Official Name: Republic of Latvia
Capital: Riga
Area: 24,938 sq mi (64,589 sq km)
Population: 2,191,580
Major Language(s): Latvian, Russian
Major Religion(s): Protestant Christianity
Currency: lati

Libya

Ar-Libya

Official Name: Libya
Capital: Tripoli
Area: 679,362 sq mi (1,759,540 sq km)
Population: 6,733,620
Major Language(s): Arabic
Major Religion(s): Sunni Islam
Currency: Libyan dinar

Lebanon

Lubnan

Official Name: Lebanese Republic
Capital: Beirut
Area: 4,015 sq mi (10,400 sq km)
Population: 4,140,289
Major Language(s): Arabic, French, English, Armenian
Major Religion(s): Islam, Christianity
Currency: Lebanese pound

Liechtenstein

Official Name: Principality of Liechtenstein
Capital: Vaduz
Area: 62 sq mi (160 sq km)
Population: 36,713
Major Language(s): German
Major Religion(s): Roman Catholic Christianity
Currency: Swiss franc

Lesotho

Official Name: Kingdom of Lesotho
Capital: Maseru
Area: 11,720 sq mi (30,355 sq km)
Population: 1,930,493
Major Language(s): Sesotho, English, Zulu, Xhosa
Major Religion(s): Christianity, Indigenous beliefs
Currency: maloti

Lithuania

Lietuva

Official Name: Republic of Lithuania
Capital: Vilnius
Area: 25,212 sq mi (65,300 sq km)
Population: 3,525,761
Major Language(s): Lithuanian, Russian
Major Religion(s): Roman Catholic Christianity
Currency: litai

Liberia

Official Name: Republic of Liberia
Capital: Monrovia
Area: 43,000 sq mi (111,370 sq km)
Population: 3,887,886
Major Language(s): English, many African languages
Major Religion(s): Christianity, Indigenous beliefs
Currency: Liberian dollar

Luxembourg

Official Name: Grand Duchy of Luxembourg
Capital: Luxembourg
Area: 998 sq mi (2,586 sq km)
Population: 509,074
Major Language(s): Luxembourgish, German, French
Major Religion(s): Roman Catholic Christianity
Currency: euro

Macedonia

Makedonija

Official Name: Republic of Macedonia
Capital: Skopje
Area: 9,781 sq mi (25,333 sq km)
Population: 2,082,370
Major Language(s): Macedonian, Albanian
Major Religion(s): Eastern Orthodox Christianity, Sunni Islam
Currency: Macedonian denar

Maldives

Dhivehi Raajje

Official Name: Republic of Maldives
Capital: Male
Area: 116 sq mi (300 sq km)
Population: 394,451
Major Language(s): Maldivian Dhivehi
Major Religion(s): Sunni Islam
Currency: rufiyaa

Madagascar

Madagascar/Madagasikara

Official Name: Republic of Madagascar
Capital: Antananarivo
Area: 226,657 sq mi (587,040 sq km)
Population: 22,585,517
Major Language(s): English, French, Malagasy
Major Religion(s): Indigenous beliefs, Christianity
Currency: ariary

Mali

Official Name: Republic of Mali
Capital: Bamako
Area: 478,767 sq mi (1,240,000 sq km)
Population: 14,533,511
Major Language(s): Bambara, French, many African languages
Major Religion(s): Sunni Islam
Currency: CFA franc

Malawi

Official Name: Republic of Malawi
Capital: Lilongwe
Area: 45,745 sq mi (118,480 sq km)
Population: 16,323,044
Major Language(s): Chichewa, Chinyanja, other African languages
Major Religion(s): Christianity, Islam
Currency: Malawian kwacha

Malta

Official Name: Republic of Malta
Capital: Valletta
Area: 122 sq mi (316 sq km)
Population: 409,836
Major Language(s): Maltese
Major Religion(s): Roman Catholic Christianity
Currency: euro

Malaysia

Official Name: Malaysia
Capital: Kuala Lumpur
Area: 127,317 sq mi (329,750 sq km)
Population: 29,179,952
Major Language(s): Bahasa Malaysia, Chinese, Tamil
Major Religion(s): Islam, Buddhism
Currency: ringgit

Marshall Islands

Official Name: Republic of the Marshall Islands
Capital: Majuro
Area: 70 sq mi (181 sq km)
Population: 68,480
Major Language(s): Marshallese
Major Religion(s): Protestant Christianity
Currency: U.S. dollar

Country and Dependency Profiles

Mauritania

Muritaniyah

Official Name: Islamic Republic of Mauritania
Capital: Nouakchott
Area: 397,955 sq mi (1,030,700 sq km)
Population: 3,359,185
Major Language(s): Arabic, Pulaar, Soninke, Wolof, French
Major Religion(s): Sunni Islam
Currency: ouguiya

Micronesia

Official Name: Federated States of Micronesia
Capital: Palikir
Area: 271 sq mi (702 sq km)
Population: 106,487
Major Language(s): English
Major Religion(s): Roman Catholic Christianity
Currency: U.S. dollar

Mauritius

Official Name: Republic of Mauritius
Capital: Port Louis
Area: 788 sq mi (2,040 sq km)
Population: 1,313,095
Major Language(s): Creole, Bhojpuri, French
Major Religion(s): Hinduism
Currency: Mauritian rupee

Moldova

Official Name: Republic of Moldova
Capital: Chisinau
Area: 13,067 sq mi (33,843 sq km)
Population: 3,656,843
Major Language(s): Moldovan, Russian
Major Religion(s): Eastern Orthodox Christianity
Currency: lei

Mayotte

(territorial overseas collectivity of France)

Official Name: Territorial Collectivity of Mayotte
Capital: Mamoutzou
Area: 144 sq mi (374 sq km)
Population: 223,765
Major Language(s): Mahorian, French
Major Religion(s): Sunni Islam
Currency: euro

Monaco

Official Name: Principality of Monaco
Capital: Monaco
Area: 1 sq mi (2 sq km)
Population: 30,510
Major Language(s): French, English, Italian, Monegasque
Major Religion(s): Roman Catholic Christianity
Currency: euro

Mexico

Official Name: United Mexican States
Capital: Mexico City
Area: 761,606 sq mi (1,972,550 sq km)
Population: 114,975,406
Major Language(s): Spanish, many Amerindian languages
Major Religion(s): Roman Catholic Christianity
Currency: Mexican peso

Mongolia

Mongol Uls

Official Name: Mongolia
Capital: Ulaanbaatar
Area: 603,909 sq mi (1,564,116 sq km)
Population: 3,179,997
Major Language(s): Khalkha Mongol
Major Religion(s): Buddhism
Currency: tögrög

Montenegro

Crna Gora

Official Name: Montenegro
Capital: Cetinje
Area: 5,415 sq mi (14,026 sq km)
Population: 657,394
Major Language(s): Montenegrin, Serbian
Major Religion(s): Eastern Orthodox Christianity, Sunni Islam
Currency: euro

Montserrat

(overseas territory of the United Kingdom)

Official Name: Montserrat
Capital: Plymouth
Area: 39 sq mi (102 sq km)
Population: 5,164
Major Language(s): English
Major Religion(s): Protestant Christianity
Currency: East Caribbean dollar

Morocco

Al Maghrib

Official Name: Kingdom of Morocco
Capital: Rabat
Area: 172,414 sq mi (446,550 sq km)
Population: 32,309,239
Major Language(s): Arabic, French, Berber dialects
Major Religion(s): Sunni Islam
Currency: Moroccan dirham

Mozambique

Moçambique

Official Name: Republic of Mozambique
Capital: Maputo
Area: 309,496 sq mi (801,590 sq km)
Population: 23,515,934
Major Language(s): Emakhuwa, Xichangana, Portuguese
Major Religion(s): Christianity, Sunni Islam
Currency: metical

Namibia

Official Name: Republic of Namibia
Capital: Windhoek
Area: 318,696 sq mi (825,418 sq km)
Population: 2,165,828
Major Language(s): Afrikaans, English, German
Major Religion(s): Christianity, Indigenous beliefs
Currency: Namibian dollar

Nepal

Official Name: Federal Democratic Republic of Nepal
Capital: Kathmandu
Area: 56,827 sq mi (147,181 sq km)
Population: 29,890,686
Major Language(s): Nepali, Maithali, Bhojpuri
Major Religion(s): Hinduism, Buddhism
Currency: Nepalese rupee

Netherlands

Nederland

Official Name: Kingdom of the Netherlands
Capital: Amsterdam
Area: 16,033 sq mi (41,526 sq km)
Population: 16,730,632
Major Language(s): Dutch, Frisian
Major Religion(s): Christianity
Currency: euro

New Caledonia

Nouvelle-Caledonie
(self-governing territory of France)

Official Name: Territory of New Caledonia and Dependencies
Capital: Noumea
Area: 7,359 sq mi (19,060 sq km)
Population: 260,166
Major Language(s): French, many Melanesian-Polynesian dialects
Major Religion(s): Roman Catholic Christianity
Currency: CFP franc

Country and Dependency Profiles

New Zealand

Official Name: New Zealand
Capital: Wellington
Area: 103,738 sq mi (268,680 sq km)
Population: 4,327,944
Major Language(s): English, Maori
Major Religion(s): Christianity
Currency: New Zealand dollar

Nicaragua

Official Name: Republic of Nicaragua
Capital: Managua
Area: 49,998 sq mi (129,494 sq km)
Population: 5,727,707
Major Language(s): Spanish
Major Religion(s): Roman Catholic Christianity
Currency: cordoba

Niger

Official Name: Republic of Niger
Capital: Niamey
Area: 489,191 sq mi (1,267,000 sq km)
Population: 17,078,839
Major Language(s): French, Hausa, Djerma
Major Religion(s): Sunni Islam
Currency: CFA franc

Nigeria

Official Name: Federal Republic of Nigeria
Capital: Abuja
Area: 356,669 sq mi (923,768 sq km)
Population: 170,123,740
Major Language(s): English, Hausa, Yoruba, Igbo, Fulani
Major Religion(s): Sunni Islam, Christianity
Currency: naira

Niue

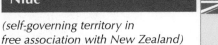

*(self-governing territory in
free association with New Zealand)*

Official Name: Niue
Capital: Alofi
Area: 100 sq mi (260 sq km)
Population: 1,269
Major Language(s): Niuean, English
Major Religion(s): Protestant Christianity
Currency: New Zealand dollar

Norfolk Island

(self-governing territory of Australia)

Official Name: Territory of Norfolk Island
Capital: Kingston
Area: 14 sq mi (35 sq km)
Population: 2,182
Major Language(s): English
Major Religion(s): Protestant Christianity
Currency: Australian dollar

Northern Mariana Islands

(commonwealth of the United States)

Official Name: Northern Mariana Islands
Capital: Saipan
Area: 179 sq mi (464 sq km)
Population: 51,395
Major Language(s): Philippine languages, Chinese
Major Religion(s): Roman Catholic Christianity
Currency: U.S. dollar

Norway

Norge

Official Name: Kingdom of Norway
Capital: Oslo
Area: 125,021 sq mi (323,802 sq km)
Population: 4,707,270
Major Language(s): Norwegian
Major Religion(s): Protestant Christianity
Currency: Norwegian kroner

Oman

Uman

Official Name: Sultanate of Oman
Capital: Muscat
Area: 82,031 sq mi (212,460 sq km)
Population: 3,090,150
Major Language(s): Arabic
Major Religion(s): Sunni Islam
Currency: Omani rial

Papua New Guinea

Papuaniugini

Official Name: Independent State of Papua New Guinea
Capital: Port Moresby
Area: 178,704 sq mi (462,840 sq km)
Population: 6,310,129
Major Language(s): Tok Pisin, English, Hiri Motu
Major Religion(s): Protestant Christianity
Currency: kina

Pakistan

Official Name: Islamic Republic of Pakistan
Capital: Islamabad
Area: 310,403 sq mi (803,940 sq km)
Population: 190,291,129
Major Language(s): Punjabi, Sindhi, Siraiki, Pashtu, Urdu
Major Religion(s): Sunni Islam
Currency: Pakistani rupee

Paraguay

Official Name: Republic of Paraguay
Capital: Asuncion
Area: 157,047 sq mi (406,750 sq km)
Population: 6,541,591
Major Language(s): Spanish, Guarani
Major Religion(s): Roman Catholic Christianity
Currency: guaraní

Palau

Belau

Official Name: Republic of Palau
Capital: Koror
Area: 177 sq mi (458 sq km)
Population: 21,032
Major Language(s): Palauan, Tobi, English
Major Religion(s): Christianity
Currency: U.S. dollar

Peru

Official Name: Republic of Peru
Capital: Lima
Area: 496,226 sq mi (1,285,220 sq km)
Population: 29,549,517
Major Language(s): Spanish, Quechua, Aymara
Major Religion(s): Roman Catholic Christianity
Currency: nuevo sol

Panama

Official Name: Republic of Panama
Capital: Panama
Area: 30,193 sq mi (78,200 sq km)
Population: 3,510,045
Major Language(s): Spanish, English
Major Religion(s): Roman Catholic Christianity
Currency: balboa

Philippines

Pilipinas

Official Name: Republic of the Philippines
Capital: Manila
Area: 115,831 sq mi (300,000 sq km)
Population: 103,775,002
Major Language(s): Filipino, Tagalog, English
Major Religion(s): Roman Catholic Christianity
Currency: Philippine peso

Country and Dependency Profiles

Pitcairn Islands

(overseas territory of the United Kingdom)

Official Name: Pitcairn, Henderson, Ducie, and Oeno Islands
Capital: Adamstown
Area: 18 sq mi (47 sq km)
Population: 48
Major Language(s): English
Major Religion(s): Protestant Christianity
Currency: New Zealand dollar

Qatar

Official Name: State of Qatar
Capital: Doha
Area: 4,416 sq mi (11,437 sq km)
Population: 1,951,591
Major Language(s): Arabic
Major Religion(s): Sunni Islam
Currency: Qatari rial

Poland

Polska

Official Name: Republic of Poland
Capital: Warsaw
Area: 120,726 sq mi (312,679 sq km)
Population: 38,415,284
Major Language(s): Polish
Major Religion(s): Roman Catholic Christianity
Currency: zloty

Romania

Official Name: Romania
Capital: Bucharest
Area: 91,699 sq mi (237,500 sq km)
Population: 21,848,504
Major Language(s): Romanian, Hungarian
Major Religion(s): Eastern Orthodox Christianity
Currency: lei

Portugal

Official Name: Portuguese Republic
Capital: Lisbon
Area: 35,672 sq mi (92,391 sq km)
Population: 10,781,459
Major Language(s): Portuguese
Major Religion(s): Roman Catholic Christianity
Currency: euro

Russia

Rossiya

Official Name: Russian Federation
Capital: Moscow
Area: 6,592,772 sq mi (17,075,200 sq km)
Population: 138,082,178
Major Language(s): Russian, many minority languages
Major Religion(s): Russian Orthodox, Muslim
Currency: Russian ruble

Puerto Rico

(commonwealth of the United States)

Official Name: Commonwealth of Puerto Rico
Capital: San Juan
Area: 5,324 sq mi (13,790 sq km)
Population: 3,998,905
Major Language(s): Spanish, English
Major Religion(s): Roman Catholic Christianity
Currency: U.S. dollar

Rwanda

Official Name: Republic of Rwanda
Capital: Kigali
Area: 10,169 sq mi (26,338 sq km)
Population: 11,689,696
Major Language(s): Kinyarwanda, French, English
Major Religion(s): Roman Catholic Christianity
Currency: Rwandan franc

Saint Barthélemy

Territorial overseas collectivity of France

Official Name: Overseas Collectivity of Saint Barthélemy
Capital: Gustavia
Area: 8 sq mi (21 sq km)
Population: 7,332
Major Language(s): French, English
Major Religion(s): Roman Catholic Christianity
Currency: euro

Saint Martin

Territorial overseas collectivity of France

Official Name: Overseas Collectivity of Saint Martin
Capital: Marigot
Area: 21 sq mi (54 sq km)
Population: 30,959
Major Language(s): French, English
Major Religion(s): Roman Catholic Christianity
Currency: euro

Saint Helena

(overseas territory of the United Kingdom)

Official Name: Saint Helena
Capital: Jamestown
Area: 159 sq mi (413 sq km)
Population: 7,637
Major Language(s): English
Major Religion(s): Protestant Christianity
Currency: Saint Helenanian pound

Saint Pierre and Miquelon

Saint-Pierre et Miquelon (territorial overseas collectivity of France)

Official Name: Territorial Collectivity of Saint Pierre and Miquelon
Capital: Saint-Pierre
Area: 93 sq mi (242 sq km)
Population: 5,831
Major Language(s): French
Major Religion(s): Roman Catholic Christianity
Currency: euro

Saint Kitts and Nevis

Official Name: Federation of Saint Kitts and Nevis
Capital: Basseterre
Area: 101 sq mi (261 sq km)
Population: 50,726
Major Language(s): English
Major Religion(s): Protestant Christianity
Currency: East Caribbean dollar

Saint Vincent and the Grenadines

Official Name: Saint Vincent and the Grenadines
Capital: Kingstown
Area: 150 sq mi (389 sq km)
Population: 103,537
Major Language(s): English, French patois
Major Religion(s): Protestant Christianity
Currency: East Caribbean dollar

Saint Lucia

Official Name: Saint Lucia
Capital: Castries
Area: 238 sq mi (616 sq km)
Population: 162,178
Major Language(s): English, French patois
Major Religion(s): Roman Catholic Christianity
Currency: East Caribbean dollar

Samoa

Official Name: Independent State of Samoa
Capital: Apia
Area: 1,137 sq mi (2,944 sq km)
Population: 194,320
Major Language(s): Samoan, English
Major Religion(s): Protestant Christianity
Currency: tala

Country and Dependency Profiles

San Marino

Official Name: Republic of San Marino
Capital: San Marino
Area: 24 sq mi (61 sq km)
Population: 32,140
Major Language(s): Italian
Major Religion(s): Roman Catholic Christianity
Currency: euro

São Tomé and Príncipe

São Tomé e Príncipe

Official Name: Democratic Republic of São Tomé and Príncipe
Capital: São Tomé
Area: 386 sq mi (1,001 sq km)
Population: 183,176
Major Language(s): Portuguese
Major Religion(s): Roman Catholic Christianity
Currency: dobra

Saudi Arabia

Al Arabiyah as Suudiyah

Official Name: Kingdom of Saudi Arabia
Capital: Riyadh
Area: 830,000 sq mi (2,149,690 sq km)
Population: 26,534,504
Major Language(s): Arabic
Major Religion(s): Sunni Islam
Currency: Saudi riyal

Senegal

Official Name: Republic of Senegal
Capital: Dakar
Area: 75,749 sq mi (196,190 sq km)
Population: 12,969,606
Major Language(s): French, Wolof, Pulaar, Jola, Mandinka
Major Religion(s): Sunni Islam
Currency: CFA franc

Serbia

Srbija

Official Name: Republic of Serbia
Capital: Belgrade
Area: 29,913 sq mi (77,474 sq km)
Population: 7,276,604
Major Language(s): Serbian
Major Religion(s): Eastern Orthodox Christianity
Currency: Serbian dinal

Seychelles

Official Name: Republic of Seychelles
Capital: Victoria
Area: 176 sq mi (455 sq km)
Population: 90,024
Major Language(s): Creole
Major Religion(s): Roman Catholic Christianity
Currency: Seychelles rupee

Sierra Leone

Official Name: Republic of Sierra Leone
Capital: Freetown
Area: 27,699 sq mi (71,740 sq km)
Population: 5,485,998
Major Language(s): Mende, Temne, English, Krio
Major Religion(s): Sunni Islam
Currency: leone

Singapore

Official Name: Republic of Singapore
Capital: Singapore
Area: 268 sq mi (693 sq km)
Population: 5,353,494
Major Language(s): Chinese (Mandarin), English, Malay, Hokkien
Major Religion(s): Buddhism, Sunni Islam
Currency: Singapore dollar

Sint Maarten

Constituent country within the Kingdom of the Netherlands

Official Name: Sint Maarten
Capital: Philipsburg
Area: 13 sq mi (34 sq km)
Population: 37,429
Major Language(s): English, Spanish, Creole
Major Religion(s): Roman Catholic Christianity
Currency: Netherlands Antilles guilder

Slovakia

Slovensko

Official Name: Slovak Republic
Capital: Bratislava
Area: 18,859 sq mi (48,845 sq km)
Population: 5,483,088
Major Language(s): Slovak, Hungarian
Major Religion(s): Roman Catholic Christianity
Currency: euro

Slovenia

Slovenija

Official Name: Republic of Slovenia
Capital: Ljubljana
Area: 7,827 sq mi (20,273 sq km)
Population: 1,996,617
Major Language(s): Slovenian
Major Religion(s): Roman Catholic Christianity
Currency: euro

Solomon Islands

Official Name: Solomon Islands
Capital: Honiara
Area: 10,985 sq mi (28,450 sq km)
Population: 584,578
Major Language(s): English, many indigenous languages
Major Religion(s): Protestant Christianity
Currency: Solomon Islands dollar

Somalia

Soomaaliya

Official Name: Somalia
Capital: Mogadishu
Area: 246,201 sq mi (637,657 sq km)
Population: 10,085,638
Major Language(s): Somali, Arabic, Italian
Major Religion(s): Sunni Islam
Currency: Somali shilling

South Africa

Official Name: Republic of South Africa
Capital: Bloemfontein (judicial), Cape Town (legislative)
Area: 471,011 sq mi (1,219,912 sq km)
Population: 48,810,427
Major Language(s): Zulu, Xhosa, Afrikaans, Sepedi, English, Setswana, Sesotho
Major Religion(s): Protestant Christianity
Currency: rand

South Sudan

Official Name: Republic of South Sudan
Capital: Juba
Area: 24,838 sq mi (64,329 sq km)
Population: 10,625,176
Major Language(s): Arabic, English
Major Religion(s): Christianity, Animist
Currency: South Sudan pound

Spain

España

Official Name: Kingdom of Spain
Capital: Madrid
Area: 194,897 sq mi (504,782 sq km)
Population: 47,042,984
Major Language(s): Spanish, Catalan, Galician, Basque
Major Religion(s): Roman Catholic Christianity
Currency: euro

Country and Dependency Profiles

Sri Lanka

Shri Lamka/Ilankai

Official Name: Democratic Socialist Republic of Sri Lanka
Capital: Colombo
Area: 25,332 sq mi (65,610 sq km)
Population: 21,481,334
Major Language(s): Sinhala, Tamil
Major Religion(s): Buddhism
Currency: Sri Lankan rupee

Sweden

Sverige

Official Name: Kingdom of Sweden
Capital: Stockholm
Area: 173,732 sq mi (449,964 sq km)
Population: 9,103,788
Major Language(s): Swedish
Major Religion(s): Protestant Christianity
Currency: Swedish kronor

Sudan

As-Sudan

Official Name: Republic of the Sudan
Capital: Khartoum
Area: 718,724 sq mi (1,861,484 sq km)
Population: 25,946,220
Major Language(s): Arabic, English
Major Religion(s): Sunni Islam
Currency: Sudanese pound

Switzerland

Schweiz/Suisse/Svizzera/Svizra

Official Name: Swiss Confederation
Capital: Bern
Area: 15,942 sq mi (41,290 sq km)
Population: 7,655,628
Major Language(s): German, French, Italian, Romansch
Major Religion(s): Christianity
Currency: Swiss franc

Suriname

Official Name: Republic of Suriname
Capital: Paramaribo
Area: 63,039 sq mi (163,270 sq km)
Population: 560,157
Major Language(s): Dutch, English
Major Religion(s): Hinduism, Christianity
Currency: Surinamese dollar

Syria

Suriyah

Official Name: Syrian Arab Republic
Capital: Damascus
Area: 71,498 sq mi (185,180 sq km)
Population: 22,530,746
Major Language(s): Arabic
Major Religion(s): Sunni Islam
Currency: Syrian pound

Swaziland

eSwatini

Official Name: Kingdom of Swaziland
Capital: Mbabane
Area: 6,704 sq mi (17,363 sq km)
Population: 1,386,914
Major Language(s): English, siSwati
Major Religion(s): Christianity, Indigenous beliefs
Currency: emalangeni

Taiwan

*T'ai-wan
(unresolved status; has limited
international recognition as the
legitimate representative of China)*

Official Name: Taiwan
Capital: Taipei
Area: 13,892 sq mi (35,980 sq km)
Population: 23,113,901
Major Language(s): Chinese (Mandarin), Taiwanese
Major Religion(s): Buddhism
Currency: New Taiwan dollar

Tajikistan

Tojikiston

Official Name: Republic of Tajikistan
Capital: Dushanbe
Area: 55,251 sq mi (143,100 sq km)
Population: 7,768,385
Major Language(s): Tajik, Russian
Major Religion(s): Sunni Islam
Currency: somoni

Togo

Official Name: Togolese Republic
Capital: Lome
Area: 21,925 sq mi (56,785 sq km)
Population: 6,961,049
Major Language(s): French, many African languages
Major Religion(s): Indigenous beliefs, Christianity
Currency: CFA franc

Tanzania

Official Name: United Republic of Tanzania
Capital: Dodoma
Area: 364,900 sq mi (945,087 sq km)
Population: 43,601,796
Major Language(s): Swahili, English
Major Religion(s): Sunni Islam, Christianity
Currency: Tanzanian shilling

Tonga

Official Name: Kingdom of Tonga
Capital: Nuku'alofa
Area: 289 sq mi (748 sq km)
Population: 106,146
Major Language(s): Tongan, English
Major Religion(s): Protestant Christianity
Currency: pa'anga

Thailand

Prathet Thai

Official Name: Kingdom of Thailand
Capital: Bangkok
Area: 198,457 sq mi (514,000 sq km)
Population: 67,091,089
Major Language(s): Thai
Major Religion(s): Buddhism
Currency: baht

Trinidad and Tobago

Official Name: Republic of Trinidad and Tobago
Capital: Port-of-Spain
Area: 1,980 sq mi (5,128 sq km)
Population: 1,226,383
Major Language(s): English, French, Spanish
Major Religion(s): Roman Catholic Christianity, Hinduism
Currency: Trinidad and Tobago dollar

Timor-Leste

Timor Lorosa'e/Timor-Leste

Official Name: Democratic Republic of Timor-Leste
Capital: Dili
Area: 5,794 sq mi (15,007 sq km)
Population: 1,201,255
Major Language(s): Tetum, Portuguese, Indonesian
Major Religion(s): Roman Catholic Christianity
Currency: U.S. dollar

Tunisia

Tunis

Official Name: Tunisian Republic
Capital: Tunis
Area: 63,170 sq mi (163,610 sq km)
Population: 10,486,339
Major Language(s): Arabic, French
Major Religion(s): Sunni Islam
Currency: Tunisian dinar

Country and Dependency Profiles

Turkey

Turkiye

Official Name: Republic of Turkey
Capital: Ankara
Area: 301,384 sq mi (780,580 sq km)
Population: 79,749,461
Major Language(s): Turkish, Kurdish
Major Religion(s): Sunni Islam
Currency: lira

Uganda

Official Name: Republic of Uganda
Capital: Kampala
Area: 91,136 sq mi (236,040 sq km)
Population: 35,873,253
Major Language(s): English, Ganda, Swahili, other African languages
Major Religion(s): Christianity
Currency: Ugandan shilling

Turkmenistan

Official Name: Turkmenistan
Capital: Ashgabat
Area: 188,456 sq mi (488,100 sq km)
Population: 5,054,828
Major Language(s): Turkmen, Russian, Uzbek
Major Religion(s): Sunni Islam
Currency: manats

Ukraine

Ukrayina

Official Name: Ukraine
Capital: Kyiv
Area: 233,090 sq mi (603,700 sq km)
Population: 44,854,065
Major Language(s): Ukrainian, Russian
Major Religion(s): Eastern Orthodox Christianity
Currency: hryvnia

Turks and Caicos Islands

(overseas territory of the United Kingdom)

Official Name: Turks and Caicos Islands
Capital: Grand Turk
Area: 166 sq mi (430 sq km)
Population: 46,335
Major Language(s): English
Major Religion(s): Protestant Christianity
Currency: U.S. dollar

United Arab Emirates

Al Imarat al Arabiyah al Muttahidah

Official Name: United Arab Emirates
Capital: Abu Dhabi
Area: 32,278 sq mi (83,600 sq km)
Population: 5,314,317
Major Language(s): Arabic
Major Religion(s): Sunni Islam
Currency: Emirati dirham

Tuvalu

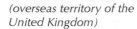

Official Name: Tuvalu
Capital: Funafuti
Area: 10 sq mi (26 sq km)
Population: 10,619
Major Language(s): Tuvaluan, English, Samoan
Major Religion(s): Protestant Christianity
Currency: Tuvaluan dollar

United Kingdom

Official Name: United Kingdom of Great Britain and Northern Ireland
Capital: London
Area: 94,526 sq mi (244,820 sq km)
Population: 63,047,162
Major Language(s): English, Welsh, Scottish Gaelic
Major Religion(s): Christianity
Currency: British pound

United States

Official Name: United States of America
Capital: Washington, D.C.
Area: 3,794,083 sq mi (9,826,630 sq km)
Population: 313,847,465
Major Language(s): English, Spanish
Major Religion(s): Christianity
Currency: U.S. dollar

Venezuela

Official Name: Bolivarian Republic of Venezuela
Capital: Caracas
Area: 352,144 sq mi (912,050 sq km)
Population: 28,047,938
Major Language(s): Spanish, many Amerindian languages
Major Religion(s): Roman Catholic Christianity
Currency: bolivar

Uruguay

Official Name: Oriental Republic of Uruguay
Capital: Montevideo
Area: 68,039 sq mi (176,220 sq km)
Population: 3,316,328
Major Language(s): Spanish
Major Religion(s): Christianity
Currency: Uruguayan peso

Vietnam

Viet Nam

Official Name: Socialist Republic of Vietnam
Capital: Hanoi
Area: 127,244 sq mi (329,560 sq km)
Population: 91,519,289
Major Language(s): Vietnamese, French, Chinese
Major Religion(s): Non-religious beliefs dominate
Currency: dong

Uzbekistan

Ozbekiston

Official Name: Republic of Uzbekistan
Capital: Tashkent
Area: 172,742 sq mi (447,400 sq km)
Population: 28,394,180
Major Language(s): Uzbek, Russian
Major Religion(s): Sunni Islam
Currency: soum

Virgin Islands

(territory of the United States)

Official Name: United States Virgin Islands
Capital: Charlotte Amalie
Area: 737 sq mi (1,910 sq km)
Population: 109,574
Major Language(s): English, Spanish
Major Religion(s): Christianity
Currency: U.S. dollar

Vanuatu

Official Name: Republic of Vanuatu
Capital: Port-Vila
Area: 4,710 sq mi (12,200 sq km)
Population: 227,574
Major Language(s): Many local languages
Major Religion(s): Protestant Christianity
Currency: vatu

Wallis and Futuna

Wallis et Futuna
(overseas collectivity of France)

Official Name: Territory of the Wallis and Futuna Islands
Capital: Mata-Utu
Area: 106 sq mi (274 sq km)
Population: 15,453
Major Language(s): Wallisian, Futunian, French
Major Religion(s): Roman Catholic Christianity
Currency: CFP franc

Country and Dependency Profiles

Yemen

Al Yaman

Official Name: Republic of Yemen
Capital: Sanaa
Area: 203,850 sq mi (527,970 sq km)
Population: 24,771,809
Major Language(s): Arabic
Major Religion(s): Sunni Islam
Currency: Yemeni rial

Zimbabwe

Official Name: Republic of Zimbabwe
Capital: Harare
Area: 150,804 sq mi (390,580 sq km)
Population: 12,619,600
Major Language(s): English, many African (Bantu) languages
Major Religion(s): Christianity, Indigenous beliefs
Currency: Zimbabwean dollar

Zambia

Official Name: Republic of Zambia
Capital: Lusaka
Area: 290,586 sq mi (752,614 sq km)
Population: 14,309,466
Major Language(s): English, many African (Bantu) languages
Major Religion(s): Christianity, Islam
Currency: Zambian kwacha

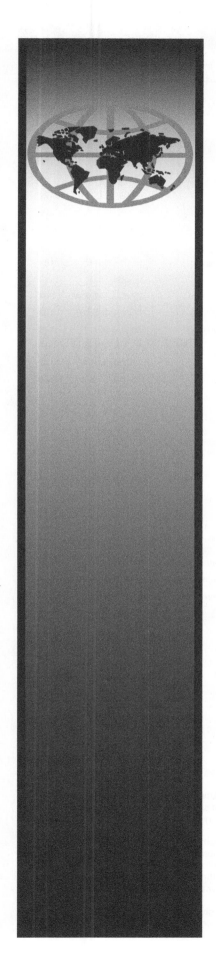

Unit IX

Geographic Index

GEOGRAPHIC INDEX

NAME/DESCRIPTION	LOCATION	PAGE	NAME/DESCRIPTION	LOCATION	PAGE
Abidjan (city, Côte d'Ivoire)	5.3°N, 4.0°W	189	Allahabad (city, India)	25.5°N, 81.9°E	197
Abu Dhabi (nat. capital, United Arab Emirates)	24.5°N, 54.4°E	196	Allier (river, Europe)	47°N, 3.1°E	182
Abuja (nat. capital, Nigeria)	9.2°N, 7.2°E	189	Almaty (city, Kazakhstan)	43.3°N, 77.0°E	195
Acapulco (city, Mexico)	16.9°N, 99.9°W	169	Alps (mountains, Europe)	46.0°N, 7.5°E	176
Accra (nat. capital, Ghana)	5.6°N, 0.2°W	189	Altiplano (plain, South America)	20.0°S, 68.0°W	172
Aconcagua 22,831 (peak, South America)	32.7°S, 70.0°W	172	Amami (island, Japan)	28.3°N, 129.4°E	198
Adana (city, Turkey)	37°N, 35.3°E	196	Amazon (river, South America)	5.0°S, 59.0°W	172
Addis Ababa (nat. capital, Ethiopia)	9.0°N, 38.7°E	189	Amazon Basin (basin, South America)	5.0°S, 65.0°W	172
Adelaide (city, Australia)	34.9°S, 138.6°E	200	Amman (nat. capital, Jordan)	32.0°N, 35.9°E	196
Aden (city, Yemen)	12.8°N, 45.0°E	196	Amritsar (city, India)	31.6°N, 74.9°E	197
Aden, Gulf of (gulf, Indian Ocean)	12.5°N, 48.0°E	196	Amsterdam (nat. capital, Netherlands)	52.4°N, 4.9°E	182
Adriatic Sea (sea, Atlantic Ocean)	42.5°N, 16.0°E	176	Amu Darya (river, Asia)	50.0°N, 65.0°E	195
Aegean Sea (sea, Atlantic Ocean)	38.5°N, 25.0°E	183	Amundsen Sea (sea, Southern Ocean)	72.5°S, 112.0°W	204
Afghanistan (country, Asia)	33.0°N, 65.0°E	196	Amur (river, Asia)	50.0°N, 125.0°E	198
Agra (city, India)	21.2°N, 78.0°E	197	Anatolian Plateau (plateau, Turkey)	39.0°N, 32.0°E	176
Aguascalientes (state capital, Mexico)	22.0°N, 102.4°W	169	Anchorage (city, U.S.)	61.2°N, 149.9°W	168
Aguascalientes (state, Mexico)	21.9°N, 102.3°W	169	Andaman Islands (islands, India)	12.0°N, 92.8°E	197
Ahaggar Mountains (mountains, Algeria)	23.3°N, 5.5°E	185	Andaman Sea (sea, Indian Ocean)	10.0°N, 95.0°E	199
Ahmedabad (city, India)	23.0°N, 72.6°E	197	Andes (mountains, South America)	15.0°S, 73.0°W	172
Ahvaz (city, Iran)	31.3°N, 48.7°E	196	Andorra (country, Europe)	42.5°N, 1.5°E	182
Alabama (river, North America)	31.1°N, 87.9°W	168	Angara (river, Asia)	58.0°N, 102.5°E	195
Alabama (state, U.S.)	33.0°N, 86.7°W	168	Angola (country, Africa)	17.0°S, 12.0°E	190
Alaska (state, U.S.)	65.0°N, 153.0°W	168	Anguilla (territory, U.K.)	18.2°N, 63.0°W	170
Alaska Range (mountains, North America)	63.1°N, 151.0°W	164	Ankara (nat. capital, Turkey)	39.9°N, 32.9°E	196
Alaska, Gulf of (gulf, Pacific Ocean)	57.0°N, 144.0°W	164	Annapolis (state capital, U.S.)	39.0°N, 76.5°W	168
Albania (country, Europe)	41.0°N, 20.0°E	182	Anshan (city, China)	41.1°N, 123.0°E	198
Albany (river, Canada)	52.3°N, 81.5°W	167	Anshun (city, China)	26.3°N, 105.9°E	198
Albany (state capital, U.S.)	42.7°N, 76.8°W	168	Antalya (city, Turkey)	36.9°N, 30.7°E	196
Albert, Lake (lake, Africa)	1.7°N, 30.9°E	185	Antananarivo (nat. capital, Madagascar)	18.9°S, 47.5°E	190
Alberta (province, Canada)	54.5°N, 115.0°W	167	Antarctic Peninsula (peninsula, Antarctica)	69.5°S, 65°W	204
Albuquerque (city, U.S.)	35.1°N, 106.6°W	168	Antigua and Barbuda (country, North America)	17.1°N, 61.9°W	170
Aldan (river, Asia)	60.0°N, 135.0°E	195	Apennines (mountains, Europe)	46.3°N, 12.6°E	176
Aleppo (city, Syria)	36.2°N, 37.2°E	196	Apia (nat. capital, Samoa)	13.8°S, 171.7°W	200
Aleutian Trench (trench, Pacific Ocean)	52.0°N, 172.0°W	4	Appalachian Mountains (mountains, North America)	40.0°N, 78.0°W	164
Alexandria (city, Egypt)	31.2°N, 29.9°E	189	Arabian Sea (sea, Indian Ocean)	15.0°N, 65.0°E	197
Algeria (country, Africa)	28.0°N, 3.0°E	189	Arafura Sea (sea, Pacific Ocean)	9.0°S, 135.0°E	201
Algiers (nat. capital, Algeria)	36.8°N, 2.1°E	189	Araguaia (river, South America)	10.0°S, 50.0°W	172
Al-Hijaz (mountains, Asia)	25.0°N, 42.0°E	192	Aral Sea (lake, Asia)	45.0°N, 60.0°E	195
Alice Springs (city, Australia)	23.7°S, 133.9°E	200			

The geographic index contains approximately 1,600 names of cities, states, countries, rivers, lakes, mountain ranges, oceans, capes, bays, and other geographic features.

The name of each geographical feature in the index is accompanied by a geographical coordinate (latitude and longitude) in degrees and by the page number of the primary map on which the geographical feature appears. Where the geographical coordinates are for specific places or points, such as a city or a mountain peak, the latitude and longitude figures give the location of the map symbol denoting that point. Thus, Los Angeles, California, is at 34°N and 118°W and the location of Mt. Everest is 28°N and 107°E.

The coordinates for political features (countries or states) or physical features (oceans, deserts) that are areas rather than points are given according to the location of the name of the feature on the map, except in those cases where the name of the feature is separated from the feature (such as a country's name appearing over an adjacent ocean area because of space requirements). In such cases, the feature's coordinates will indicate the location of the center of the feature. The coordinates for the Sahara Desert will lead the reader to the place name "Sahara Desert" on the map; the coordinates for North Carolina will show the center location of the state since the name appears over the adjacent Atlantic Ocean. Finally, the coordinates for geographical features that are lines rather than points or areas will also appear near the center of the text identifying the geographical feature.

Alphabetizing follows general conventions; the names of physical features such as lakes, rivers, mountains are given as: proper name, followed by the generic name. Thus "Mount Everest" is listed as "Everest, Mt." Where an article such as "the," "le," "al" appears in a geographic name, the name is alphabetized according to the article. Hence, "La Paz" is found under "L" and not under "P."

GEOGRAPHIC INDEX

GEOGRAPHIC INDEX

NAME/DESCRIPTION	LOCATION	PAGE
Celebes (island, Indonesia)	3.0°N, 122.0°E	201
Celebes Sea (sea, Pacific Ocean)	3.0°N, 122.0°E	199
Central African Republic (country, Africa)	7.0°N, 21.0°E	189
Central Siberian Plateau (plateau, Asia)	68.0°N, 95.0°E	192
Ceram (island, Indonesia)	3.1°S, 129.5°E	199
Ceram Sea (sea, Pacific Ocean)	2.3°S, 128°E	199
Chad (country, Africa)	15.0°N, 19.0°E	189
Chad, Lake (lake, Africa)	13.0°N, 14.0°E	185
Chandigarh (city, India)	30.7°N, 76.8°E	197
Changchun (city, China)	45.9°N, 125.3°E	198
Changde (city, China)	29°N, 111.7°E	198
Changsha (city, China)	28.2°N, 113.0°E	198
Charleston (state capital, U.S.)	38.4°N, 81.6°W	168
Charlotte (city, U.S.)	35.2°N, 80.9°W	168
Charlottetown (prov. capital, Canada)	46.2°N, 63.1°W	167
Chelyabinsk (city, Russia)	55.2°N, 61.4°E	195
Chenab (river, Asia)	31.2°N, 72.2°E	197
Chengdu (city, China)	30.7°N, 104.1°E	198
Chennai (city, India)	35.0°N, 105.0°E	197
Chetumal (state capital, Mexico)	18.5°N, 88.3°W	169
Cheyenne (state capital, U.S.)	41.1°N, 104.8°W	168
Chiapas (state, Mexico)	16.4°N, 92.4°W	169
Chicago (city, U.S.)	41.9°N, 87.6°W	168
Chifeng (city, China)	42.3°N, 118.9°E	198
Chihuahua (state capital, Mexico)	28.6°N, 106.1°W	169
Chihuahua (state, Mexico)	28.8°N, 106.4°W	169
Chihuahuan Desert (desert, North America)	30.5°N, 103.8°W	164
Chile (country, South America)	28.0°S, 70.0°W	171
Chilpancingo (state capital, Mexico)	17.6°N, 99.5°W	169
Chimborazo 20,702 (peak, Ecuador)	1.5°S, 78.8°W	172
China (country, Asia)	35.0°N, 105.0°E	198
Chisinau (nat. capital, Moldova)	47.0°N, 28.8°E	183
Chittagong (city, Bangladesh)	22.4°N, 91.8°E	197
Chongqing (city, China)	29.6°N, 106.6°E	198
Christchurch (city, New Zealand)	43.5°S, 172.6°E	200
Chukchi Sea (sea, Arctic Ocean)	69.0°N, 171.0°W	204
Churchill (city, Canada)	48.8°N, 94.2°W	167
Churchill (river, Canada)	57.8°N, 94.2°W	167
Churchill Lake (lake, Canada)	55.9°N, 108.3°W	167
Cincinnati (city, U.S.)	39.1°N, 84.5°W	168
Ciudad Juárez (city, Mexico)	31.7°N, 106.5°W	169
Ciudad Victoria (state capital, Mexico)	23.7°N, 99.1°W	169
Cleveland (city, U.S.)	41.5°N, 81.7°W	168
Coahuila (state, Mexico)	27.3°N, 102.1°W	169
Coast Mountains (mountains, North America)	51.4°N, 125.3°W	164
Coast Ranges (mountains, North America)	38.3°N, 122.6°W	164
Coastal Plain (plain, North America)	32.0°N, 90.0°W	164
Coats Land (region, Antarctica)	77.0°S, 27.5°W	204

NAME/DESCRIPTION	LOCATION	PAGE
Cod, Cape (cape, U.S.)	41.7°N, 70.3°W	164
Coimbatore (city, India)	11.3°N, 77°E	197
Colima (state capital, Mexico)	19.2°N, 103.7°W	169
Colima (state, Mexico)	19.1°N, 104.0°W	169
Cologne (city, Germany)	51.0°N, 7.0°E	182
Colombia (country, South America)	5.0°N, 73.0°W	171
Colombo (nat. capital, Sri Lanka)	6.9°N, 79.9°E	197
Colorado (river, North America)	39.0°N, 105.5°W	168
Colorado (river, South America)	36.2°S, 70.4°W	172
Colorado (state, U.S.)	39.0°N, 105.5°W	168
Colorado Plateau (plateau, U.S.)	37.0°N, 110.0°W	164
Columbia (river, North America)	46.2°N, 124.1°W	168
Columbia (state capital, U.S.)	34.0°N, 81.1°W	168
Columbus (state capital, U.S.)	40.0°N, 83.0°W	168
Comoros (country, Africa)	12.2°S, 44.3°E	190
Conakry (nat. capital, Guinea)	9.5°N, 13.7°W	189
Conception, Point (cape, U.S.)	34.5°N, 120.5°W	164
Concord (state capital, U.S.)	43.2°N, 71.5°W	168
Congo (river, Africa)	1.0°N, 24.0°E	190
Congo, Democratic Republic of (country, Africa)	5.0°S, 25.0°E	190
Congo, Republic of (country, Africa)	0.7°S, 15.0°E	190
Connecticut (state, U.S.)	41.6°N, 72.7°W	168
Cook Strait (strait, New Zealand)	41.2°S, 174.5°E	201
Copenhagen (nat. capital, Denmark)	55.7°N, 12.6°E	182
Coral Sea (sea, Pacific Ocean)	15.0°S, 150.0°E	201
Cordoba (city, Argentina)	31.4°S, 63.2°W	171
Corsica (island, France)	0.9°N, 9.3°E	182
Costa Rica (country, North America)	10.0°N, 84.1°W	169
Côte d'Ivoire (country, Africa)	8.0°N, 5.0°W	189
Crete (island, Greece)	35.3°N, 24.8°E	183
Croatia (country, Europe)	45.2°N, 15.5°E	182
Cuando (river, Africa)	18°S, 24.3°E	190
Cuanza (river, Africa)	9.4°S, 13.2°E	190
Cuba (country, North America)	22.0°N, 79.5°W	170
Cuernavaca (state capital, Mexico)	18.9°N, 99.2°W	169
Cuiabá (city, Brazil)	15.6°S, 56.1°W	171
Culiacan (state capital, Mexico)	24.8°N, 107.4°W	169
Cunene (river, Africa)	17.3°S, 11.8°E	190
Curaçao (territory, Netherlands)	12.2°N, 69°W	170
Curitiba (city, Brazil)	25.4°S, 49.3°W	171
Cuyo (island, Philippines)	10.9°N, 121°E	199
Cyprus (country, Europe)	35.0°N, 33.0°E	196
Czechia (country, Europe)	49.8°N, 15.5°E	182
Dadu (river, China)	29.6°N, 103.8°E	198
Daegu (city, South Korea)	35.9°N, 128.6°E	198
Dakar (nat. capital, Senegal)	14.7°N, 17.4°W	189
Dalian (city, China)	38.9°N, 121.6°E	198
Dallas (city, U.S.)	32.8°N, 96.8°W	168

GEOGRAPHIC INDEX

GEOGRAPHIC INDEX

NAME/DESCRIPTION	LOCATION	PAGE	NAME/DESCRIPTION	LOCATION	PAGE
Fredericton (prov. capital, Canada)	46.0°N, 66.7°W	167	Great Victoria Desert (desert, Australia)	29.0°S, 129.0°E	201
Freetown (nat. capital, Sierra Leone)	8.5°N, 13.3°W	189	Greater Kingan Range (mountains, Asia)	49.4°N, 123.2°E	192
French Guiana (department, France)	4.0°N, 53.0°W	171	Greece (country, Europe)	38.0°N, 22.0°E	182
Fuji, Mt. 12,388 (peak, Japan)	35.4°N, 138.7°E	192	Green (river, North America)	35.2°N, 109.9°W	168
Fukuoka (city, China)	33.6°N, 130.4°E	198	Greenland (island, North America)	72.0°N, 40.0°W	164
Funafuti (nat. capital, Tuvalu)	8.5°S, 179.2°E	200	Greenland Sea (sea, Atlantic Ocean)	79.0°N, 5.0°W	204
Fundy, Bay of (bay, Atlantic Ocean)	45°N, 65.8°W	167	Grenada (country, North America)	12.1°N, 61.8°W	170
Fushun (city, China)	41.9°N, 123.9°E	198	Grootfontein (city, Namibia)	19.6°S, 18.1°E	190
Fuzhou (city, China)	26.1°N, 119.3°E	198	Guadalajara (state capital, Mexico)	20.7°N, 103.4°W	169
Gabon (country, Africa)	1.0°S, 11.8°E	189	Guadeloupe (territory, France)	16.3°N, 61.6°W	170
Gaborone (nat. capital, Botswana)	24.8°S, 25.9°E	190	Guadiana (river, Europe)	37.2°N, 7.4°W	182
Galapagos Islands (islands, Ecuador)	0.7°S, 90.6°W	172	Guam (territory, U.S.)	13.5°N, 144.8°E	199
Gambia (country, Africa)	13.5°N, 15.5°W	189	Guanajuato (state capital, Mexico)	21.0°N, 101.3°W	169
Gan (river, China)	29.8°N, 116.2°E	198	Guanajuato (state, Mexico)	21.0°N, 101.3°W	169
Ganges (river, Asia)	25.0°N, 81.0°E	197	Guangzhou (city, China)	23.2°N, 113.4°E	198
Garonne (river, Europe)	45°N, 0.6°W	182	Guatemala (country, North America)	14.6°N, 90.5°W	169
Gaza (city, Palestinian Territories)	31.5°N, 34.5°E	196	Guatemala (nat. capital, Guatemala)	14.6°N, 90.5°W	169
Gaziantep (city, Turkey)	37.1°N, 37.4°E	196	Guayaquil (city, Ecuador)	2.2°S, 79.9°W	171
Geographe Bay (bay, Australia)	33.6°S, 115.3°E	201	Guerrero (state, Mexico)	17.6°N, 100.0°W	169
Georgetown (nat. capital, Guyana)	6.8°N, 58.2°W	171	Guiana Shield (plateau, South America)	4.0°N, 62.0°W	172
Georgia (country, Asia)	42.0°N, 43.5°E	195	Guilin (city, China)	25.3°N, 110.3°E	198
Georgia (state, U.S.)	33.0°N, 83.5°W	168	Guinea (country, Africa)	11.0°N, 10.0°W	189
Germany (country, Europe)	51.0°N, 9.0°E	182	Guinea, Gulf of (gulf, Atlantic Ocean)	3.0°N, 2.5°E	189
Ghana (country, Africa)	8.0°N, 2.0°W	189	Guinea-Bissau (country, Africa)	12.0°N, 15.0°W	189
Ghardaia (city, Algeria)	32.5°N, 3.7°E	189	Guiyang (city, China)	26.7°N, 106.6°E	198
Gibraltar, Strait of (strait, Atlantic Ocean)	36.0°N, 5.6°W	176	Gujranwala (city, Pakistan)	32.2°N, 74.2°E	197
Gibson Desert (desert, Australia)	23.0°S, 125.0°E	201	Guwahati (city, India)	26.2°N, 91.7°E	197
Gila (river, North America)	32.7°N, 114.6°W	168	Guyana (country, South America)	5.0°N, 59.0°W	171
Gilbert Islands (islands, Oceania)	1.5°N, 170.0°W	5	Gwalior (city, India)	26.2°N, 78.2°E	197
Glasgow (city, U.K.)	55.9°N, 4.3°E	182	Gwangju (city, South Korea)	35.2°N, 126.9°E	198
Gobi Desert (desert, Asia)	42.5°N, 107.0°E	192	Haifa (city, Israel)	32.8°N, 35°E	196
Godavari (river, Asia)	20.0°N, 77.0°E	197	Haikou (city, China)	20.1°N, 110.3°E	198
Goiánia (city, Brazil)	16.7°S, 49.3°W	171	Hainan (island, China)	19.0°N, 109.5°E	198
Gotland (island, Sweden)	57.5°N, 18.6°E	182	Haiphong (city, Vietnam)	20.9°N, 106.7°E	199
Gran Chaco (plain, South America)	24.0°S, 60.0°W	172	Haiti (country, North America)	18.5°N, 72.0°W	170
Grand Erg Occidental (desert, Africa)	30.7°N, 0.0°W	185	Halifax (prov. capital, Canada)	44.7°N, 63.6°W	167
Grand Erg Oriental (desert, Africa)	29.0°N, 8.0°E	185	Halmahera (island, Indonesia)	0.6°N, 127.9°E	199
Great Artesian Basin (basin Australia)	25.0°S, 144.0°E	201	Hamburg (city, Germany)	53.6°N, 10.0°E	182
Great Australian Bight (bay, Indian Ocean)	35.0°S, 130.0°E	201	Han (river, China)	30.6°N, 114.3°E	198
Great Barrier Reef (reef Australia)	18.0°S, 148.0°E	201	Handan (city, China)	36.6°N, 114.5°E	198
Great Basin (basin, North America)	40.7°N, 117.7°W	164	Hangzhou (city, China)	30.3°N, 120.2°E	198
Great Bear Lake (lake, Canada)	66.0°N, 121.0°W	167	Hanoi (nat. capital, Vietnam)	21.0°N, 105.9°E	199
Great Dividing Range (mountains, Australia)	31.0°S, 151.0°E	201	Harare (nat. capital, Zimbabwe)	17.8°S, 31.1°E	190
Great Indian Desert (desert, Asia)	27.0°N, 71.0°E	192	Harbin (city, China)	45.8°N, 126.6°E	198
Great Plains (plains, North America)	37.0°N, 97.0°W	164	Harrisburg (state capital, U.S.)	40.3°N, 76.9°W	168
Great Salt Lake (lake, U.S.)	41.2°N, 112.6°W	168	Hartford (state capital, U.S.)	41.8°N, 72.7°W	168
Great Sandy Desert (desert, Australia)	20.0°S, 125.0°E	201	Hatteras, Cape (cape, U.S.)	35.3°N, 75.5°W	164
Great Slave Lake (lake, Canada)	61.7°N, 114.0°W	167	Havana (nat. capital, Cuba)	23.1°N, 82.4°W	170

GEOGRAPHIC INDEX

GEOGRAPHIC INDEX

NAME/DESCRIPTION	LOCATION	PAGE
Jodhpur (city, India)	26.3°N, 73.0°E	197
Johannesburg (city, South Africa)	26.2°S, 28.1°E	190
Jordan (country, Asia)	31.0°N, 36.0°E	196
Juba (nat. capital, South Sudan)	4.9°N, 31.6°E	189
Jujuy (city, Argentina)	24.2°S, 65.3°W	171
Juneau (state capital, U.S.)	58.5°N, 134.2°W	168
K2 28,251 (peak, Asia)	35.9°N, 76.5°E	192
Kabul (nat. capital, Afghanistan)	34.5°N, 69.2°E	196
Kaduna (city, Nigeria)	10.6°N, 7.5°E	189
Kafue (river, Africa)	16°S, 28°E	190
Kalahari Desert (desert, Africa)	24.5°S, 21.0°E	185
Kamchatka Peninsula (peninsula, Asia)	56.0°N, 160.0°E	192
Kampala (nat. capital, Uganda)	0.3°N, 32.6°E	189
Kananga (city, D.R. Congo)	5.9°S, 22.5°E	190
Kangaroo Island (island, Australia)	35.8°S, 137.3°E	201
Kano (city, Nigeria)	12.0°N, 8.5°E	189
Kanpur (city, India)	26.5°N, 80.3°E	197
Kansas (state, U.S.)	38.5°N, 98.0°W	168
Kansas City (city, U.S.)	39.1°N, 94.6°W	168
Kaohsiung (city, Taiwan)	22.6°N, 120.3°E	198
Kapuas (river, Indonesia)	0.1°N, 110.0°E	199
Kara Sea (sea, Arctic Ocean)	76.0°N, 80.0°E	204
Karachi (city, Pakistan)	24.9°N, 67.1°E	197
Karaj (city, Iran)	35.8°N, 51°E	196
Kasai (river, Africa)	4.0°S, 19.6°E	190
Kathmandu (nat. capital, Nepal)	27.7°N, 85.3°E	197
Kattegat (strait, Atlantic Ocean)	57.0°N, 11.0°E	176
Kaua'i (island, U.S.)	22.1°N, 159.5°W	168
Kayan (river, Indonesia)	2.5°N, 116.0°E	199
Kazakhstan (country, Asia)	48.0°N, 68.0°E	195
Kazan (city, Russia)	55.8°N, 49.2°E	195
Kentucky (state, U.S.)	37.5°N, 85.0°W	168
Kenya (country, Africa)	1.0°N, 38.0°E	189
Kenya, Mt. 17,058 (peak, Kenya)	0.2°S, 37.3°E	185
Kerguelen Islands (islands, Indian Ocean)	49.3°S, 69.0°E	5
Kermandec Trench (trench, Pacific Ocean)	30.0°S, 178.0°E	5
Kermanshah (city, Iran)	34.3°N, 47.1°E	196
Kharkiv (city, Ukraine)	48.9°N, 36.3°E	183
Khartoum (nat. capital, Sudan)	15.6°N, 32.5°E	189
Kheta (river, Asia)	71.9°N, 102.1°E	195
Khulna (city, Bangladesh)	22.8°N, 89.6°E	197
Kiev (nat. capital, Ukraine)	50.4°N, 30.5°E	183
Kigali (nat. capital, Rwanda)	2.0°S, 30.1°E	190
Kilamanjaro, Mt. 19,340 (peak, Tanzania)	3.1°S, 37.4°E	185
Kingston (nat. capital, Jamaica)	18.0°N, 76.8°W	170
Kingstown (nat. capital, St. Vincent & the Grenadines)	13.2°N, 61.2°W	170
Kinshasa (nat. capital, D.R. Congo)	4.3°S, 15.3°E	190

NAME/DESCRIPTION	LOCATION	PAGE
Kiribati (country, Oceania)	1.0°N, 173.0°E	200
Kochi (city, India)	10.0°N, 76.3°E	197
Kodiak Island (island, North America)	57.8°N, 152.4°W	164
Kola Peninsula (peninsula, Europe)	67.7°N, 36°E	176
Kolkata (city, India)	22.5°N, 88.4°E	197
Kolyma (river, Asia)	67.5°N, 156.3°E	195
Konqi (river, China)	42.1°N, 86.6°E	198
Konya (city, Turkey)	37.9°N, 32.5°E	196
Korea Strait (strait, Pacific Ocean)	34.0°N, 129.0°E	198
Kosciuszko, Mt. 7,310 (peak, Australia)	36.5°S, 15.3°E	201
Kosovo (country, Europe)	42.5°N, 21.0°E	182
Kota (city, India)	25.2°N, 75.8°E	197
Kotto (river, Africa)	6.8°N, 22.5°E	189
Kozhikode (city, India)	11.3°N, 75.8°E	197
Kraków (city, Poland)	50.1°N, 19.9°E	182
Krasnoyarsk (city, Russia)	56.0°N, 93.1°E	195
Krishna (river, Asia)	16°N, 81°E	197
Kuala Lumpur (nat. capital, Malaysia)	3.2°N, 101.7°E	199
Kuban (river, Russia)	45.3°N, 37.4°E	183
Kunlun Shan (mountains, Asia)	36.0°N, 84.0°E	192
Kunming (city, China)	25.0°N, 102.7°E	198
Kuril Islands (islands, Asia)	46.2°N, 152.0°E	192
Kurile Trench (trench, Pacific Ocean)	50.0°N, 160.0°E	5
Kuskokwim (river, North America)	60.1°N, 162.3°W	168
Kuwait (country, Asia)	29.5°N, 45.8°E	196
Kuwait (nat. capital, Kuwait)	29.4°N, 48°E	196
Kwango (river, Africa)	3.3°S, 17.4°E	190
Kyoto (city, Japan)	35°N, 135.8°E	198
Kyrgyzstan (country, Asia)	41.0°N, 75.0°E	195
Kyushu (island, Japan)	33.0°N, 131.0°E	198
La Paz (nat. capital, Bolivia)	16.5°S, 68.2°W	171
La Paz (state capital, Mexico)	24.1°N, 110.3°W	169
La Plata (city, Argentina)	34.9°S, 58.0°W	171
Labrador Sea (sea, Atlantic Ocean)	60.0°N, 55.0°W	167
Ladoga, Lake (lake, Russia)	61.0°N, 31.5°E	195
Lagos (city, Nigeria)	6.5°N, 3.4°E	189
Lahore (city, Pakistan)	31.6°N, 74.4°E	197
Lake of the Woods (lake, North America)	49.3°N, 94.8°W	168
Lanai (island, U.S.)	20.8°N, 156.9°W	168
Lansing (state capital, U.S.)	42.7°N, 84.6°W	168
Lanzhou (city, China)	36.1°N, 103.8°E	198
Laos (country, Asia)	18.0°N, 105.0°E	199
Lapland (region, Europe)	65.9°N, 16.8°E	176
Laptev Sea (sea, Arctic Ocean)	76.0°N, 126.0°E	204
Las Vegas (city, U.S.)	36.2°N, 115.1°W	168
Latvia (country, Europe)	57.0°N, 25.0°E	182
Laurentian Highlands (plateau, Canada)	50.0°N, 70.0°W	167
Lebanon (country, Asia)	33.8°N, 36.8°E	196

GEOGRAPHIC INDEX

GEOGRAPHIC INDEX

NAME/DESCRIPTION	LOCATION	PAGE
Mato Grosso, Planalto do (plateau, Brazil)	15.0°S, 52.0°W	172
Maui (island, U.S.)	20.8°N, 156.3°W	168
Mauritania (country, Africa)	20.0°N, 12.0°W	189
Mauritius (country, Indian Ocean)	20.2°S, 57.5°E	5
Mbabane (nat. capital, Swaziland)	26.3°S, 31.1°E	190
Mbuji-Mayi (city, D.R. Congo)	6.2°S, 23.6°E	190
McClintock Channel (channel, Atlantic Ocean)	72°N, 102°W	167
McKinley, Mt. 20,327 (peak, U.S.)	63.1°N, 151.0°W	164
McMurdo Station (research station, Antarctica)	77.9°S, 166.7°E	204
Mecca (city, Saudi Arabia)	21.5°N, 39.8°E	196
Medan (city, Indonesia)	3.6°N, 98.7°E	199
Medellin (city, Colombia)	6.2°N, 75.6°W	171
Medina (city, Saudi Arabia)	24.5°N, 39.6°E	196
Mediterranean Sea (sea, Atlantic Ocean)	36.0°N, 15.0°E	176
Meerut (city, India)	30.0°N, 77.7°E	197
Mekong (river, Asia)	22.0°N, 100.0°E	199
Melbourne (city, Australia)	37.8°S, 145.0°E	200
Melekeok (nat. capital, Palau)	7.5°N, 134.6°E	200
Memphis (city, U.S.)	35.1°N, 90.0°W	168
Mendocino, Cape (cape, U.S.)	40.4°N, 124.4°W	164
Mendoza (city, Argentina)	32.9°S, 68.8°W	171
Merida (state capital, Mexico)	21.0°N, 89.6°W	169
Meuse (river, Europe)	50.5°N, 5.0°E	182
Mexicali (state capital, Mexico)	32.7°N, 115.5°W	169
Mexico (country, North America)	20.0°N, 100.0°W	163
Mexico (state, Mexico)	19.4°N, 99.6°W	169
Mexico City (nat. capital, Mexico)	19.4°N, 99.2°W	169
Mexico, Gulf of (gulf, Atlantic Ocean)	25.0°N, 90.0°W	168
Miami (city, U.S.)	25.8°N, 80.2°W	168
Michigan (state, U.S.)	44.3°N, 85.6°W	168
Michigan, Lake (lake, North America)	43.5°N, 87.5°W	168
Michoacán (state, Mexico)	19.2°N, 101.9°W	169
Micronesia, Federated States of (country, Oceania)	7.0°N, 158.0°E	200
Mid-Atlantic Ridge (ridge, Atlantic Ocean)	10.0°N, 40.0°W	4
Middle America Trench (trench, Pacific Ocean)	15.0°N, 100.0°W	4
Mid-Indian Ridge (ridge, Indian Ocean)	5.0°S, 65.0°E	5
Milan (city, Italy)	45.5°N, 9.2°E	182
Milwaukee (city, U.S.)	43.1°N, 88.0°W	168
Mindanao (island, Philippines)	8.0°N, 125.0°E	201
Mindanao (island, Philippoines)	8°N, 125°E	199
Mindoro (island, Philippines)	12.9°N, 121.1°E	199
Minneapolis (city, U.S.)	45.0°N, 93.3°W	168
Minnesota (state, U.S.)	46.0°N, 94.0°W	168
Minsk (nat. capital, Belarus)	53.9°N, 27.6°E	183
Mississippi (river, North America)	29.2°N, 89.3°W	168
Mississippi (state, U.S.)	33.0°N, 90.0°W	168
Missouri (river, North America)	38.8°N, 90.1°W	168

NAME/DESCRIPTION	LOCATION	PAGE
Missouri (state, U.S.)	38.5°N, 92.5°W	168
Mistassini, Lake (lake, Canada)	51°N, 73.7°W	167
Mitchell River (river, Australia)	15.2°S, 141.6°E	201
Mogadishu (nat. capital, Somalia)	2.1°N, 45.4°E	189
Moldova (country, Europe)	49.0°N, 29.0°E	183
Molokai (island, U.S.)	21.1°N, 157.0°W	168
Molucca Sea (sea, Pacific Ocean)	0.4°S, 125.4°E	199
Monaco (country, Europe)	43.7°N, 7.4°E	182
Mongol Altai (mountains, Asia)	49.0°N, 89.0°E	192
Mongolia (country, Asia)	46.0°N, 105.0°E	198
Monrovia (nat. capital, Liberia)	6.3°N, 10.8°W	189
Montana (state, U.S.)	47.0°N, 110.0°W	168
Montenegro (country, Europe)	42.8°N, 19.5°E	182
Monterrey (state capital, Mexico)	25.7°N, 100.3°W	169
Montevideo (nat. capital, Uruguay)	34.9°S, 56.2°W	171
Montgomery (state capital, U.S.)	32.4°N, 86.3°W	168
Montpelier (state capital, U.S.)	44.3°N, 72.6°W	168
Montréal (city, Canada)	45.5°N, 73.6°W	167
Montserrat (territory, U.K.)	16.8°N, 62.2°W	170
Morelia (state capital, Mexico)	19.8°N, 101.2°W	169
Morelos (state, Mexico)	18.8°N, 99.1°W	169
Morocco (country, Africa)	32.0°N, 5.0°W	189
Moroni (nat. capital, Comoros)	11.7°S, 43.3°E	184
Moscow (nat. capital, Russia)	55.8°N, 37.6°E	195
Mosel (river, Europe)	50.4°N, 7.6°E	182
Mosul (city, Iraq)	36.3°N, 43.1°E	196
Mozambique (country, Africa)	18.3°S, 35.0°E	190
Mozambique Channel (channel, Indian Ocean)	19.0°S, 41.0°E	185
Multan (city, Pakistan)	30.2°N, 71.5°E	197
Mumbai (city, India)	19.0°N, 72.8°E	197
Munich (city, Germany)	48.1°N, 11.6°E	182
Murmansk (city, Russia)	69°N, 33.1°E	195
Murray (river, Australia)	36.0°S, 145.0°E	201
Muscat (nat. capital, Oman)	23.6°N, 58.6°E	196
Myanmar (Burma) (country, Asia)	22.0°N, 98.0°E	199
Mysore (city, India)	12.3°N, 76.7°E	197
Nagoya (city, Japan)	35.2°N, 136.9°E	198
Nagpur (city, India)	21.2°N, 79.1°E	197
Nairobi (nat. capital, Kenya)	1.3°S, 36.8°E	189
Namib Desert (desert, Africa)	24.0°S, 15.0°E	185
Namibia (country, Africa)	22.0°S, 17.0°E	190
Nampo (city, North Korea)	38.7°N, 125.4°E	198
Nanchang (city, China)	28.7°N, 115.9°E	198
Nanjing (city, China)	32.1°N, 118.8°E	198
Nanning (city, China)	22.8°N, 108.3°E	198
Nantong (city, China)	32°N, 120.9°E	198
Nanyang (city, China)	33.0°N, 112.5°E	198
Naples (city, Italy)	40.9°N, 14.3°E	182

GEOGRAPHIC INDEX

GEOGRAPHIC INDEX

NAME/DESCRIPTION	LOCATION	PAGE
Ohio (state, U.S.)	40.5°N, 82.5°W	168
Oka (river, Russia)	56.3°N, 44°E	183
Okavango (river, Africa)	19°S, 22.5°E	190
Okeechobee, Lake (lake, U.S.)	26.9°N, 80.8°W	168
Okhotsk, Sea of (sea, Pacific Ocean)	53.0°N, 150.0°E	195
Okinawa (island, Japan)	26.5°N, 127.9°E	198
Oklahoma (state, U.S.)	65.5°N, 98.0°W	168
Oklahoma City (state capital, U.S.)	35.5°N, 97.5°W	168
Olenyok (river, Asia)	70.5°N, 121.0°E	195
Olympia (state capital, U.S.)	47.0°N, 122.9°W	168
Omaha (city, U.S.)	41.3°N, 96.0°W	168
Oman (country, Asia)	21.0°N, 57.0°E	196
Oman, Gulf of (gulf, Indian Ocean)	24.5°N, 58.5°E	196
Omsk (city, Russia)	55.0°N, 73.4°E	195
Onega, Lake (lake, Russia)	61.7°N, 35.7°E	195
Ontario (province, Canada)	50.7°N, 86.1°W	167
Ontario, Lake (lake, North America)	43.5°N, 78.0°W	168
Oran (city, Algeria)	35.7°N, 0.6°W	189
Orange (river, Africa)	29.6°S, 22.7°E	190
Ord River (river, Australia)	15.5°S, 128.4°E	201
Oregon (state, U.S.)	44.0°N, 120.5°W	168
Orinoco (river, South America)	8.0°N, 63.0°W	172
Orizaba 18,491 (peak, Mexico)	19.0°N, 97.3°W	164
Orlando (city, U.S.)	28.5°N, 81.4°W	168
Osaka (city, Japan)	34.7°N, 135.5°E	198
Oslo (nat. capital, Norway)	59.9°N, 10.8°E	182
Ostrov Vrangelya (islands, Russia)	71.2°N, 179.4°W	204
Ottawa (nat. capital, Canada)	45.4°N, 75.7°W	167
Ottawa (river, Canada)	45.4°N, 75.7°W	167
Ouagadougou (nat. capital, Burkina Faso)	12.4°N, 1.5°W	189
Ozark Plateau (plateau, U.S.)	37.2°N, 92.5°W	164
P'yongyang (nat. capital, North Korea)	39.0°N, 125.8°E	198
Pachuca (state capital, Mexico)	20.1°N, 98.7°W	169
Padang (city, Indonesia)	1°S, 100.4°E	199
Pakistan (country, Asia)	30.0°N, 70.0°E	197
Palana (city, Russia)	59.1°N, 159.9°E	195
Palau (country, Oceania)	7.5°N, 134.5°E	200
Palawan (island, Philippines)	10.1°N, 118.9°E	199
Palembang (city, Indonesia)	3.0°S, 104.8°E	199
Palermo (city, Italy)	38.1°N, 13.4°E	182
Palikir (nat. capital, Micronesia)	6.9°N, 158.1°E	200
Palk Strait (strait, Indian Ocean)	10°N, 79.8°E	197
Pamir (river, Asia)	37°N, 72.7°E	197
Pampas (plain, South America)	35.0°S, 63.0°W	172
Panama (country, North America)	9.0°N, 79.5°W	169
Panama Canal (canal, Panama)	9.1°N, 79.7°W	169
Panama, Isthmus of (isthmus, North America)	9.0°N, 80.0°W	169
Panama City (nat. capital, Panama)	9.0°N, 79.5°W	169

NAME/DESCRIPTION	LOCATION	PAGE
Pangani (river, Africa)	5.4°S, 39°E	190
Papua New Guinea (country, Asia)	6°S, 147°E	199
Papua New Guinea (country, Oceania)	6.0°S, 144.0°E	200
Papua, Gulf of (gulf, Papua New Guinea)	9.0°S, 145.0°E	199
Paraguay (country, South America)	23.0°S, 58.0°W	171
Paraguay (river, South America)	23.0°S, 61.0°W	172
Paramaribo (nat. capital, Suriname)	5.8°N, 55.2°W	171
Paraná (river, South America)	21.0°S, 52.0°W	172
Paris (nat. capital, France)	48.9°N, 2.3°E	182
Patagonia (region, South America)	48.0°S, 61.0°W	172
Patna (city, India)	25.6°N, 85.1°E	197
Peace (river, Canada)	59.0°N, 111.4°W	167
Pechora (river, Asia)	65.5°N, 52.0°E	195
Pecos (river, North America)	29.7°N, 101.4°W	168
Pennsylvania (state, U.S.)	41.0°N, 77.5°W	168
Perm (city, Russia)	58.0°N, 56.3°E	195
Persian Gulf (gulf, Indian Ocean)	27.0°N, 51.0°E	196
Perth (city, Australia)	31.9°S, 115.8°E	200
Peru (country, South America)	8.0°S, 76.0°W	171
Peru-Chile Trench (trench, Pacific Ocean)	23.0°S, 73.0°W	4
Peshawar (city, Pakistan)	34°N, 71.6°E	197
Philadelphia (city, U.S.)	40.0°N, 75.2°W	168
Philippine Sea (sea, Pacific Ocean)	20.0°N, 134.0°E	201
Philippines (country, Asia)	13.0°N, 122.0°E	199
Phnom Penh (nat. capital, Cambodia)	11.6°N, 104.9°E	199
Phoenix (state capital, U.S.)	33.5°N, 112.1°W	168
Pierre (state capital, U.S.)	44.4°N, 100.3°W	168
Pittsburgh (city, U.S.)	40.4°N, 80.0°W	168
Plateau of Tibet (plateau, Asia)	33°N, 88°E	192
Platte (river, North America)	41.1°N, 95.9°W	168
Plenty, Bay of (bay, New Zealand)	37.3°S, 176.8°E	201
Po (river, Europe)	45.0°N, 10.0°E	182
Podgorica (nat. capital, Montenegro)	42.4°N, 19.3°E	182
Poland (country, Europe)	52.0°N, 20.0°E	182
Port Elizabeth (city, South Africa)	34.0°S, 25.6°E	190
Port Moresby (nat. capital, Papua New Guinea)	9.5°S, 147.2°E	200
Port Sudan (city, Sudan)	19.6°N, 37.2°E	189
Port-au-Prince (nat. capital, Haiti)	18.5°N, 72.3°W	170
Portland (city, U.S.)	45.5°N, 122.7°W	168
Porto (city, Portugal)	41.2°N, 8.6°W	182
Porto Allegre (city, Brazil)	30.0°S, 51.2°W	171
Porto Novo (nat. capital, Benin)	6.5°N, 2.6°E	189
Port-of-Spain (nat. capital, Trinidad & Tobago)	10.7°N, 61.5°W	170
Portugal (country, Europe)	40.0°N, 8.0°E	182
Port-Vila (nat. capital, Vanuatu)	17.7°S, 168.3°E	200
Prague (nat. capital, Czechia)	50.1°N, 14.5°E	182
Pretoria (nat. capital, South Africa)	25.7°S, 28.2°E	190

GEOGRAPHIC INDEX

NAME/DESCRIPTION	LOCATION	PAGE
Prince Edward Island (province, Canada)	46.3°N, 63.3°W	167
Pripyat (river, Europe)	52.1°N, 28.0°E	183
Pristina (nat. capital, Kosovo)	42.7°N, 21.2°E	175
Providence (state capital, U.S.)	41.8°N, 71.4°W	168
Puebla (state capital, Mexico)	19.1°N, 98.2°W	169
Puebla (state, Mexico)	19.0°N, 97.9°W	169
Puerto Rico (commonwealth, U.S.)	18.5°N, 66.1°W	170
Pune (city, India)	18.5°N, 73.9°E	197
Punta Lavapié (cape, South America)	38.0°S, 73.0°W	172
Punta Negra (cape, South America)	7.0°S, 81.5°W	172
Purus (river, South America)	3.4°S, 61.5°W	172
Putumayo (river, South America)	3.1°S, 68.0°W	172
Pyrenees (mountains, Europe)	42.5°N, 0.8°E	176
Qaraghandy (city, Kazakhstan)	49.8°N, 73.2°E	195
Qatar (country, Asia)	25.3°N, 51.5°E	196
Qingdao (city, China)	36.1°N, 120.4°E	198
Qiqihar (city, China)	47.4°N, 123.9°E	198
Qom (city, Iran)	34.6°N, 50.9°E	196
Quanzhou (city, China)	24.9°N, 118.6°E	198
Québec (prov. capital, Canada)	46.8°N, 71.3°W	167
Québec (province, Canada)	53.8°N, 72.0°W	167
Queen Charlotte Islands (islands, Canada)	53.0°N, 132.0°W	167
Queen Elizabeth Islands (islands, Canada)	78.0°N, 95.0°W	167
Queen Maud Land (region, Antarctica)	73.5°S, 12.0°E	204
Queensland (state, Australia)	23.0°S, 143.0°E	200
Queretaro (state capital, Mexico)	20.6°N, 100.4°W	169
Queretaro (state, Mexico)	20.8°N, 99.9°W	169
Quetta (city, Pakistan)	30.2°N, 67.3°E	197
Quintana Roo (state, Mexico)	19.6°N, 87.9°W	169
Quito (nat. capital, Ecuador)	0.2°S, 78.5°W	171
Rabat (nat. capital, Morocco)	34.0°N, 6.9°W	189
Rainier, Mt. 14,411 (peak, U.S.)	46.9°N, 121.8°W	164
Raipur (city, India)	21.1°N, 81.4°E	197
Rajkot (city, India)	22.3°N, 70.8°E	197
Raleigh (state capital, U.S.)	35.8°N, 78.7°W	168
Ranchi (city, India)	23.4°N, 85.3°E	197
Rawalpindi (city, Pakistan)	33.6°N, 73°E	197
Recife (city, Brazil)	8.1°S, 34.9°W	171
Red (river, North America)	31.0°N, 91.8°W	168
Red Sea (sea, Indian Ocean)	20.0°N, 38.0°E	196
Regina (prov. capital, Canada)	50.5°N, 104.6°W	167
Reindeer Lake (lake, Canada)	57.6°N, 102.3°W	167
Réunion (territory, France)	21.0°S, 55.5°E	5
Reykjanes Ridge (ridge, Atlantic Ocean)	60.0°N, 30.0°W	4
Reykjavík (nat. capital, Iceland)	64.2°N, 22.0°W	182
Rhine (river, Europe)	50.4°N, 7.6°E	182
Rhode Island (state, U.S.)	41.7°N, 71.5°W	168
Rhone (river, Europe)	44.8°N, 4.8°E	182

NAME/DESCRIPTION	LOCATION	PAGE
Richmond (state capital, U.S.)	37.5°N, 77.4°W	168
Riga (nat. capital, Latvia)	57.0°N, 24.1°E	182
Riga, Gulf of (gulf, Baltic Sea)	57.6°N, 23.6°E	183
Rio Branco (city, Brazil)	10.0°S, 67.8°W	171
Rio de Janeiro (city, Brazil)	22.9°S, 43.3°W	171
Rio Gallegos (city, Argentina)	51.6°S, 69.2°W	171
Rio Grande (river, North America)	26.0°N, 97.2°W	169
Riyadh (nat. capital, Saudi Arabia)	24.6°N, 46.7°E	196
Rocky Mountains (mountains, North America)	52.0°N, 110.0°W	164
Romania (country, Europe)	47.0°N, 25.0°E	183
Rome (nat. capital, Italy)	41.9°N, 12.5°E	182
Ronne Ice Shelf (ice shelf, Antarctica)	77.9°S, 61.3°W	204
Rosario (city, Brazil)	33°S, 60.7°W	171
Roseau (nat. capital, Dominica)	15.3°N, 61.4°W	170
Ross Ice Shelf (ice shelf, Antarctica)	81.5°S, 175.0°W	204
Ross Sea (sea, Southern Ocean)	76.0°S, 175.0°W	204
Rostov (city, Russia)	57.2°N, 39.4°E	195
Rotterdam (city, Netherlands)	51.9°N, 4.5°E	182
Rub Al Khali (desert, Asia)	19.5°N, 49.0°E	192
Rufiji (river, Africa)	8°S, 39.3°E	190
Rupert (river, Canada)	51.5°N, 78.8°W	167
Russia (country, Asia)	60.0°N, 100.0°E	195
Ruvuma (river, Africa)	10.5°S, 40.4°E	190
Rwanda (country, Africa)	2.0°S, 30.0°E	190
Rybinsk Reservoir (lake, Russia)	58.4°N, 38.4°E	183
Ryukyu Islands (islands, Asia)	26.5°N, 128.0°E	192
Sable, Cape (cape, U.S.)	43.5°N, 63.6°W	164
Sacramento (state capital, U.S.)	38.6°N, 121.5°W	168
Sahara (desert, Africa)	23.0°N, 13.0°E	185
Sahel, The (region, Africa)	15.0°N, 8.0°W	185
Saint George's (nat. capital, Grenada)	12.1°N, 61.8°W	170
Saint John's (prov. capital, Canada)	47.6°N, 52.7°W	167
Saint John's (nat. capital, Antiqua & Barbuda)	17.1°N, 61.9°W	170
Saint Kitts and Nevis (country, North America)	17.3°N, 62.7°W	170
Saint Lucia (country, North America)	14°N, 61°W	170
Saint Vincent & the Grenadines (country, North America)	13.2°N, 61.2°W	170
Sakhalin (island, Russia)	51.0°N, 143.0°E	195
Salado (river, South America)	31.7°S, 60.7°W	172
Salem (city, India)	11.6°N, 78.2°E	197
Salem (state capital, U.S.)	44.9°N, 123.0°W	168
Salt Lake City (state capital, U.S.)	40.8°N, 111.9°W	168
Salta (city, Argentina)	24.8°S, 65.4°W	171
Saltillo (state capital, Mexico)	25.4°N, 101.0°W	169
Salton Sea (lake, U.S.)	33.3°N, 115.8°W	168
Salvador (city, Brazil)	13.0°S, 38.5°W	171
Salween (river, Myanmar)	17.2°N, 97.6°E	199
Samar (island, Philippines)	12.0°N, 125.0°E	199

GEOGRAPHIC INDEX

GEOGRAPHIC INDEX

NAME/DESCRIPTION	LOCATION	PAGE
Tianjin (city, China)	39.1°N, 117.2°E	198
Tiber (river, Europe)	42.5°N, 12.3°E	182
Tibesti Massif (mountains, Chad)	20.8°N, 18.1°E	185
Tibet, Plateau of (plateau, China)	32.0°N, 90.0°E	192
Tien Shan (mountains, Asia)	42.0°N, 80.0°E	192
Tierra del Fuego (archipelago, South America)	54.0°S, 69.0°W	172
Tigris (river, Asia)	36.5°N, 43.0°E	196
Tijuana (city, Mexico)	32.5°N, 117.0°W	169
Timbuktu (city, Mali)	16.8°N, 3°W	189
Timor (island, Asia)	9.0°S, 125.0°E	199
Timor Sea (sea, Pacific Ocean)	11.0°S, 128.0°E	201
Timor-Leste (country, Asia)	9.0°S, 126.0°E	199
Tirana (nat. capital, Albania)	41.3°N, 19.8°E	182
Tiruchirappalli (city, India)	10.8°N, 78.7°E	197
Titicaca, Lake (lake, South America)	15.8°S, 69.4°W	172
Tlaxcala (state capital, Mexico)	19.5°N, 98.4°W	169
Tlaxcala (state, Mexico)	19.4°N, 98.2°W	169
Tobol (river, Asia)	58.2°N, 68.2°E	195
Tocantins (river, South America)	1.8°S, 49.2°W	172
Togo (country, Africa)	8.0°N, 1.2°E	189
Tokyo (nat. capital, Japan)	35.7°N, 139.8°E	198
Toluca (state capital, Mexico)	19.3°N, 99.7°W	169
Tomsk (city, Russia)	56.5°N, 85°E	195
Tonga (country, Oceania)	21.1°S, 175.2°W	200
Tonkin, Gulf of (gulf, Pacific Ocean)	20.0°N, 108.0°E	199
Tonle Sap (river, Asia)	12.9°N, 104.1°E	199
Topeka (state capital, U.S.)	39.1°N, 95.7°W	168
Toronto (prov. capital, Canada)	43.7°N, 79.4°W	167
Torreon (city, Mexico)	25.5°N, 103.5°W	169
Torres Strait (strait, Australia)	9.8°S, 142.5°E	201
Toulouse (city, France)	43.6°N, 1.5°E	182
Trenton (state capital, U.S.)	40.2°N, 74.8°W	168
Trinidad and Tobago (country, North America)	10.4°N, 61.3°W	170
Tripoli (nat. capital, Libya)	32.9°N, 13.2°E	189
Trujillo (city, Peru)	8.1°S, 79.0°W	171
Tuamotu Archipelago (islands, Pacific Ocean)	18.0°S, 141.4°W	4
Tucson (city, U.S.)	32.2°N, 110.9°W	168
Tucumán (city, Argentina)	29.8°S, 65.2°W	171
Tugela (river, Africa)	28.8°S, 28.9°E	190
Tunis (nat. capital, Tunisia)	36.8°N, 10.2°E	189
Tunisia (country, Africa)	35.0°N, 9.2°E	189
Turin (city, Italy)	45.1°N, 7.7°E	182
Turkana, Lake (lake, Africa)	3.6°N, 36.1°E	185
Turkey (country, Asia)	39.0°N, 35.0°E	196
Turkmenistan (country, Asia)	40.0°N, 60.0°E	195
Turks and Caicos (territory, U.K.)	21.5°N, 71.8°W	170
Tuvalu (country, Oceania)	8.5°S, 179.2°E	200
Tuxtla Gutierrez (state capital, Mexico)	16.8°N, 93.1°W	169

NAME/DESCRIPTION	LOCATION	PAGE
Tyrrhenian Sea (sea, Atlantic Ocean)	40.0°N, 12.0°E	176
Ubangi (river, Africa)	4.8°N, 20.2°E	190
Ucayali (river, South America)	8.6°N, 74.4°W	172
Uganda (country, Africa)	1.5°N, 32.0°E	189
Ujungpandang (city, Indonesia)	5.1°S, 119.4°E	199
Ukraine (country, Europe)	49.0°N, 32.0°E	183
Ulaanbaatar (nat. capital, Mongolia)	47.9°N, 106.9°E	198
Ulsan (city, South Korea)	35.6°N, 129.3°E	198
Uluru 2,831 (peak, Australia)	25.3°S, 131.0°E	201
Ungava Bay (bay, Atlantic Ocean)	59.5°N, 67.3°W	167
Ungava Peninsula (peninsula, Canada)	59.0°N, 74.0°W	167
United Arab Emirates (country, Asia)	24.0°N, 54.0°E	196
United Kingdom (country, Europe)	54.0°N, 2.0°W	182
United States (country, North America)	40.0°N, 100.0°W	163
Ural (river, Asia)	52.0°N, 55.0°E	195
Ural Mountains (mountains, Asia)	60.0°N, 60.0°E	192
Uruguay (country, South America)	33.0°S, 56.0°W	171
Uruguay (river, South America)	34.2°S, 58.3°W	172
Ürümqi (city, China)	43.7°N, 87.6°E	198
Utah (state, U.S.)	39.5°N, 111.5°W	168
Uzbekistan (country, Asia)	41.0°N, 64.0°E	195
Vaal (river, Africa)	29.1°S, 23.6°E	190
Vadodara (city, India)	22.3°N, 73.2°E	197
Valdez (city, U.S.)	61.1°N, 146.4°W	168
Valencia (city, Spain)	39.5°N, 0.4°W	182
Valencia (city, Venezuela)	10.2°N, 68°W	171
Valparaiso (city, Chile)	33.1°S, 71.6°W	171
Vancouver (city, Canada)	49.3°N, 123.1°W	167
Vancouver Island (island, Canada)	49.8°N, 126.0°W	167
Vanua Levu (island, Fiji)	16.6°S, 179.2°E	201
Vanuatu (country, Oceania)	16.0°S, 167.0°E	200
Varanasi (city, India)	25.3°N, 83.0°E	197
Vatican City (country, Europe)	41.9°N, 12.5°E	182
Venezuela (country, South America)	7.0°N, 67.0°W	171
Veracruz (state, Mexico)	19.4°N, 96.4°W	169
Verkhoyansk Range (mountains, Asia)	67.0°N, 129.0°E	192
Vermont (state, U.S.)	44.0°N, 72.7°W	168
Victoria (prov. capital, Canada)	48.4°N, 123.4°W	167
Victoria (state, Australia)	37.0°S, 144.0°E	200
Victoria Island (island, Canada)	71.0°N, 110.0°W	204
Victoria Land (region, Antarctica)	72.0°S, 155.0°E	204
Victoria, Lake (lake, Africa)	1.0°S, 33.0°E	185
Viedma (city, Argentina)	40.8°S, 63.0°W	171
Vienna (nat. capital, Austria)	48.2°N, 16.4°E	182
Vientiane (nat. capital, Laos)	18.0°N, 102.6°E	199
Vietnam (country, Asia)	16.0°N, 107.5°E	199
Vijayawada (city, India)	16.5°N, 80.6°E	197
Villahermosa (state capital, Mexico)	18.0°N, 92.9°W	169

GEOGRAPHIC INDEX

Sources

Maps

Base maps created using Natural Earth Data. Available online at http://www.naturalearthdata.com/

Map 3a   Fekete, B.M. 2002. Water Systems Analysis Group, University of New Hampshire, NH.

Map 6a   World Clim; http://www.wordclim.org/

Map 6b   World Clim; http://www.wordclim.org/

Map 11   Olson, D. M, E. Dinerstein, E. D. Wikramanayake, N. D. Burgess, G.V.N. Powell, E.C. Underwood, J.A. D'amico, I. Itoua, H.E. Strand, J.C. Morrison, C.J. Loucks, T.F. Allnutt, T.H. Ricketts, Y. Kura, J.F. Lamoreux, W.W. Wettengel, P. Hedao, & K.R. Kassem. 2001. Terrestrial Olson, D. M, E. Dinerstein, E.D. Wikramanayake, N.D. Burgess, G.V.N. Powell, E.C. Underwood, J.A. D'amico, I. Itoua, H.E. Strand, J.C. Morrison, C.J. Loucks, T.F. Allnutt, T.H. Ricketts, Y. Kura, J.F. Lamoreux, W.W. Wettengel, P. Hedao, & K.R. Kassem. 2001. Terrestrial Olson, D. M, E. Dinerstein, E.D. Wikramanayake, N.D. Burgess, G.V.N. Powell, E.C. Underwood, J.A. D'amico, I. Itoua, H.E. Strand, J.C. Morrison, C.J. Loucks, T.F. Allnutt, T.H. Ricketts, Y. Kura, J.F. Lamoreux, W.W. Wettengel, P. Hedao, & K.R. Kassem. 2001. Terrestrial Ecoregions of the World: A New Map of Life on Earth. BioScience 51:933–938.

Map 18   Central Intelligence Agency. 2012. The World Factbook [Online]. Available online at https://www.cia.gov/library/publications/the-world-factbook/UN Department of Economic and Social Affairs, Population Division. World Urbanization Prospects: The 2011 Revision [Online]. Available online at http://esa.un.org/unup/

Map 22   Barrett, David B., George T. Kurian, Todd M. Johnson. 2001. *World Christian Encyclopedia: A Comparative Survey of Churches and Religions in the Modern World*, 2nd edition. Oxford; New York : Oxford University Press.

Map 23   Lewis, M. Paul, ed. 2009. *Ethnologue: Languages of the World*, Sixteenth edition. Dallas, Tex.: SIL International [Online]. Available online at http://www.ethnologue.com/

Map 24   Institute For Endangered Languages. 2007. Living Tongues [Online]. Available online at http://www.livingtongues.org/

Map 26   Central Intelligence Agency. 2012. The World Factbook [Online]. Available online at https://www.cia.gov/library/publications/the-world-factbook/

Map 27   World Bank. 2012. World Development Indicators [Online]. Available online at http://data.worldbank.org/

Map 28   Central Intelligence Agency. 2012. The World Factbook [Online]. Available online at https://www.cia.gov/library/publications/the-world-factbook/

Map 29a   World Bank. 2012. World Development Indicators [Online]. Available online at http://data.worldbank.org/

Map 29b   Central Intelligence Agency. 2012. The World Factbook [Online]. Available online at https://www.cia.gov/library/publications/the-world-factbook/

Map 30   Central Intelligence Agency. 2012. The World Factbook [Online]. Available online at https://www.cia.gov/library/publications/the-world-factbook/

Map 31   World Bank. 2012. World Development Indicators [Online]. Available online at http://data.worldbank.org/

Map 32   Central Intelligence Agency. 2012. The World Factbook [Online]. Available online at https://www.cia.gov/library/publications/the-world-factbook/

Map 33   Central Intelligence Agency. 2012. The World Factbook [Online]. Available online at https://www.cia.gov/library/publications/the-world-factbook/

Map 34   Food and Agriculture Organization of the United Nations. 2012. FAO Statistical Yearbook 2012 [Online]. Available online at http://faostat.fao.org/

Map 35   International Food Policy Research Institute. 2011. 2011 Global Hunger Index [Online]. Available online at http://www.ifpri.org/publication/2011-global-hunger-index

Map 36   World Bank. 2012. World Development Indicators [Online]. Available online at http://data.worldbank.org/

Map 37   Food and Agriculture Organization of the United Nations. 2012. FAO Statistical Yearbook 2012 [Online]. Available online at http://faostat.fao.org/

Map 38   World Health Organization. 2012. Global Burden of Disease [Online]. Available online at http://www.who.int/healthinfo/global_burden_disease/en/

Map 39a   World Health Organization. 2012. Global Burden of Disease [Online]. Available online at http://www.who.int/healthinfo/global_burden_disease/en/

Map 39b   World Health Organization. 2012. Global Burden of Disease [Online]. Available online at http://www.who.int/healthinfo/global_burden_disease/en/

Map 39c	World Health Organization. 2012. Global Burden of Disease [Online]. Available online at http://www.who.int/healthinfo/global_burden_disease/en/
Map 39d	World Health Organization. 2012. Global Burden of Disease [Online]. Available online at http://www.who.int/healthinfo/global_burden_disease/en/
Map 40	Central Intelligence Agency. 2012. The World Factbook [Online]. Available online at https://www.cia.gov/library/publications/the-world-factbook/World Bank. 2012. World Development Indicators [Online]. Available online at http://data.worldbank.org/
Map 41	UNICEF. 2012. Childinfo [Online]. Available online at http://www.childinfo.org/World Bank. 2012. *World Development Indicators* [Online]. Available online at http://data.worldbank.org/
Map 42	United Nations Development Programme. 2011. Human Development Report 2011 [Online]. Available online at http://hdr.undp.org/en/
Map 43	UN Department of Economic and Social Affairs, Population Division. World Urbanization Prospects: The 2011 Revision [Online]. Available online at http://esa.un.org/unup/
Map 44	Central Intelligence Agency. 2012. *The World Factbook* [Online]. Available online at https://www.cia.gov/library/publications/the-world-factbook/
Map 45	World Bank. 2012. *World Development Indicators* [Online]. Available online at http://data.worldbank.org/
Map 46	World Bank. 2012. *World Development Indicators* [Online]. Available online at http://data.worldbank.org/
Map 47	World Bank. 2012. *World Development Indicators* [Online]. Available online at http://data.worldbank.org/
Map 48	World Bank. 2012. *World Development Indicators* [Online]. Available online at http://data.worldbank.org/
Map 49	World Bank. 2012. *World Development Indicators* [Online]. Available online at http://data.worldbank.org/
Map 50	Central Intelligence Agency. 2012. *The World Factbook* [Online]. Available online at https://www.cia.gov/library/publications/the-world-factbook/
Map 51	World Bank. 2012. *World Development Indicators* [Online]. Available online at http://data.worldbank.org/
Map 52	United Nations. 2012. *Millennium Development Goals Indicators* [Online]. Available online at http://mdgs.un.org/unsd/mdg/Default.aspx
Map 53	World Bank. 2012. *World Development Indicators* [Online]. Available online at http://data.worldbank.org/
Map 54	World Bank. 2012. *World Development Indicators* [Online]. Available online at http://data.worldbank.org/
Map 55	Central Intelligence Agency. 2012. *The World Factbook* [Online]. Available online at https://www.cia.gov/library/publications/the-world-factbook/
Map 56	Central Intelligence Agency. 2012. *The World Factbook* [Online]. Available online at https://www.cia.gov/library/publications/the-world-factbook/
Map 57	Food and Agriculture Organization of the United Nations. 2012. *FAO Statistical Yearbook 2012* [Online]. Available online at http://faostat.fao.org/
Map 58	World Trade Organization. 2011. *International Trade Statistics, 2011* [Online]. Available online at http://www.wto.org/english/res_e/statis_e/its2011_e/its11_toc_e.htm
Map 59	World Bank, Development Prospects Group. 2011. *Remittance Flows in 2011: An Update* [Online]. Available online at http://go.worldbank.org/A8EKPX2IA0
Map 61	World Bank. 2012. *World Development Indicators* [Online]. Available online at http://data.worldbank.org/
Map 62	Central Intelligence Agency. 2012. *The World Factbook* [Online]. Available online at https://www.cia.gov/library/publications/the-world-factbook/
Map 63	World Bank. 2012. *World Development Indicators* [Online]. Available online at http://data.worldbank.org/
Map 64	World Bank. 2012. *World Development Indicators* [Online]. Available online at http://data.worldbank.org/
Map 65	Deutsche Gesellschaft für Technische Zusammenarbeit (GTZ). *International Fuel Prices 2009, 6th Edition* [Online]. Available online at http://www.gtz.de/fuelprices
Map 66	World Bank. 2012. *World Development Indicators* [Online]. Available online at http://data.worldbank.org/
Map 67	World Bank. 2012. *World Development Indicators* [Online]. Available online at http://data.worldbank.org/

Map 68	World Bank. 2012. *World Development Indicators* [Online]. Available online at http://data.worldbank.org/
Map 69	British Petroleum. 2012. *Statistical Review of World Energy 2012* [Online]. Available online at http://www.bp.com/sectionbodycopy.do?categoryId=7500&contentId=7068481
Map 70	World Bank. 2012. *World Development Indicators* [Online]. Available online at http://data.worldbank.org/
Map 71	Central Intelligence Agency. 2012. *The World Factbook* [Online]. Available online at https://www.cia.gov/library/publications/the-world-factbook/
Map 74	World Bank. 2012. *World Development Indicators* [Online]. Available online at http://data.worldbank.org/
Map 79	World Bank. 2012. *World Development Indicators* [Online]. Available online at http://data.worldbank.org/
Map 80	United Nations. 2012. *Millennium Development Goals Indicators* [Online]. Available online at http://mdgs.un.org/unsd/mdg/Default.aspx
Map 83	McGuire, V.L., 2009, Water-level changes in the High Plains aquifer, predevelopment to 2007, 2005–06, and 2006–07: U.S. Geological Survey Scientific Investigations Report 2009–5019, 9 p [Online]. Available online at: http://pubs.usgs.gov/sir/2009/5019/.
Map 85	Food and Agriculture Organization of the United Nations. 2012. *FAO Statistical Yearbook 2012* [Online]. Available online at http://faostat.fao.org/
Map 86	World Bank. 2012. *World Development Indicators* [Online]. Available online at http://data.worldbank.org/
Map 88	Food and Agriculture Organization of the United Nations. 2012. *FAO Statistical Yearbook 2012* [Online]. Available online at http://faostat.fao.org/
Map 89	World Bank. 2012. *World Development Indicators* [Online]. Available online at http://data.worldbank.org/
Map 90	Food and Agriculture Organization of the United Nations. 2012. *FAO Statistical Yearbook 2012* [Online]. Available online at http://faostat.fao.org/
Map 91	Food and Agriculture Organization of the United Nations. 2012. *FAO Statistical Yearbook 2012* [Online]. Available online at http://faostat.fao.org/
Map 92	World Bank. 2012. *World Development Indicators* [Online]. Available online at http://data.worldbank.org/
Map 94	World Bank. 2012. *World Development Indicators* [Online]. Available online at http://data.worldbank.org/
Map 95	Myers, N., R. A. Mittermeier, C. G. Mittermeier, G. A. B. da Fonseca, and J. Kent. 2000. Biodiversity hotspots for conservation priorities. *Nature* 403:853–858
Map 100	Central Intelligence Agency. 2012. *The World Factbook* [Online]. Available online at https://www.cia.gov/library/publications/the-world-factbook/
Map 104	Central Intelligence Agency. 2012. *The World Factbook* [Online]. Available online at https://www.cia.gov/library/publications/the-world-factbook/
Map 109	United Nations. *Member States of the United Nations* [Online]. Available online at http://www.un.org/en/members/
Map 110	Central Intelligence Agency. 2012. *The World Factbook* [Online]. Available online at https://www.cia.gov/library/publications/the-world-factbook/
Map 111	United Nations. 2012 [Online]. Available online at http://unstats.un.org/unsd/methods/m49/m49regin.htm
Map 112	Federation of American Scientists. 2012. *Status of World Nuclear Forces* [Online]. Available online at http://www.fas.org/programs/ssp/nukes/nuclearweapons/nukestatus.html
Map 113	World Bank. 2012. *World Development Indicators* [Online]. Available online at http://data.worldbank.org/
Map 114	Central Intelligence Agency. 2012. *The World Factbook* [Online]. Available online at https://www.cia.gov/library/publications/the-world-factbook/
Map 116	Central Intelligence Agency. 2012. *The World Factbook* [Online]. Available online at https://www.cia.gov/library/publications/the-world-factbook/
Map 117	United Nations High Commissioner for Refugees. 2011. *A Year of Crises: UNHCR Global Trends 2011* [Online]. Available online at http://www.unhcr.org/pages/49c3646c4d6.html
Map 118	United Nations High Commissioner for Refugees. 2011. *A Year of Crises: UNHCR Global Trends 2011* [Online]. Available online at http://www.unhcr.org/pages/49c3646c4d6.html
Map 119	United Nations High Commissioner for Refugees. 2011. *A Year of Crises: UNHCR Global Trends 2011* [Online]. Available online at http://www.unhcr.org/pages/49c3646c4d6.html

Map 120   World Bank. 2012. *Worldwide Governance Indicators* [Online]. Available online at http://info
           .worldbank.org/governance/wgi/index.asp

Map 121   Freedom House. 2012. *2012 Freedom in the World* [Online]. Available online at http://www
           .freedomhouse.org/report-types/freedom-world

Map 122   Gibney, M., Cornett, L., & Wood, R. 2012. *Political Terror Scale 1976–2008* [Online].
           Available online http://www.politicalterrorscale.org/

Map 123   Central Intelligence Agency. 2012. *The World Factbook* [Online]. Available online at https://
           www.cia.gov/library/publications/the-world-factbook/

Map 124   Amnesty International. 2012. *Abolitionist and retentionist countries* [Online]. Available online
           at http://www.amnesty.org/en/death-penalty/abolitionist-and-retentionist-countries

Map 125   Molzahn, Cory, Viridiana Ríos, and David A. Shirk. 2012. *Drug Violence in Mexico: Data
           and Analysis Through 2011*. Trans-Border Institute [Online]. Available online at http://
           justiceinmexico.files.wordpress.com/2012/03/2012-tbi-drugviolence.pdf

Map 130   Center for International Earth Science Information Network (CIESIN)/Columbia University,
           International Food Policy Research Institute (IFPRI), The World Bank, and Centro Internacional
           de Agricultura Tropical (CIAT). 2011. Global Rural-Urban Mapping Project, Version 1
           (GRUMPv1): Urban Extents Grid. Palisades, NY: NASA Socioeconomic Data and Applications
           Center (SEDAC). http://sedac.ciesin.columbia.edu/data/set/grump-v1-urban-extents.

Map 130   Center for International Earth Science Information Network (CIESIN)/Columbia University,
           International Food Policy Research Institute (IFPRI), The World Bank, and Centro Internacional
           de Agricultura Tropical (CIAT). 2011. Global Rural-Urban Mapping Project, Version 1
           (GRUMPv1): Urban Extents Grid. Palisades, NY: NASA Socioeconomic Data and Applications
           Center (SEDAC). http://sedac.ciesin.columbia.edu/data/set/grump-v1-urban-extents.

Map 138   Center for International Earth Science Information Network (CIESIN)/Columbia University,
           International Food Policy Research Institute (IFPRI), The World Bank, and Centro Internacional
           de Agricultura Tropical (CIAT). 2011. Global Rural-Urban Mapping Project, Version 1
           (GRUMPv1): Urban Extents Grid. Palisades, NY: NASA Socioeconomic Data and Applications
           Center (SEDAC). http://sedac.ciesin.columbia.edu/data/set/grump-v1-urban-extents.

Map 142   Center for International Earth Science Information Network (CIESIN)/Columbia University,
           International Food Policy Research Institute (IFPRI), The World Bank, and Centro Internacional
           de Agricultura Tropical (CIAT). 2011. Global Rural-Urban Mapping Project, Version 1
           (GRUMPv1): Urban Extents Grid. Palisades, NY: NASA Socioeconomic Data and Applications
           Center (SEDAC). http://sedac.ciesin.columbia.edu/data/set/grump-v1-urban-extents.

Map 149   Center for International Earth Science Information Network (CIESIN)/Columbia University,
           International Food Policy Research Institute (IFPRI), The World Bank, and Centro Internacional
           de Agricultura Tropical (CIAT). 2011. Global Rural-Urban Mapping Project, Version 1
           (GRUMPv1): Urban Extents Grid. Palisades, NY: NASA Socioeconomic Data and Applications
           Center (SEDAC). http://sedac.ciesin.columbia.edu/data/set/grump-v1-urban-extents.

Map 157   Center for International Earth Science Information Network (CIESIN)/Columbia University,
           International Food Policy Research Institute (IFPRI), The World Bank, and Centro Internacional
           de Agricultura Tropical (CIAT). 2011. Global Rural-Urban Mapping Project, Version 1
           (GRUMPv1): Urban Extents Grid. Palisades, NY: NASA Socioeconomic Data and Applications
           Center (SEDAC). http://sedac.ciesin.columbia.edu/data/set/grump-v1-urban-extents.

Map 166   Center for International Earth Science Information Network (CIESIN)/Columbia University,
           International Food Policy Research Institute (IFPRI), The World Bank, and Centro Internacional
           de Agricultura Tropical (CIAT). 2011. Global Rural-Urban Mapping Project, Version 1
           (GRUMPv1): Urban Extents Grid. Palisades, NY: NASA Socioeconomic Data and Applications
           Center (SEDAC). http://sedac.ciesin.columbia.edu/data/set/grump-v1-urban-extents.